Stereodynamics of
Molecular Systems

Pergamon Titles of Related Interest

Norwich: MOLECULAR DYNAMICS IN BIOSYSTEMS
Sarma: GEOMETRY AND DYNAMICS OF NUCLEIC ACIDS
Srinivasan: BIOMOLECULAR STRUCTURE, CONFORMATION,
FUNCTION AND EVOLUTION

Stereodynamics of Molecular Systems

Proceedings of a symposium held at the
State University of New York at Albany
23-24 April 1979

Edited by

Ramaswamy H. Sarma

Director
Institute of Biomolecular Stereodynamics
State University of New York at Albany

Pergamon Press
New York ☐ Oxford ☐ Toronto ☐ Sydney ☐ Frankfurt ☐ Paris

Pergamon Press Offices:

U.S.A.   Pergamon Press Inc., Maxwell House, Fairview Park,
                         Elmsford, New York 10523, U.S.A.

U.K.   Pergamon Press Ltd., Headington Hill Hall,
                         Oxford OX3 0BW, England

CANADA   Pergamon of Canada, Ltd., 150 Consumers Road,
                         Willowdale, Ontario M2J, 1P9, Canada

AUSTRALIA   Pergamon Press (Aust) Pty. Ltd., P O Box 544,
                         Potts Point, NSW 2011, Australia

FRANCE   Pergamon Press SARL, 24 rue des Ecoles,
                         75240 Paris, Cedex 05, France

FEDERAL REPUBLIC   Pergamon Press GmbH, 6242 Kronberg/Taunus,
OF GERMANY   Pferdstrasse 1, Federal Republic of Germany

Cover illustration of one complete turn of B-DNA
© copyright by Irving Geis.

Library of Congress Cataloging in Publication Data
Main entry under title:

Stereodynamics of molecular systems.

Sponsored by the Dept. of Chemistry and Institute
of Biomolecular Stereodynamics of the State University
of New York at Albany.
    Includes index.
    1. Nucleic acids--Spectra--Congresses. 2. Nuclear
magnetic resonance spectroscopy--Congresses. 3. Stereo-
chemistry--Congresses. 4. Molecules--Spectra--
Congresses. I. Sarma, Ramaswamy H., 1939-
II. New York (State). State University, Albany. Dept.
of Chemistry. III. New York (State). State University,
Albany. Institute of Biomolecular Sterodynamics.

QP620.S73 1979   574.1'9283   79-19669
ISBN 0-08-024629-X

Printed in the United States of America

Preface

These are the proceedings of the Conversation in the Discipline, "Stereodynamics of Molecular Systems," held at the State University of New York at Albany, April 23-24, 1979 under the auspices of the Department of Chemistry and organized by the University's Institute of Biomolecular Stereodynamics. The Conversation attempted to cover an exploding area in chemistry and biology today viz the geometry and shape of molecules and their dynamics. Within the limited time available we have attempted to cover a large area with no particular distinction between chemical and biological systems. The contributions deal with stereodynamics of small molecules and ions, followed by medium size systems, then larger and complex biomolecules and toward the end take up intact organs and organisms. To provide some depth an area of current interest *viz.* nucleic acid structure was given intense examination.

From the conference it emerged that nucleic acids are considerably more flexible than previously believed and that this flexibility plays a dominant role in their biological function. It is gratifying to note that the deliberations on the nucleic acid statics and dynamics received extensive report in the News and Views Section of *Nature* (*279*, 474, 1979).

The text opens with a section on methodologies which are employed in later parts of the book. NMR spectroscopy and its use for determining geometry and dynamics is a science unto itself that it became necessary to have three separate contributions to cover the methodology of NMR. I thank Professor C. Hackett Bushweller for "NMR Spectroscopy and Stereodynamics of Small Molecules." I am deeply indebted to Dr. Frank A. Bovey for his contribution "NMR Spectroscopy and Solution Dynamics of Polymer Chains" and making a complex methodology understandable and interesting to knowledgeable readers who may not be specialists. I also thank Professor Nadrian C. Seeman for his contribution "Single Crystal Crystallography" in the methodology section.

I thank President Vincent O'Leary for welcoming the delegates. I am deeply indebted to Nobel Laureate Ivar Giaever for inaugurating the symposium and providing an auspicious beginning. My grateful thanks are due to Professor Henry M. Sobell and Leonard Lerman for helping in arranging the public lecture by Nobel Laureate Max Perutz. I acknowledge the assistance and help from our colleagues, particularly Nadrian C. Seeman, Shelton Bank, M. M. Dhingra and C. K. Mitra.

I take this opportunity to thank the State University of New York and its Conversation in the Discipline Series, President Vincent O'Leary, Vice President for Research Louis Salkaver, General Electric Company and Merck and Co. for financially supporting the symposium.

This book was set in type by a high speed, third generation CRT typesetter—the University's new Compugraphic Videosetter Universal, driven by the UNIVAC 1110. I thank Mr. Steve Rogowski and Ms. Katie Huxford of the Computing Center for their efficient services in typesetting. I greatly appreciate the services of Ms. Barbara Hale who patiently fed the manuscript through the terminal to the UNIVAC.

Above all, I thank the speakers and delegates for their enthusiastic response to our invitations and delivering the manuscripts in time for publication.

Ramaswamy H. Sarma
June 30, 1979

Contents

Part IV: Systems of Higher Order. Nucleic Acid Statics and Dynamics

Part V: Intact Biological Systems

*President Vincent O'Leary
welcomes the speakers and delegates*

The State University of New York at Albany

Nobel Laureates Ivar Giaever and Max Perutz at the
State University of New York at Albany

Inaugural Address by Nobel Laureate Ivar Giaever

Good morning,

It is my pleasure to add to President O'Leary's welcome and welcome you all here to the Conference on Stereodynamics of Molecular Systems. This conference is sponsored by the State University of New York and the General Electric Company and since I work for the latter, I'll try to represent General Electric Company. General Electric Company, of course, is a very large organization. I'm from Norway and I've been told that the sales of General Electric Company are greater than the Gross National Product of Norway. Perhaps that tells you something about Norway anyway. The business of General Electric Company and many of the products of General Electric Company had their start in the humble beginnings of basic research. I am very confident that the leadership of General Electric Company recognizes that. This was particularly true in the past for Chemistry and Physics and not so much for Biology, but who knows what the future will bring. Because of that, General Electric Company has chosen to sponsor selective conferences, and I'm very happy that they have chosen to sponsor this one, because after reading the abstract I find that I'm really trying to work in the particular area. I've changed from Physics into Biology in a certain sense. As a matter of fact, Professor Sarma was kind enough to invite me to contribute scientifically to this conference but I wasn't quite sure what he meant by Stereodynamics of Molecular Systems so I wisely declined that particular offer. After looking at Professor Sobell's molecular models here on the first row, the subject appears even more complicated than I thought and I anticipate being educated by you all in the next two days. Therefore I find myself as an inauguration speaker instead of contributing scientifically. I've been to many conferences in my life and I can only remember one inauguration speech and let me tell you about that one because it relates to the weather which was mentioned here. This was a conference on super-conductivity and a very famous man whose name I will not reveal to you inaugurated the conference. He went up and said he was very happy the weather was so nice and he looked forward to meeting all his friends out in the golf course, and the chairman did not look very happy with this remark, and I see Professor Sarma is a little nervous up there now. Of course, we all recognize that the informal part of a scientific conference very often is as important as the formal programs, and I went to the wine and cheese party yesterday and I noticed that the scientific discussions and conversations had already started. Some people actually were dedicated enough to discuss the things up on the blackboard, forgetting about the wine and the cheese which shows you they were really dedicated scientists. One thing I like particularly here is the name of the conference. It is called a conversation in the discipline of Stereodynamics and Molecular Systems, and I think we should do just that. For the next two days, let's have a conversation in this particular discipline. I think the subject of the discipline is very timely and I think Professor Sarma has gotten the right people together, we have an exciting book of abstracts so let's all put our heart and mind together and make this conference a successful one, one of those we'll look back and remember with excitement and pleasure. Let me finally close by saying I wish you a good time in the Tri-City Area and I hope you'll postpone your golf until Wednesday morning.

Part I
Methodology
NMR Spectroscopy and
Crystallography

Nuclear Magnetic Resonance Spectroscopy
Chemical Shifts, Coupling Constants and Molecular Geometry

M. M. Dhingra[98] **and Ramaswamy H. Sarma**
Institute of Biomolecular Stereodynamics
and
Department of Chemistry
State University of New York at Albany
Albany, New York 12222

Introduction

High resolution nuclear magnetic resonance spectroscopy provides information about the geometry and dynamics of molecular systems in solution. The methodology involved in the determination of the dynamics of small and macromolecules are discussed in the succeeding chapters and here we address the use of the NMR parameters, the chemical shifts (δ) and coupling constants (J) to derive molecular geometry. An elementary knowledge of NMR spectroscopy is assumed and the discussion is mostly limited to proton magnetic resonance spectroscopy. In order to make the presentation well focused we discuss the use of chemical shifts and coupling constants to derive the conformational features of one molecular system viz nucleic acid structures. However, the same methodology can be applied to any other system which has a reasonable number of coupled nuclei.

Constitutional Features of Nucleic Acids

Nucleic acids have three main structural features. i) a five-membered sugar ring which is ribose for RNA and deoxyribose for DNA; ii) the heterocyclic bases, adenine, guanine, cytosine, uracil/thymine attached to the C1′ of the sugar ring in the β-configuration; iii) the 3′5′phosphodiester linkage joining the two sugar rings. In Figure 1 is shown the structure of the nucleic acid segment Adenylyl (3′5′) adenosine i.e. 3′5′ ApA. The numbering scheme for various atoms and the nomenclature for various torsion angles are also shown. The different torsion angles are defined as:

O1′-C1′-N9-C8	χ (purines)
O1′-C1′-N1-C6	χ (pyrimidines)
C4′-C3′-O3′-P	α
C3′-O3′-P-O5′	β
O3′-P-O5′-C5′	γ
P-O5′-C5′-C4′	δ
O5′-C5′-C4′-C3′	ϵ
C5′-C4′-C3′-O3′	ζ

3

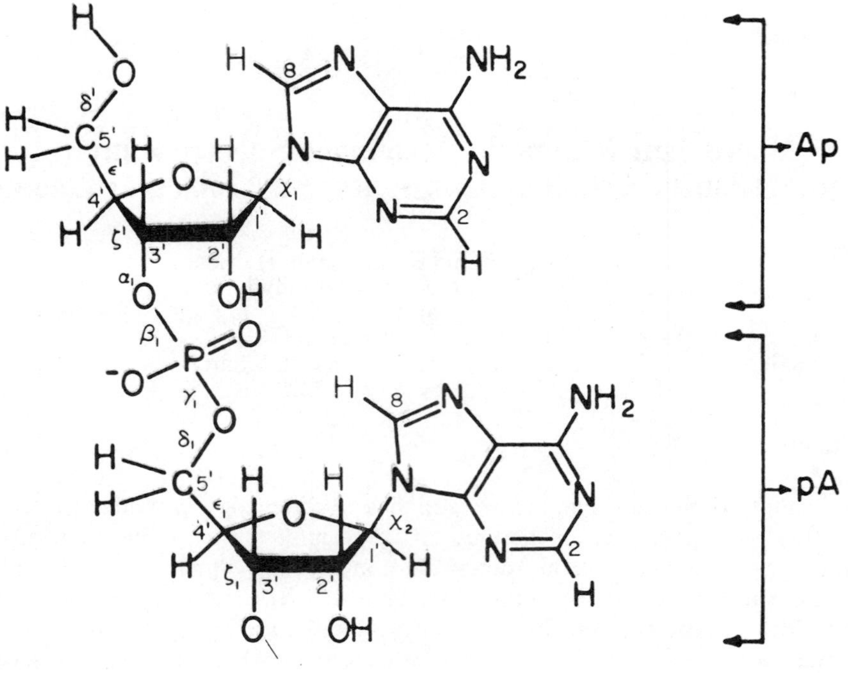

Figure 1. Structure of 3′5′ ApA, the numbering scheme and the proposed IUPAC-IUB nomenclature for the various torsion angles. See Figure 3 for the old nomenclature.

The value of the torsion angle is zero for a cis planar arrangement and clockwise rotation results in a positive value (see later Figures 9-14).

Nucleic acid structures are fairly flexible due to the possibility of internal rotation about the various single bonds. A complete conformational analysis involves the determination of the preferred orientation or the minimum energy conformation about these bonds. For a dinucleoside monophosphate (Figure 1) this includes the determination of (i) the two sugar base torsions, i.e., glycosidic torsions χ_1 and χ_2, (ii) the mode of pucker of the two sugar rings; (iii) torsion angles α_1, β_1, γ_1, δ_1, ϵ_1 and (iv) torsion angles of the free exocyclic CH_2OH group, i.e., ϵ' and δ'. In the case of a trimer, the situation is more complex, i.e., three glycosidic torsions, mode of pucker of three sugar rings and twelve other torsion angles. The determination of these conformational features essentially involves assignment and analysis of NMR spectra, accurate extraction of chemical shifts and coupling constants and translation of these NMR parameters to conformational angles. The methodology is extremely time consuming, complex and difficult, sometimes bordering on the torturous. However, a knowledge of the approaches to solve the NMR spectra of nucleic acid systems should make such studies of organic and inorganic molecules very easy.

Assignments and Analysis of the NMR Spectra of Nucleic Acids

The application of NMR spectroscopy to delineate the spatial configuration of nucleic acid structures involves a step-by-step analysis of the spectra. The foremost step is the unambiguous assignment of the various resonance lines which becomes

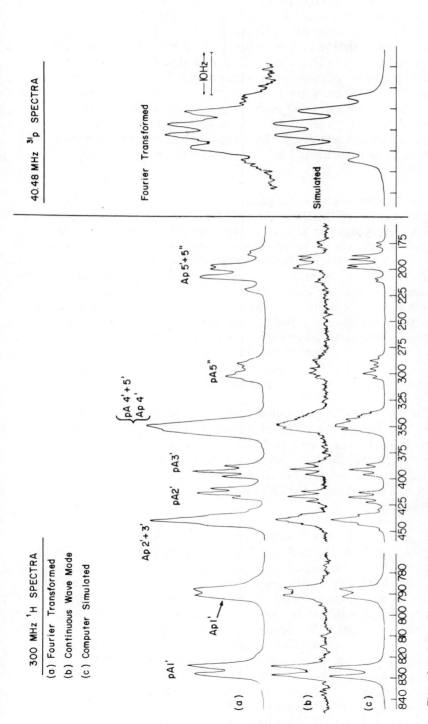

Figure 2. *Left* The diagrams a and b respectively represent the 300 MHz ¹HNMR spectra of ApA (0.05 M, pD 7.5, temp. 27°C) obtained by Fourier transform (FT) methods, (a) and in continuous wave (CW) modes (b). The diagram c is the computer simulated one. Only the ribose region is shown. The FT spectra were superior in S/N ratio but suffered from a poor resolution of 0.7 Hz because only 8K of the total 16K memory was available to transform a band width of 2500 Hz. The CW spectra was superior in resolution (0.2 Hz) but had only fair S/N ratio. By carefully studying both FT and CW spectra peaks were identified and finally the spectra were simulated. The chemical shifts are expressed in Hz, 300 MHz system, upfield from internal tetramethyl ammonium chloride. *Right* The Fourier transformed (top) and computer simulated (bottom) ³¹P NMR spectra (40.48 MHz) of dApdA 0.05 M, pD 7.5 and temp. 27°C. From references 9 and 10.

complex as the number of nucleotidyl units increase. One may succeed to assign the spectral lines which originate due to couplings by appropriate homo and heteronuclear decoupling experiments (*vide infra*). However, the unambiguous assignment of the various single resonances (chemical shifted) observed particularly in oligonucleotides is still a formidable task. This is because one has to determine from which protons the lines originate and to which residue the protons belong. For example in ApA (Figure 1), there are four well separated downfield resonances which arise from a pair of H8 and a pair of H2. The distinction between H8 and H2 is achieved on the basis that H8's are relatively more acidic and hence exchange with deuterium in D_2O at high temperatures.[1/2] They can also be distinguished from spin lattice relaxation time measurements,[3] H2 having a much longer relaxation time than H8.

There have been several approaches to assign the H8 to individual segments. Ts'o, *et al.*[4] achieved the assignment of the H8 to the 5' or 3' nucleotidyl unit by comparing the chemical shifts in ApA and pApA. They showed that the phosphate group at the 5' position will deshield the H8 of -pA in ApA and hence assigned[5] the lower field resonance to H8 of -pA and the higher field one to the H8 of Ap- in ApA. Chan and Nelson[6] assigned the H8 resonances based on Mn^{++} ion binding to the phosphate groups. The paramagnetic ions are known to broaden proton resonances because of the electronic spin nuclear spin dipole dipole interactions. The effectiveness of a paramagnetic center in broadening a nuclear resonance varies as the square of the paramagnetic moment and as the inverse sixth power of the separation between the nucleus and the paramagnetic center. Hence the methodology enables to locate the proton relative to the site of Mn^{++} binding, i.e., the phosphate backbone. Their assignments[6] in ApA were identical to that proposed by Ts'o, *et al.*[4]

The assignment of H2 from individual segments is a difficult task because of its lack of sensitivity to the nature of the ribose phosphate backbone. Chan and Nelson[6] have proposed their assignments based on their sensitivity to the temperature induced destacking of ApA, and consequent reduction in the ring current shielding. These methods of assignments of H8 and H2 to individual segments are by no means unambiguous because they assume special geometrical features for the molecules to arrive at the assignments.

At the level of a dinucleoside monophosphate such as ApA (Figure 1), the assignments of the sugar protons can be achieved unambiguously by a series of homo and heteronuclear decouplings. For example, Ap- H5' and H5'' of ApA occurs at the highest field (Figure 2), decoupling of which will enable to locate Ap- H4'. Then this can be decoupled to locate Ap- H3'. This location can be further independently verified by phosphorus decoupling. Once H3' is assigned, the H2' and H1' of the Ap- residue is assigned by appropriate decouplings. Once the H1' of Ap- is assigned, the H1' of -pA is automatically assigned because the H1's usually appear in a region well separated from other sugar protons. The decoupling of H1' of -pA will assign H2' and which in turn can be used to assign H3', H4', etc. of the -pA unit. The series of decouplings described here have been employed by Cheng and Sarma[7] to arrive at the unambiguous assignments in 2'-0-methyl CpC. No individual unambiguous assignments of H5' and H5'' have been made so far. Generally, the low field

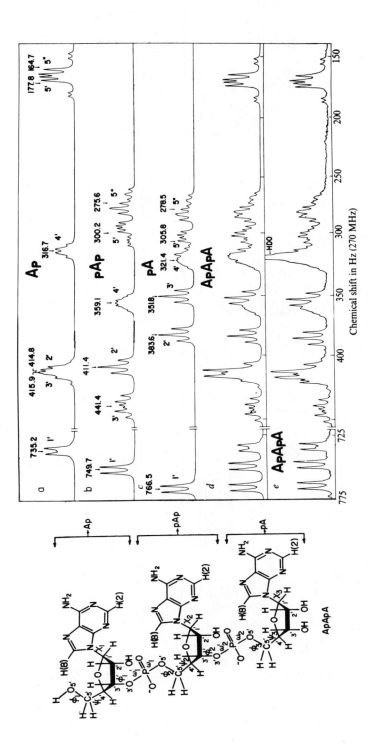

Figure 3. *Left* Structure of ApApA. Note that the old IUPAC-IUB nomenclature is used. Compare this nomenclature with the proposed IUPAC-IUB nomenclature in Figure 1 for easy conversions. *Right* Computer simulation of the Ap- (a), - pAp- (b) and -pA (c) parts of ApApA; (d) Combination of the above three parts into one to produce ApApA simulation; (e) The 270 MHz ¹H experimental NMR spectrum of ApApA at 72°C, pD 7.0, 0.02 M, shifts are up-field from tetramethylammonium chloride. From reference 12 with slight modification.

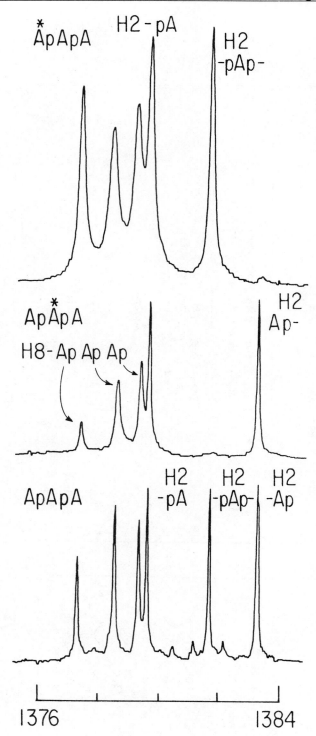

Figure 4. The base proton region of A*pApA, ApA*pA and ApApA, 270 MHz ¹HNMR system, the chemical shifts are upfield from interal tetramethylammonium chloride, 19° C. Unpublished data. (Sarma, Danyluk, et al).

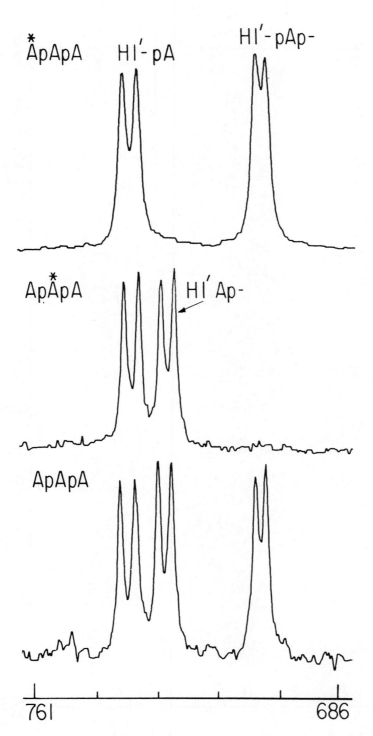

Figure 5. The H1′ region of A*pApA, ApA*pA and ApApA. The rest of the details as in Figure 4. Unpublished data. (Sarma, Danyluk, et al).

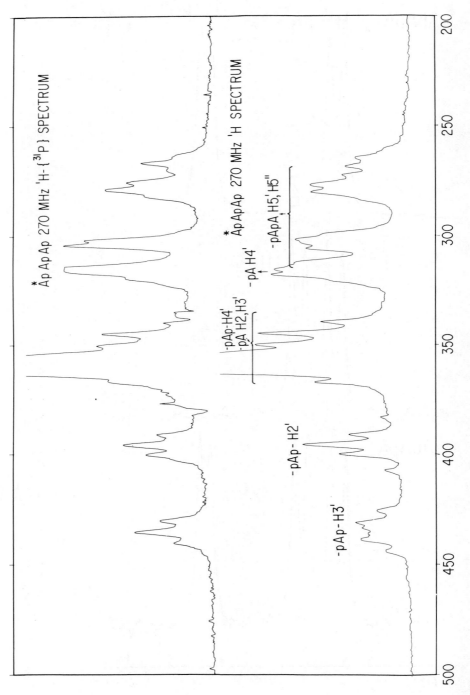

Figure 6. 270 MHz ¹HNMR spectra of A*pApA taken under conditions in which phosphorus-31 was decoupled (top) and coupled (bottom). The shifts are upfield from tetramethylammonium chloride and the employed temperature is 50°C. Comparison of spectra *top* and *bottom* immediately allows recognition of signals from protons coupled to the backbone phosphorus atoms. Unpublished data. (Sarma, Danyluk, et al).

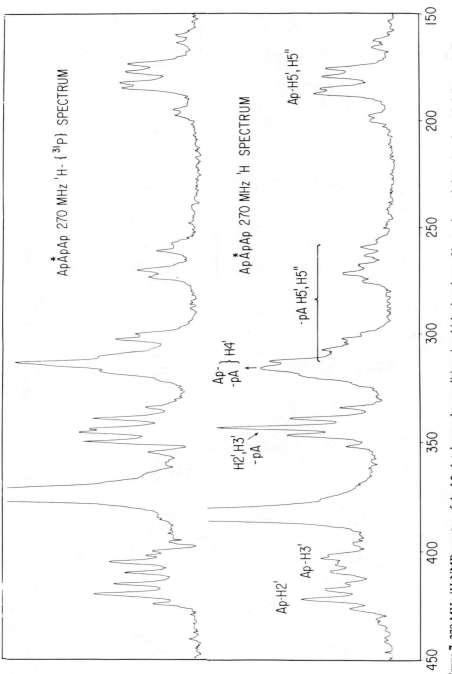

Figure 7. 270 MHz ^{1}H NMR spectra of ApA*pA taken under conditions in which phosphorus-31 was decoupled (top) and coupled (bottom). Temperature, 40°C and compensation for heating due to decoupling was not made and hence HDO peak position is affected. Comparison of spectra in Figures 6 and 7 makes possible the assignment of protons coupled to the backbone phosphorus -31 as well as protons from the common residue - pA, the resonances of which (for example, 2' and 3') are highly temperature sensitive. Unpublished data. (Sarma, Danyluk, et al).

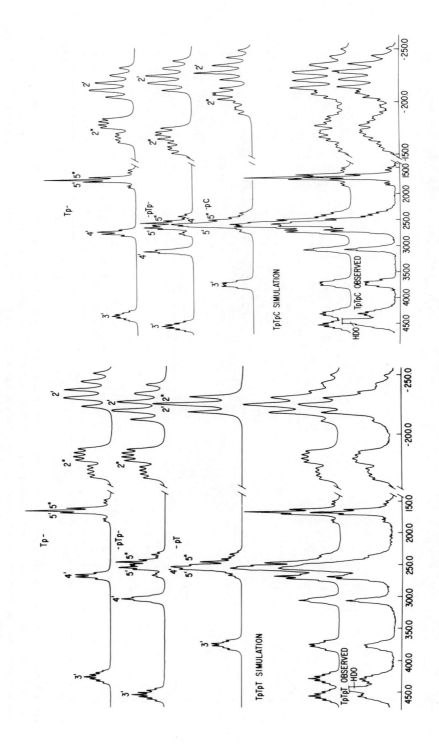

Figure 8. The observed and computer simulated 270 MHz ^{1}H NMR spectra of d-TpTpT (left) and d-TpTpC (right) at 20°C. As is clear the regions from the three nucleotidyl units were separately simulated and then combined to produce the total spectra of the trimers. The shifts are in Hz from tetramethyl ammonium chloride.

hydrogen from the ABX system has been assigned[8] to H5′ and the higher field to H5″. According to this assignment, the H5′ refers to that geminal hydrogen which is *gauche* to H4′ and the ether oxygen of the ribose.

Once the assignments are done, the next step involves the extraction of the accurate NMR parameters, i.e., chemical shifts and coupling constants from the observed spectrum. This is achieved by computer synthesizing the observed spectrum using various programs such as LAOCN III, LAME and BIGNMR. By simulation techniques the chemical shifts and coupling constants can be obtained with an accuracy of ± 0.005 ppm and ± 0.1 Hz respectively. The first complete assignment and simulation of a regular dinucleoside monophosphate was reported from this laboratory.[9/10] One of these original spectra along with the simulation is shown in Figure 2.

From the discussion presented above, it appears that analysis and assignments of NMR spectra of dimer segments of nucleic acids are fairly straightforward. However, it becomes a difficult problem in the case of a trimer. From the inspection of the structure of the trimer ApApA (Figure 3, left), it is apparent that one is facing a formidable task to completely analyze the spectrum of ApApA. Nucleic acid NMR spectroscopists have overcome this problem by synthesizing ApApA in which certain residues are completely deuterated.[11] Thus A*pApA and ApA*pA in which the residue marked * has been fully deuterated have been synthesized.[11] In Figures 4 and 5, we illustrate the base proton and the H1′ region of the 270 ¹H NMR spectra of A*pApA, ApA*pA and ApApA and they immediately provide unambiguous information about the assignments of the base protons and H1′ region in ApApA. In Figures 6 and 7 are illustrated the high field 270 MHz spectra of A*pApA and ApA*pA taken under conditions in which phosphorus-31 was coupled and decoupled. The assignments indicated in the figures are derived from phosphorus decouplings as well as from a series of step-by-step closely interconnected homonuclear decoupling experiments. Once the assignments of the protons of the individual residues at a convenient temperature is achieved, the complete spectra can be computer synthesized by separately synthesizing spectra for the Ap-, -pAp- and -pA residues and thus adding them together. Complex simulation of such magnitude was achieved for the first time in this laboratory for ApApA and was reported[12] in 1976 and this is reproduced in Figure 3 with a small change. This small change involves the exchange of the Ap- and pAp- parts of H1′, H2′ and H3′. This is because in the original report[12] there was a computer mix-up of the H1′, H2′ and H3′ region of Ap- and -pAp- regions; however, this does not affect any interpretations.[12] It may be noted that phosphorus-31 and homonuclear decouplings were carried out at 40°C for ApA*pA and 50°C for A*pApA. This was the lowest temperature at which the resonances began to separate so that decoupling experiments could be carried out meaningfully. Once the assignments are made, by following the shift trends with temperature carefully, complete computer simulation at most temperatures can be achieved. Comparison of the spectra of the H1′ at 19°C (Figure 5) with that of 72°C (Figure 3) indicates that H1′ of Ap- and -pAp- cross over with temperature. Also a comparison of spectra of A*pApA (50°C), ApA*pA (40°C) and ApApA (72°C) clearly shows that elevation of temperature causes significant changes in chemical shifts for several protons; particularly noteworthy are the changes on -pA H3′, H2′ as well as -pApA H5′, H5″ region. The computer simulation provides accurate values for coupling constants and chemical shifts which can then be translated into conformational parameters. In the case of deoxy nucleic acid systems assignments and analysis of spectra even at the level of a trimer can be done without selective deuteration. This is because the complex set of resonances from the 2'2'' region shift upfield from the envelope containing the remaining sugar protons. Here again the assignments are based on careful step-wise decoupling experiments and computer simulation. Details on this has been presented by Cheng, *et al.*[35] In Figure 8 are shown the experimentally recorded and the completely analyzed spectra of the deoxy trimers d-TpTpT and d-TpTpC.

The Vicinal Coupling Constants

The three bond vicinal coupling constants between various nuclei (^{1}H-^{1}H, ^{1}H-^{31}P, ^{1}H-^{13}C, ^{13}C-^{31}P) provide information about the conformational features about the various single bonds in nucleic acid structures. Karplus[13] has related the observed vicinal coupling constants to the dihedral angle ϕ between vicinal nuclei. The analytical expression for ^{1}H-^{1}H vicinal coupling is given by Equation 1:

$$J_{HH} = A \cos^2\phi + B \tag{1}$$

where A and B are constants and A has different magnitude for the ranges $0° < \phi < 90°$ and $90° < \phi < 180°$. Similar equations for other vicinal couplings J_{HCOP}, J_{CCOP}, J_{HCCC} has been derived (*vide infra*). One uses all these expressions to derive geometry information about nucleic acid structures.

Conformation of the Sugar Rings

It is well known that the ribose and deoxyribose rings exhibit a non-planar puckered conformation which can be described by ring torsion angles, by least square planes[14] and by pseudo rotation parameters.[15] Single crystal data[16/17] on several nucleic acid components show that the furanose ring exists in a majority of cases in a conformation in which either C2' or C3' atom is displaced toward C5' from the plane of the other ring atoms of the sugar moiety. The conformation in which C2' is displaced is

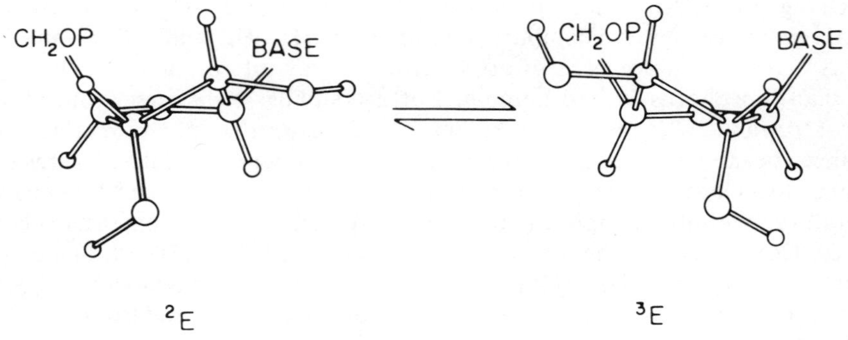

Figure 9. The two major conformations of the sugar ring.

designated as ^2E ($C2'$-endo) and the one in which $C3'$ is displaced as ^3E ($C3'$-endo). These conformations are shown in Figure 9. In addition a few other twist and envelope conformations[16/17] have been reported for pentose rings in nucleic acids.

These solid state findings are helpful to determine the solution stereodynamics of pentose rings in nucleic acids. The dihedral angles between the C-H bonds on adjacent carbons in a furanose ring are determined by the mode and the extent of the ring puckering and thus a knowledge of the vicinal H-H coupling constants will permit determination of its stereochemistry,[18/19] i.e., the observed coupling constants can be related via Karplus equation to apparent values of $\phi_{1'2'}$ and $\phi_{3'4'}$ in the relevant H-C-C-H fragments. However, there are difficulties in using Karplus equation to distinguish between the small differences in dihedral angles that exist among the various possible furanose ring conformations. This has been discussed elsewhere[20] in detail.

It has been proposed[21] that the best way to handle the pentose coupling constants is to treat them as arising from an equilibrium blend of ^2E and ^3E conformations. The concentration dependence of the coupling constants $J_{1'2'}$, $J_{2'3'}$, and $J_{3'4'}$ in 5'AMP indicates that $J_{1'2'} + J_{3'4'}$ essentially remains constant.[21] This pattern of variation in the endocyclic proton-proton coupling constants is consistent with the existence of ^2E and ^3E conformations in equilibrium, the relative population of these states being determined by the magnitude of $J_{1'2'}$ and $J_{3'4'}$. To illustrate this point in Table I are listed the projected coupling constants for various envelope conformations. These are calculated using the modified Karplus equation[15] $J_{HH} = 10.5 \cos^2\phi_{HH} - 1.2 \cos\phi_{HH}$. The data show that for ^2E and ^3E the sum $J_{1'2'} + J_{3'4'}$ has a value of 10.3 to 10.4 Hz and $J_{2'3'}$ varies from 5.5 to 4.9. For a pure ^2E state in solution the value of $J_{1'2'}$ would be about 10.1 Hz for a pure ^3E state $J_{3'4'}$ would be about 10.3 Hz. Note that $J_{3'4'}$ for ^2E and $J_{1'2'}$ for ^3E are practically zero. Any deviation from these pattern of coupling constants suggest deviation from the ideal $^2E \rightleftharpoons {}^3E$ equilibrium such as changes in the phase angle of pseudorotation as well as in the amplitude of pucker.[15] The Karplus approach cannot be used to determine precisely the conformational features of the pentose ring,[21] and we believe that the treatment of the endocyclic coupling constants on the basis of a $^2E \rightleftharpoons {}^3E$ or a $^2T_3 \rightleftharpoons {}^3T_2$ equilibrium (where T = twist chair) gives reasonably satisfactory results. Theoretical calculations[22/23] do also support such an approach. Operationally ratio of $J_{1'2'}$ to the sum $J_{1'2'} + J_{3'4'}$ gives the fractional population of ^2E conformers; 100 – percent ^2E = percent ^3E.

Table I

Calculated Coupling Constants in Hz
in a Pentose Ring for Various Envelope Conformations

	$J_{1'2'}$	$J_{1''2'}$	$J_{2'3'}$	$J_{2''3'}$	$J_{3'4'}$
C2'-exo	0.2	5.0	4.9	11.0	8.2
C3'-endo	0.1	7.4	4.9	11.0	10.3
C4'-exo	2.4	9.1	7.0	9.0	10.3
C1'-exo	10.1	5.5	7.6	0.1	2.9
C2'-endo	10.1	5.5	5.5	0.1	0.2
C3'-exo	7.9	7.6	5.5	0.1	0.0

Conformation About the C4'-C5' (ε) Bond

The symbol ε defines the torsion about C4'-C5' and ε is zero when C4'-C3' is cis planar to the C5'-O5' bond, i.e., the fully eclipsed conformation. The minimum energy rotamers about C4'-C5' are shown in Figure 10. They are the *gauche gauche* (*gg* ε = 60°, g⁺) *gauche trans* (*gt* ε = 180°, t) and *trans gauche* (*tg*, ε = 300°, g⁻) conformers. In aqueous solution there is rapid interconversion among these conformers and the relative population is influenced by a large number of factors including the nature of the base and the value of χ, furanose ring puckering and solvent properties.[24-26] Usually, distribution of conformers about the C4'-C5' bond is calculated from the experimental sum $J_{4'5'}$ + $J_{4'5''}$ and this allows the determination

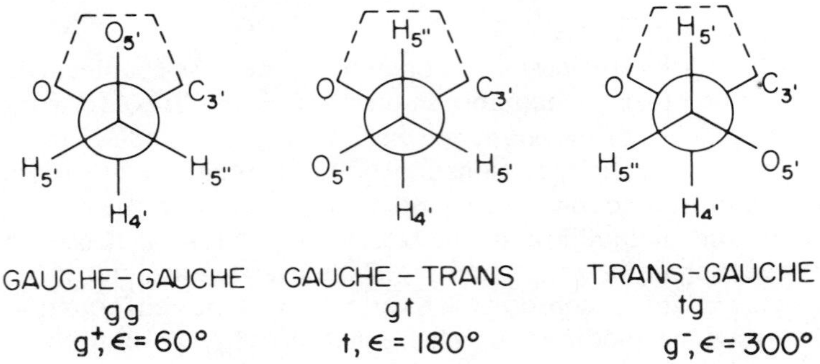

GAUCHE - GAUCHE   GAUCHE - TRANS   TRANS - GAUCHE
        gg   gt   tg
   g⁺, ε = 60°   t, ε = 180°   g⁻, ε = 300°

Figure 10. Classical staggered rotamers about the C4'-C5' bond.

of the population of *gg* conformers and the combined population of *gt* and *tg*. This approach is used because the evaluation of the individual contributions of *gt* and *tg* requires a knowledge of the absolute assignments of H5' and H5'', and this is not known. The use of the experimental sum is adequate for monitoring perturbations in the time averaged conformation about C4'-C5'. Any perturbation resulting in the increase of the *gg* rotamer populations will result in the decrease in the magnitude of $J_{4'5'}$ + $J_{4'5''}$ where as an increase in *gt* and *tg* populations at the expense of *gg* rotamers will result in an increase in the observed sum.

The empirical equations developed earlier[27-29] for the calculation of rotamer distributions about C4'-C5' have been modified by Lee and Sarma[30] to reflect the more reasonable values of J_t = 11.7 Hz and J_g = 2.0 Hz. The modified expression is:

$$\text{percent } gg = (13.7 - \Sigma) \, 100/9.7 \qquad (2)$$

where $\Sigma = J_{4'5'} + J_{4'5''}$. The percentage (*gt* + *tg*) = 100 – percent *gg*.

This modified expression has been recently used to evaluate the population distribution of conformers about the C4'-C5' bond in 3'5'ribodinucleoside monophosphates,[31/32] 3'5' deoxyribodinucleoside monophosphates,[33] 2'5' ribodimers[34] and trinucleoside diphosphates.[35]

Conformation About The C5′-O5′ (δ) Bond

The torsion about C5′-O5′ bond is denoted by δ and the minimum energy conformers are shown in Figure 11. Population distribution of conformers about this

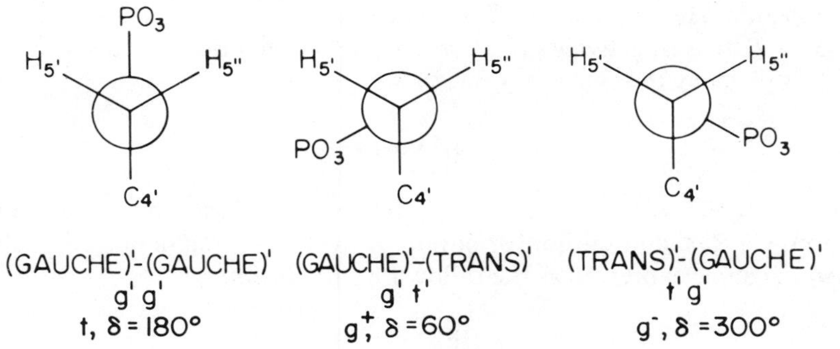

(GAUCHE)′-(GAUCHE)′   (GAUCHE)′–(TRANS)′   (TRANS)′-(GAUCHE)′
g′ g′   g′ t′   t′ g′
t, δ = 180°   g$^+$, δ = 60°   g⁻, δ = 300°

Figure 11. Staggered rotamers about the C5′-O5′.

bond can be obtained from the magnitude of the vicinal coupling $J_{5'P}$ and $J_{5''P}$ and is given by the expression

$$\text{percent } g'g' = (25-\Sigma') \, 100/20.8 \tag{3}$$

where Σ' is the sum $J_{5'P} + J_{5''P}$. It may be noted that when C4′-C5′ and C5′-O5′ rotamers exist in the *gauche gauche* arrangement the H4′ and P in a 5′-mononucleotide lie in an inplane 'w' path and this will result in a long-range four bond coupling (*vide infra*).

Conformation About the C3′-O3′ (α) Bond

Theoretically, there are three staggered conformations possible about C3′-O3′ (Figure 12), and the three bond HCOP coupling between H3′ and the phosphorus

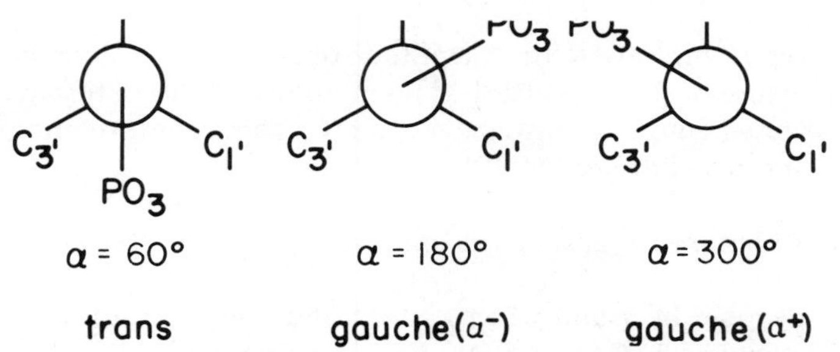

α = 60°   α = 180°   α = 300°

trans   gauche (a⁻)   gauche (a⁺)

Figure 12. Staggered rotamers about the C3′-O3′ bond.

should yield information about C3'-O3' torsion. However, analysis of rotation about C3'-O3' in terms of the three staggered forms leads to conflicting results.

The problem is rendered simple if one assumes that the *trans ($\alpha = 60°$)* conformers does not contribute to the conformational blend. The experimental and theoretical studies which support this assumption are detailed elsewhere.[7] We have shown[7] that simple interpretation of the coupling constants in terms of a $\alpha^+ \rightleftharpoons \alpha^-$ equilibrium is a reasonable one. Here (+) and (-) denotes conformations in the range of 270° – 285° and 195° – 210°, respectively. The correlation between the vicinal HP coupling and the ϕPH is given by the expression

$$J_{HP} = 18.1 \cos^2\phi - 4.8 \cos\phi \tag{4}$$

The value of ϕPH = 0 when $\alpha = 240°$. A close correlation exists between the torsion about C3'-O3' and the conformation of the ribose ring in nucleic acids and this correlation can be expressed as the following equilibrium

$$^2E\alpha^+ \rightleftharpoons {}^3E\alpha^- \tag{Figure 13}$$

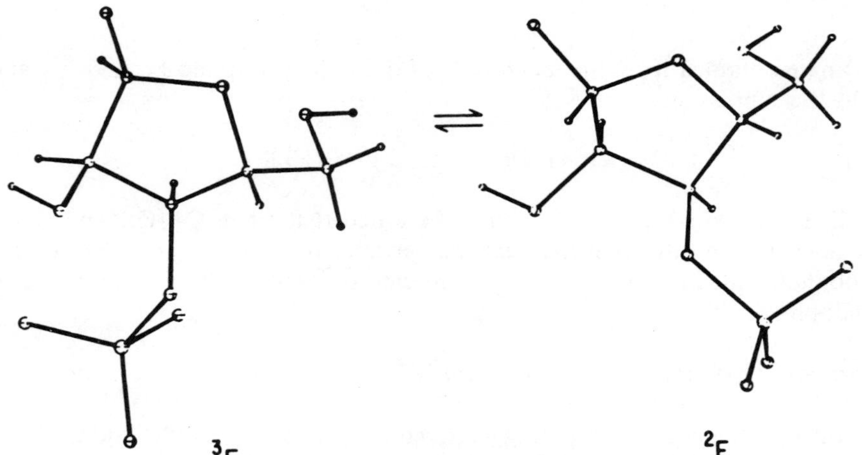

Figure 13. Conformational interrelationships between sugar pucker and C3'-O3' torsion.

Destacking causes an increase in the magnitude of χ in the anti domain (*vide infra*) and this in turn causes a shift toward $^2E\alpha^+$ population and this in turn manifests in a long range four bond coupling between H2' and P. These conformational interrelationships are described elsewhere.[31/32/36]

Conformation About the Glycosidic Linkage (χ)

This torsion is denoted by χ and when $\chi = 0°$, it corresponds to the eclipsed conformation of bonds O1'-C1' and N9-C8 (purines) and O1'-C1' and N1-C6 (pyrimidines). Clockwise rotation of N9-C8/N1-C6 bonds relative to O1'-C1' when looking along the C1'-N9/C1'-N6 results in positive values. The domain of $\chi \simeq 0 \pm 90°$ is *anti* and $\chi \simeq 180° \pm 90°$ is *syn* (Figure 14).

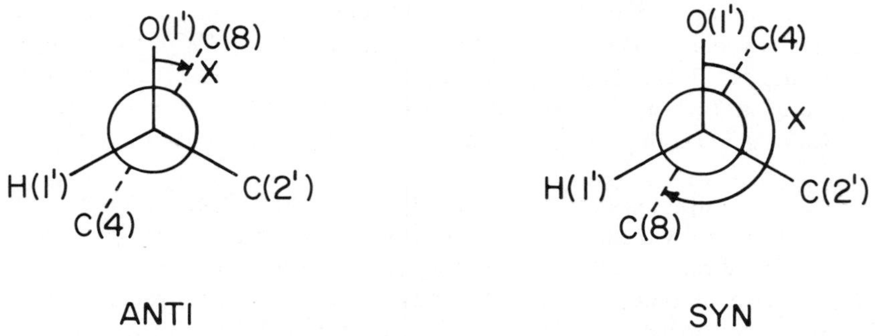

Figure 14. Definition of sugar-base torsion (see also Figure 18).

Employment of vicinal coupling constants to determine χ has been very difficult. This is because there is no observable proton-proton vicinal coupling that will provide information about χ. NMR methods based on chemical shifts, pH and Mn^{++} ion effects, NOE and T_2 measurements have provided qualitative information about the magnitude of χ in nucleic acid structures.[6-37-40]

Theoretically, the glycosidic torsion should manifest in the vicinal coupling constant $J_{H1'-C1'-N9-C8}$ and $J_{H1'-C1'-N9-C4}$ for purines and $J_{H1'-C1'-N1-C6}$ and $J_{H1'-C1'-N1-C2}$ for pyrimidines. Determination of these coupling constants have been very difficult because of the poor abundance of ^{13}C nuclei. The natural abundance proton coupled ^{13}C Fourier transform NMR spectra of suitably concentrated solutions has made possible the determination of such couplings.

To translate the observed vicinal coupling constants into conformational parameters one requires a Karplus type equation correlating J_{HCNC} with the dihedral angle. Lemieux and coworkers[41-43] from a study of ^{13}C enriched pyrimidine derivatives showed that the vicinal proton-carbon coupling between H1' and C2 exhibited a Karplus type dependence in which the magnitude of J varied from 0 to 8 Hz and is governed by the equation:

$$J_{HC} = 6.7 \cos^2\phi + 1.3 \cos\phi \qquad (5)$$

Application of this equation to the $J_{C4H1'}$ and $J_{C8H1'}$ in cyclic purine derivatives[44/45] gave unsatisfactory fitting. This may be due to the strain in the cyclo systems and it is possible that the above relation is applicable only for the pyrimidines.

The Four Bond Coupling Constants

With the availability of high frequency FT NMR spectrometers which provide well resolved and dispersed proton spectra, a number of long range coupling constants have been measured and they can provide information about the spatial arrangement in many a nucleic acid systems. Long range coupling constants in both saturated and unsaturated systems involving hetero atoms have been well reviewed.[46-48]

Four bond couplings like three bond vicinal couplings are sensitive to stereochemistry. When the coupling path is a planar zig-zag 'W' observably large couplings results.[46/47] Four bond J_{HCCCH} couplings have been observed for the sugar ring protons of a few nucleic acid derivatives,[49-51] but they are of such small magnitude (0.3 - 0.7 Hz) that they contain little geometry information.

Four bond J_{HCCOP} couplings of magnitude 2.4 - 2.7 Hz have been observed between the ^{31}P and H4′ in many a nucleic acid systems. Their presence were discovered in this laboratory[52] by recording spectra in which phosphorus-31 was decoupled. Inspection of spectra in Figures 15a and c makes it unmistakably clear that H4′ is coupled to the phorphorus in 5′-AMP and the accurate value of the coupling can be obtained from computer simulation (Figure 15b and d). Significantly large coupling between H4′ and phosphorus atoms in nucleic acids results when the C4′-C5′ and C5′-O5′ are predominantly oriented in gg and g′g′ conformation creating a zig-zag 'W' inplane path between H4′ an P (Figure 16). Hall and coworkers[53-55] have observed similar four bond coupling in many non-nucleic acid systems. Sarma, et al.[56] have observed linear relations between J_{HCCOP} and the gg and g′g′ populations about the C4′-C5′ and C5′-O5′ bonds. Thus the four bond coupling J_{HCCOP} has been shown to be a simultaneous measure of the orientation about the C4′-C5′ and C5′-O5′ bonds.

Four bond couplings between H2′ and the phosphorus atom have also been observed.[32/36] Inspection of Figure 13 indicates that the geometric relationship between H2′ and ^{31}P is an inplane 'W' in $^2E\alpha^+$ conformation. As is seen in Figure 17, elevation of temperature causes the appearance of fine structure in the H2′ region of Ap- of ApU clearly indicating $J_{H2'-P}$ couplings. This is due to a shift in the $^3E\alpha^- \rightleftharpoons {}^2E\alpha^+$ equilibrium toward the right with increasing temperature. Elevation of temperature causes unstacking of the bases, which results in an increase in the magnitude of χ_1 and a coupled shift of $^3E\alpha^-$ to $^2E\alpha +$ conformation.

Four bond couplings J_{HCCCH} between sugar H1′ and the base protons have been observed in a number of C-nucleosides, i.e., α-pseudouridine,[57] β-pseudouridine

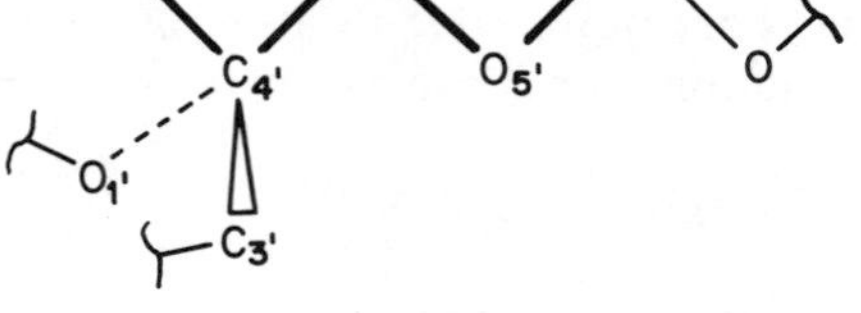

Figure 16. The inplane zig-zag 'W' path between H4′ and P when C4′-C5′ and C5′-O5′ are in gg and g′g′ conformations (see Figures 10, 11 and 18).

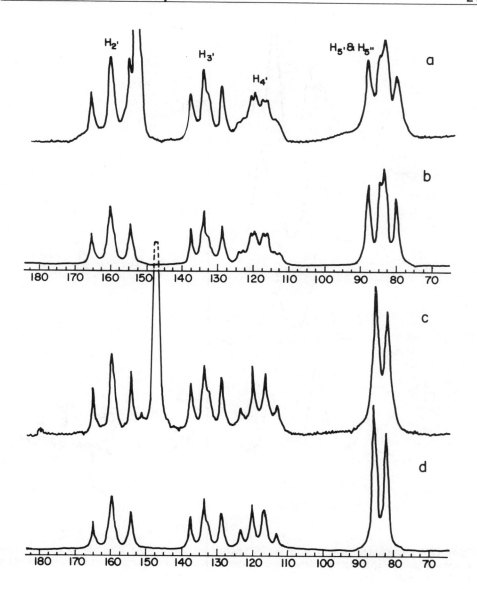

Figure 15. 100 MHz ¹H NMR spectra of 5′AMP under conditions in which phosphorus-31 was coupled (a) and decoupled (c). Compare the H4′ and H5′, H5″ regions in both spectra. The corresponding simulations are in b and d.

and its 3′-monophosphate,[58] showdomycin[59/60] and oxazinomycine.[61] These couplings will be useful in determining glycosidic torsion once a relationship between χ and the magnitude of these couplings is established.

The Five Bond Coupling Constants

Five bond couplings of the type J_{HCCNCH} have been used to investigate glycosidic torsion in certain pyrimidine and triazole nucleosides.[62/63] Examination of the observed

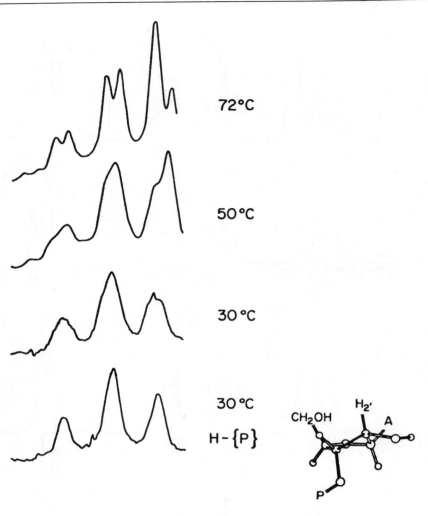

Figure 17. Temperature dependence of the H2′ signals of Ap- of ApU. The bottom spectrm is the phosphorus decoupled spectrum. Elevation of temperature causes the emergence of fine structure due to H2′-P four bond coupling.

magnitude of five bond couplings in various systems[48] shows a maximum value of 1.5 Hz. These couplings are examples of m-benzylic interactions[64/65] in which the maximum couplings are favored by an extended zig-zag planar path. Evans and Sarma[66] have observed five bond coupling of about 0.5 Hz between H1′ and H7 in tubercidin-5′-phosphate which indicates detectable populations of conformers in which χ is in the anti domain. (Figure 18)

The Chemical Shifts

The computer simulated spectra provide the chemical shifts with a high degree of accuracy. Among the various factors which influence the position of a proton signal, the important ones are (a) the electron density at various carbon atoms bearing the

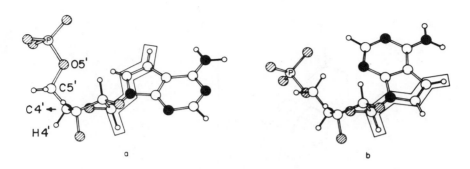

Figure 18. Perspective drawing of the *anti* (a) and *syn* (b) orientations about the glycosyl bond. In the *anti* conformation the bond system between H1′ and H7 has an in-plane zig-zag geometry (boxed regions). In the *syn* conformation, the zig-zag pattern is destroyed. and the C4′-C5′ is *gt*. In the *anti* conformation note the in-plane 'W' path between phosphorus and H4′ which is absent in the *syn* conformation.

proton, (b) the diamagnetic and paramagnetic susceptibility anisotropy of the hetero atoms, (c) the ring current effects associated with mobile π electron cloud, (d) the electric field polarization effects, (e) solute-solute and solute-solvent interactions, (f) hydrogen bonding effects, etc. The theory of chemical shifts and the contribution from these interaction have been discussed in several books and monographs, for example, references 67 to 70. In order to arrive at the spatial configuration of molecular systems by NMR methods one indeed exploits the effect of these factors on molecular geometry.

Thus information about glycosidic torsion in nucleic acid structures have been derived from the changes in base proton chemical shifts due to the selective field effect of a charged phosphate group,[38/71] effect of substituents,[72-74] and effect of metal cations.[75] The anisotropic effect of the hetero atom and the ring current field effects of the heterocycle on the sugar proton chemical shifts have also been exploited to arrive at information about sugar-base torsion.[24/25/75-79]

At the level of oligonucleotides th application of the above approaches are complicated, and in order to simplify matters chemical shifts are treated in a relative manner, i.e., what are the changes from monomer to dimer, from dimer to trimer, etc. And further shift trends are used to obtain information which cannot usually be obtained from coupling constants. For example, for a dimer, torsional preferences about α, δ, ϵ and about the sugar pucker can be obtained from coupling constants, but information about χ_1, χ_2, β and γ do not follow from coupling measurements because of the lack of desirable nuclei which can couple and produce a discernable coupling constant.

Cheng and Sarma[33] have developed a method based on dimerization shift data (i.e., difference in chemical shifts between monomers and dimers) to determine χ_1, χ_2, β and γ based on ring current theory. Recently, the method was improved by Sarma, *et al.*[80] to take into consideration the contribution to shielding/deshielding from diamagnetic and paramagnetic anisotropy contributions. The method essentially involves a conformational search in the normally expected domains of χ_1, χ_2, β and γ

$g^- g^+$

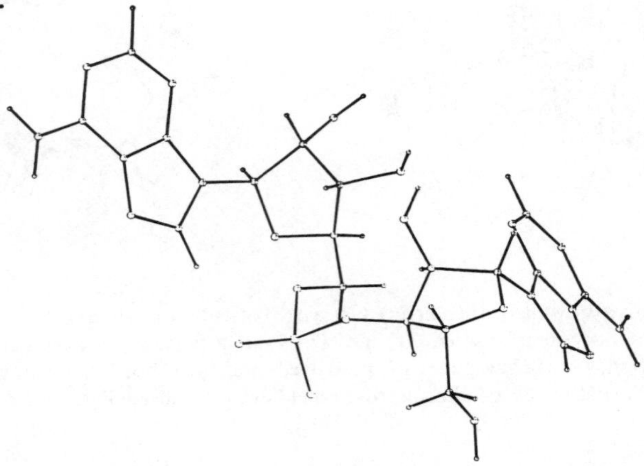

$g^+ g^-$

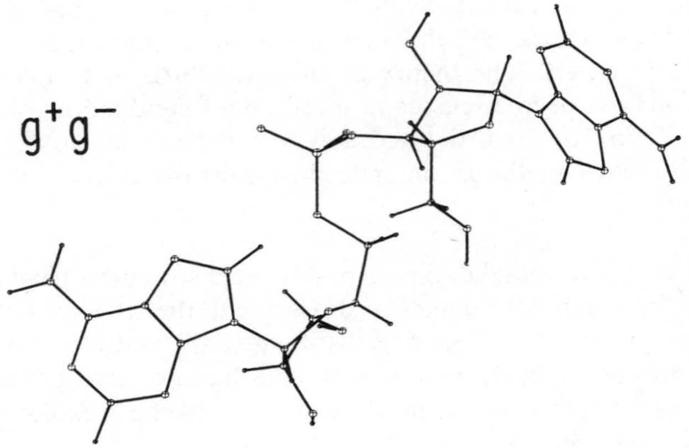

$g^- g^-$

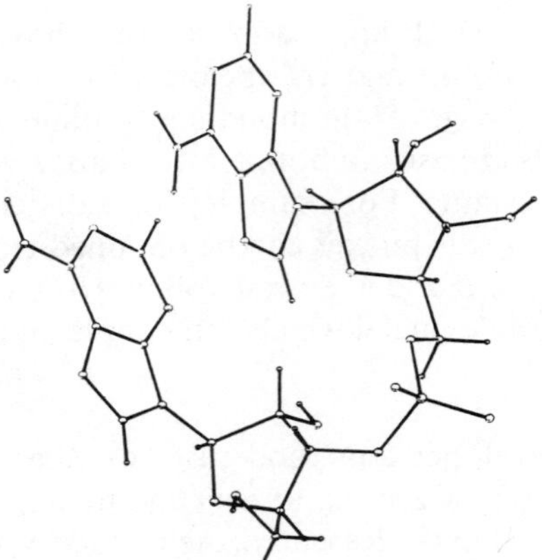

t t

t g⁻

g⁻ t

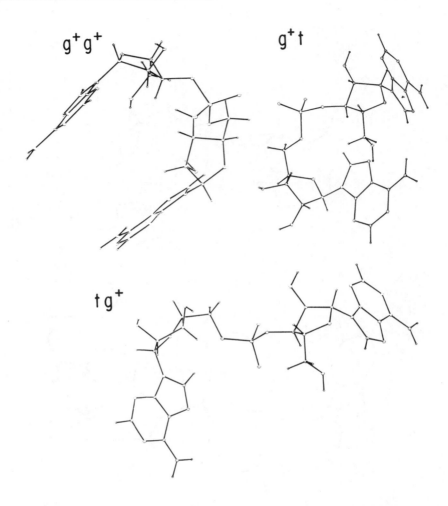

Figure 19. The nine possible conformations about the β, γ phosphodiester bonds. Note that in the g^-g^- and g^+g^+ arrays base stacking is possible. For all diagrams sugar pucker is ^3E, $\chi_1 = \chi_2 = 0°$, $\alpha = 206°$, $\delta = 180°$, ϵ and $\epsilon' = 60°$.

to duplicate the experimentally observed dimerization data. The coupling constant data already provide information about the preferred torsional isomers about C3'-O3', O5'-C5', and C5'-C4' and about the sugar ring conformations. It is generally known that χ_1 and χ_2 are in the anti domain, and that the phosphodiester torsions (β,γ) may display any of the possible nine conformations (Figure 19). Inspection of these projections clearly reveals that in the g^-g^- conformation there is substantial base base stacking interactions. In fact, the dimerization data show that the base protons and a few protons of the sugar ring in, for example ApA, are shifted upfield compared to the monomers. This suggests the possible presence of g^-g^- arrays. It is important to emphasize that only in the g^-g^- stacked array the *base overlap* and *base separation* permits the base protons to experience ring current upfield shifts, i.e., the observed upfield shifts of the base protons essentially reflect the presence of g^-g^- stacks. In the g^+g^+ loop stack, the base separation is too large to result in any ring-current shifts for the base protons. So the x, y, z coordinates for the hydrogens are generated for conformational arrays using experimentally observed torsion angles α, δ, ϵ, ϵ' and sugar pucker and varying β, γ, χ_1, χ_2 torsion angles initially at intervals of 20° and finally at 5° in the g^-g^-, and *anti* domains. From these coordinates contributions to shielding/deshielding from ring current fields, diamagnetic and paramagnetic anisotropy were determined using principles discussed in references 80 and 81 and references therein. The method essentially is an iterative procedure where the iterations are carried out on β, γ, χ_1 and χ_2 till a good fit is obtained between the observed dimerization shifts and those calculated for protons from NMR theory for a specific conformation. In Figure 20, the preferred spatial configuration of ApA so derived is shown. In Table II are listed the final torsion angles and the x, y, and z of the protons. Table II also contains the theoretically projected and experimentally observed dimerization data for the -pA segment of ApA. Except for H5′ and H3′, there is reasonable agreement between the calculated and observed values. The large deshielding of the H5′ of -pA originate from the juxtaposition between this H5′ and 2′OH of Ap- in the g^-g^- conformation (Figure 19, g^-g^-) and provide further evidence for the presence of g^-g^- arrays as has been discussed in extenso elsewhere.[31/32] The H3′ is affected because of the shift toward ^3E in the $^2E \rightleftharpoons {}^3E$ equilibrium upon dimerization.

Even though an important component of the procedure elaborated above is the effect of intramolecular base stacking on chemical shifts, the same approach can be employed to determine whether base stacked and non-base stacked arrays contribute significantly to the conformational blend, and even to determine the extent of

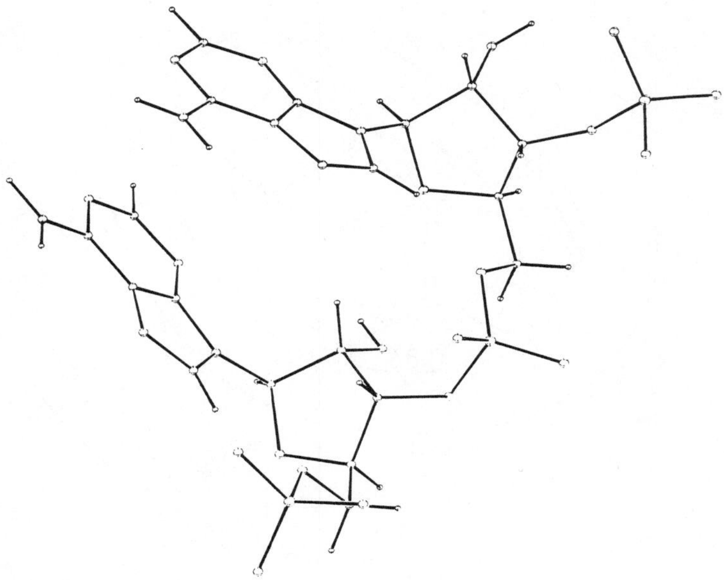

Figure 20. The preferred spatial configuration of ApA determined by the procedure outlined in the text.

Table II

Conformational Angles for the Derived Geometry

of ApA are $\alpha = 206°$, $\beta = 285°$, $\gamma = 290°$, $\delta = 180°$, $\epsilon = 60°$, $\chi_1 = 25°$, $\chi_2 = 50°$, ^3E, ^3E.

Calculated Shieldings for the Effect of Ap- Residue on -pA Residue Protons.

Protons of -pA residue	x	y	z	Ring Current	Diamag Aniso	Paramag Aniso	δ Total	δ obs
H - 1'	-0.077	3.065	5.470	0.127	0.000	0.037	0.164	0.180
H - 2'	-1.730	4.921	4.633	0.118	-0.001	0.030	0.147	0.170
H - 3'	-2.797	3.709	2.991	0.093	-0.001	0.027	0.119	0.000
H - 4'	-2.735	1.323	4.690	0.018	-0.001	-0.003	0.014	0.015
H - 5'	-2.404	0.224	2.643	-0.013	-0.001	-0.012	-0.026	-0.245
H - 5''	-3.701	1.202	2.480	0.018	-0.001	0.005	0.024	-0.051
H - 2	3.917	5.287	5.789	0.090	0.002	0.055	0.147	0.201
H - 8	-0.797	3.701	1.855	0.262	-0.004	0.070	0.328	0.250

oscillations about the various bonds. A case in point is provided by the dinucleoside monophosphates in which one or both of the bases are replaced by the modified Y base ϵA (Figure 21). The dimerization data for $\epsilon Ap\epsilon A$ indicated substantial

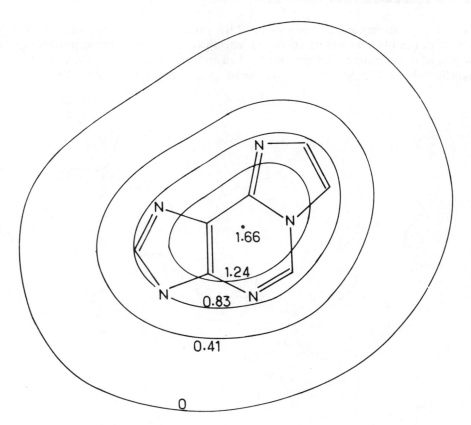

Figure 21. The structure of the modified Y base ϵA and its shielding values due to the sum of the ring current effect and of the atomic diamagnetic susceptibility anisotropy in a plane 3.4 Å distant from the molecular surface. From reference 82.

deshielding[82] of H5′ of -pϵA suggetisting significant populations of g^-g^- arrays. This was further reflected in the upfield shifts of the base protons in ϵApϵA. A search in the g^-g^- *anti* conformations space using the methodology discussed above indicated the best fit happens when $\beta = 285°$, $\gamma = 295°$, $\chi_1 = 25°$, $\chi_2 = 40°$ and the remaining torsion angles: $\alpha = 205°$, $\delta = 180°$, $\epsilon = 60°$, $\epsilon' = 60°$ and sugar puckers ^3E, ^3E. A perspective of this conformation is shown in Figure 22. The data for this con-

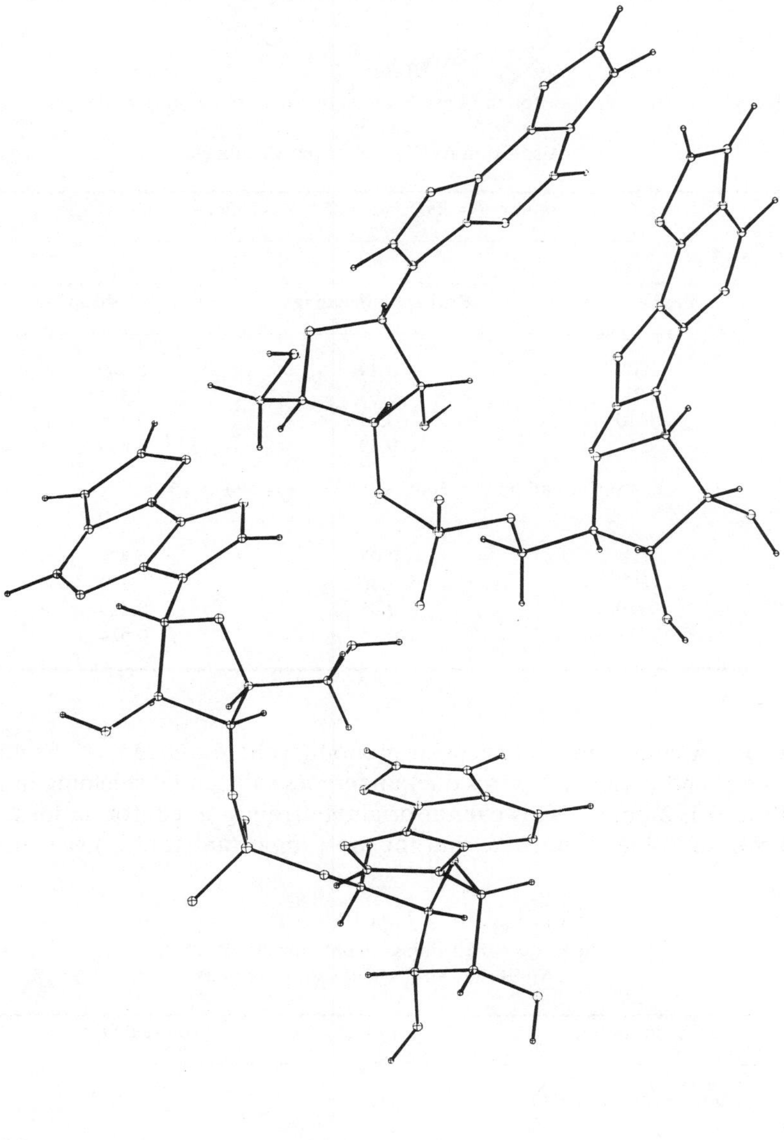

Figure 22. *Right* A perspective of ϵApϵA in the g^-g^- conformation. The sugar pucker is ^3E and the torsion angles are: $\alpha = 205°$, $\beta = 285°$, $\gamma = 295°$, $\delta = 180°$, $\epsilon = 60°$ $\chi_1 = 25°$ and $\chi_2 = 40°$. *Left* A perspective of ϵApϵA in the g^+t conformation. Except for $\beta = 110°$, $\gamma = 215°$, $\chi_1 = 20°$, and $\chi_2 = 50°$, rest of the geometrical details same as the g^-g^- situation.

formation in Table III indicate that there is an agreement between the projected and observed shifts, except for H8. In order to account for the shift of H8 an intensive search in the g^-g^- conformation space was made and it was found that when β was changed from 285° to 300°, γ from 295° to 290° and χ_2 from 40° to 45°, the projection for H8 agreed with the observed value and under these slightly different torsion angles, the remaining base protons experience very little shielding (Table III). Thus the data clearly demonstrate that ϵApϵA populate significantly in the g^-g^- conformation space and that the O3'-P torsion (β) has a local flexible domain (285°-300°) in the g^-g^- array.

Table III

The Projected Shieldings in ppm for the Base Protons of the ϵAp- Residue in the g^-g^- Conformation Space.
Also Given Are The Observed Shieldings

Conformation: $\beta = 285°$, $\chi_1 = 25°$, $\gamma = 295°$, $\chi_2 = 40°$		
Base Protons	Projected Shieldings	Observed Shieldings
H2	0.39	0.409
H8	0.06	0.324
H10	0.45	0.449
H11	0.49	0.534
Conformation: $\beta = 300°$, $\chi_1 = 25°$, $\gamma = 290°$, $\chi_2 = 45°$		
H2	0.08	0.409
H8	0.30	0.324
H10	0.05	0.449
H11	0.15	0.534

In many an ϵA containing dinucleoside monophosphates the H5' of the nucleotidyl unit at the 5' end (i.e., the Np- residue) undergoes substantial shielding in the range of 0.440 to 0.192 ppm.[82] This cannot originate from g^-g^- conformations. Calculations shown in Table IV indicate that the most important conformer which can in-

Table IV

The Projected Shieldings in ppm for the H5' of the
ϵAp- Residue in Various Conformations

Conformation	Proton	Projected Shielding
$g^+g^+(\beta = 80°, \gamma = 80°)$	H5'	0.12
$g^+t\ (\beta = 110°, \gamma = 215°)$	H5'	0.56
$tg^+\ (\beta = 180°, \gamma = 80°)$	H5'	0.03

fluence the chemical shift of ϵAp- H5′ is g⁺t (Figure 22) and hence dinucleoside monophosphates which contain ϵA will have a significant proportion of g^+t conformers. Because of the large number of variables and observables our procedure may not give a unique solution for β, γ, χ_1 and χ_2 but definitely will predict the correct domains for these torsion angles. The use of chemical shifts to delineate the molecular details of bulged configurations and miniature double helices have been described elsewhere.[35/80]

Recent Developments

During this decade, in addition to the development of high frequency Fourier transform NMR spectrometers, various spectral analysis aids and structural and conformational probing aids have been developed. These techniques have been developed from different angles but the final goal is to derive information about the three dimensional geometry of molecular systems. A brief sketch of these are presented below.

Heteronuclear Two-Dimensional NMR Spectroscopy

The most simple and familiar example of two dimensional NMR spectrum is the stacking in a two dimensional manner a set of spectra as a function of time as is usually done in the spin-lattice relaxation time measurements.[83] The two dimensional perspective is due to time-space displacements. A conventional NMR spectrum is a plot of frequency *versus* the intensity of the signals. However, when all the variables of the plotted function are frequencies, then a two dimensional spectrum results. For example, for the correlation of proton and phosphorus chemical shifts, the experiment consists of two rf pulses applied at the proton frequency, spearated by an evolution period t_1 and followed by a normal phosphorus observation pulse with subsequent acquisition of the free induction decay. The signal is then subjected to a Fourier transformation and this yields the phorphorus spectrum. The pulse sequence and the first Fourier transformation are repeated for some 300 regular increments of t_1. The result is a stacked plot of phosphorus frequency (F_2) versus t_1 where one sees the modulation of the phosphorus signals. The modulation is then analyzed by a second Fourier transformation which substitutes the proton frequency (F_1) for the time variable, and this results in the two dimensional NMR spectrum. A two dimensional NMR spectrum of 2′GMP so obtained[84] is illustrated in Figure 23. Note that in the proton dimension, only the H2′ which is coupled to the phosphorus gives rise to signals, thus enabling the assignments straight forward. For details of the two dimensional heteronuclear spectroscopy, consult references 84-90.

J Resolved Two Dimensional NMR Spectroscopy

This version of the two dimensional spectroscopy extends NMR spectral-information into a new frequency dimension where the scalar spin-spin coupling is used as the spreading parameter;[87-90] along one axis chemical shifts in frequency units and along the other axis the scalar coupling constants. This kind of resolution helps in studying the proton coupled spectra of large molecules that otherwise would have been complex by overlaps.[91] One can also improve resolution in

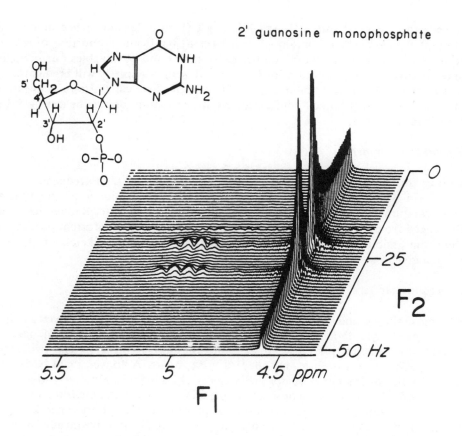

Figure 23. The heteronuclear two dimensional spectrum of 2'-GMP. The F_1 dimension reflects the chemical shifts of protons and F_2 that of phosphorus. We thank Dr. P. H. Bolton for this Figure.

macromolecular spectra by completely eliminating proton couplings and thus present a single resonance for each proton.[91/92] It relies on the fact that all multiplets lie on a 45° line with respect to chemical shift and coupling constant axes. Projection along this axis to zero J eliminates all couplings.

The two dimensional J resolved NMR technique is based on the spin echo experiment where the echo amplitudes are not affected by chemical shifts but are modulated by J. The 90°-τ-180°-τ echo pulse sequence (Figure 24) can be used to

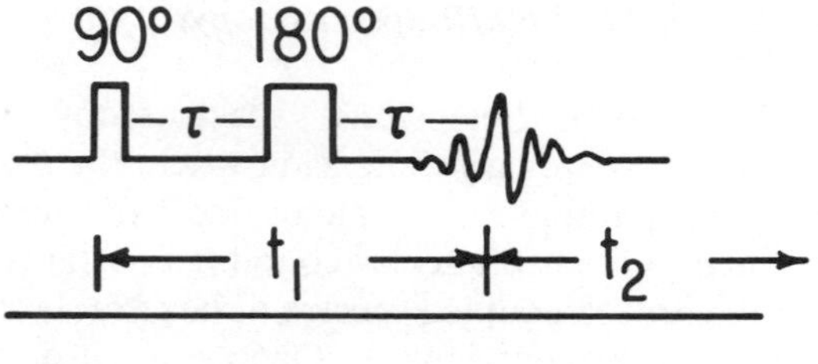

Figure 24. The 90°-τ-180°-τ spin echo pulse sequence for 2D spectroscopy.

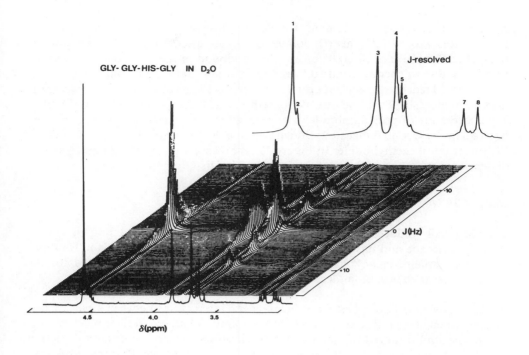

GLY- GLY- HIS-GLY   IN D₂O

J-resolved

cross-sections in the J direction for each of the numbered peaks

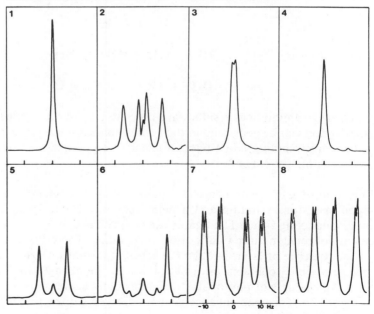

Figure 25. The J resolved two dimensional 400 MHz proton spectrum of Gly-Gly-His-Gly. Note that the spectrum along the chemical shift axis (shown in right hand top corner in line drawing for clarity) is like a broad band decoupled spectrum and instantly provides chemical shift data. The bottom part provides cross-sections in the J direction for each of the numbered peaks. Spectrum, courtesy of Bruker Instruments, Inc.

generate the necessary two domains for 2D spectroscopy. A series of half-echos are accumulated with $\sim$1024 time (t_2) domain points for a series of $\sim$512 t_1 domain settings of τ. The effect of the 180° pulse at the midpoint of t_1 is to change the sense of precession due to chemical shift differences, but not that due to coupling constant differences. Hence at 2τ all lines differing only in chemical shift fall in phase, but lines influenced by J will be out of phase by an amount proportional to J and τ. Transformation of peak amplitudes in $t_1 = 2\tau$ gives a resonance displaced in proportion to J in the second dimension. Bodenhausen et al.[93] have published the J resolved ^{13}C spectrum of methyliodide. In Figure 25 is shown a J-resolved homonuclear two dimensional spectrum of the tetrapeptide Gly-Gly-His-Gly.

Photo-CIDNP

The chemically induced dynamic nuclear polarization by photo ractions is known for over a decade and has been used to study the organic free radicals in solution. Recently, Kaptein, et al.[94] have developed a laser photo-CIDNP method for the study of interactions in biomolecular systems.

Photo-CIDNP method is based on the magnetic interactions between a pair of free radicals which are produced by means of laser and is observed in their reaction products. The application of this method to proteins rests on a reversible photo-reaction of a dye (D) such as flavin with the amino acid residues tyrosine, histidine, and tryptophan.[95] The photo reaction that takes place may be represented by

$$D \xrightarrow{h\nu} {}^1D \rightarrow {}^3D \tag{6}$$

$$D^3 + TyrH \rightarrow DH\bullet + Tyr\bullet \tag{7}$$

$$DH\bullet + Tyr\bullet \rightarrow D + Tyr\,H \tag{8}$$

When these amino acid residues are accessible to photo-excited dye, nuclear spin polarization can be generated in their side chains resulting in selective enhancement of the corresponding NMR lines. Nuclear spin polarization arises from the radical pair recombination step 8.

Detection of CIDNP in complex NMR spectra of proteins requires a high power argon laser for excitation and a high frequency NMR spectrometer operating at superconducting fields. The sample can be irradiated in the probe and the NMR data can be collected in pulse Fourier transform mode. The subtraction of light spectrum (in the presence of laser) and dark spectrum (normal spectrum) gives the difference spectrum which contains information about the changes in the intensities of the lines. An example of photo-CIDNP in the amino acid N-acetyl tryptophan is shown in Figure 26. The peak enhancement is observed for the ring protons and emission from the β-CH_2 group. This indicates the interaction of the dye with the amino acid. It appears that the photo-CIDNP method will be an important one in the future to study enzyme-inhibitor, protein-nucleic acid as well as nucleic acid drug interactions.

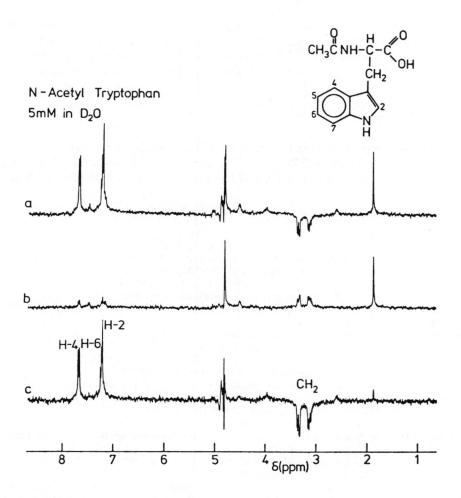

Figure 26. 360 MHz ¹H FT NMR spectra of 5 × 10⁻³M N-acetyl tryptophan and 2 × 10⁻⁴ M flavin in D₂O. (a) light spectrum (b) dark spectrum (c) difference spectrum. Reprinted with permission of D. Reidel Publishing Company, reference 95.

NMR Zeugmatography

See contribution by Paul C. Lauterbur in this volume.[96]

Cross-Polarization, Magic Angle NMR

See contribution by Steven S. Danyluk in this volume.[97]

Summary

In these pages, we have attempted to summarize the use of coupling constants and chemical shifts to derive molecular geometry with nucleic acid structures as examples as well as to provide a brief sketch of some of the recent developments in NMR spectroscopy. Because of the limitations of space, we have not provided details or for that matter the limitations and pitfalls of the methods. For this we recommend the readers consult the original literature which are referred to in the article.

Acknowledgement

This research was supported by Grant CA12462 from National Cancer Institute of NIH and Grant PCM-7822531 from National Science Foundation. This research was also supported by Grant 1-P07-PR-PR00798 from the Division of Research Resources, NIH.

References and Footnotes

1. Schweizer, M. P., Chan, S. I., Helmkamp, G. K., Ts′o, P. O. P., *J. Am. Chem. Soc. 86*, 696 (1966).
2. Bullock, F. J., and Jardetzky, O., *J. Org. Chem. 29*, 1988 (1964).
3. Akasaka, K., Imoto, T., and Hatano, H., *Chem. Phys. Lett. 21*, 398 (1973).
4. Ts′o, P. O. P., Kondo, N. S., Schweizer, M. P., and Hollis, D. P., *Biochemistry 8*, 997 (1969).
5. Schweizer, M. P., Broom, A. D., Ts′o, P. O. P., and Hollis, D. P., *J. Am. Chem. Soc. 90*, 1042 (1968).
6. Chan, S. I., and Nelson, J. H., *J. Am. Chem. Soc. 91*, 168 (1969).
7. Cheng, D. M., and Sarma, R. H., *Biopolymers 16*, 1687 (1977).
8. Remin, M., and Shugar, D., *Biochem. Biophys. Res. Commun. 48*, 636 (1972).
9. Lee, C. H., Evans, F. E., and Sarma, R. H., *FEBS Letters 51*, 73 (1975).
10. Evans, F. E., Lee, C. H., and Sarma, R. H., *Biochem. Biophys. Res. Commun. 63*, 106 (1975).
11. Kondo, N. S., Ezra, F., and Danyluk, S. S., *FEBS Letters 53*, 213 (1975).
12. Evans, F. E., and Sarma, R. H., *Nature 263*, 567 (1976).
13. Karplus, M., *J. Chem. Phys. 33*, 1842 (1969).
14. Davies D. B. in *Progress in Nuclear Magnetic Resonance Spectroscopy* (Emsley, J. W., Feeney, J., and Sutcliffe, L. H., Eds.), *12*, 135 (1978).
15. Altona, C., and Sundaralingam, M., *J. Am. Chem. Soc. 94*, 8205 (1972); *95*, 2333 (1973).
16. Sundaralingam, M., *Biopolymers 7*, 821 (1969).
17. Sundaralingam, M., in *Conformations of Biological Molecules and Polymers. The Jerusalem Symposium on Quantum Chemistry and Biochemistry V*. Editors Bergman, E., and Pullman, B., p. 417 (1973).
18. Jardetzky, C. D., *J. Am. Chem. Soc. 82*, 229 (1960).
19. Jardetzky, C. D., *J. Am. Chem. Soc., 84*, 62 (1962).
20. Sarma, R. H., and Mynott, R. J., *J. Am. Chem. Soc. 95*, 1641 (1973).
21. Evans, F. E., and Sarma, R. H., *J. Biol. Chem. 249*, 4754 (1974).
22. Sasisekharan, V., in *Conformations of Biological Molecules and Polymers. The Jerusalem Symposium on Quantum Chemistry and Biochemistry V*. Editors Bergman, E., and Pullman, B., p. 247 (1973).
23. Saran, A. Perahia, D., and Pullman, B., *Theor. Chim. Acta 30*, 31 (1973).
24. Lee, C. H., Evans, F. E., and Sarma, R. H., *J. Biol. Chem. 250*, 1290 (1975).
25. Sarma, R. H., Lee, C. H., Evans, F. E., Yathindra, N., and Sundaralingam, M., *J. Am. Chem. Soc. 96*, 7337 (1974).
26. Hruksa, F. E., in *Conformation of Biological Molecules and Polymers. The Jerusalem Symposium on Quantum Chemistry and Biochemistry V*. Editors Bergman, E., and Pullman, B., p. 345 (1973).
27. Hruska, F. E., Wood, D. J., Mynott, R. J. and Sarma, R. H., *FEBS Letters 31*, 153 (1973).
28. Wood, D. J., Mynott, R. J., Hruska, F. E., and Sarma, R. H., *FEBS Letters 34*, 323 (1973).
29. Wood, D. J., Hruska, F. E., Mynott, R. J., and Sarma, R. H., *Can. J. Chem. 51*, 2571 (1973).
30. Lee, C. H., and Sarma, R. H., *J. Am. Chem. Soc. 98*, 3541 (1976).
31. Lee, C. H., Ezra, F. S., Kondo, N. S., Sarma, R. H., and Danyluk, S. S., *Biochemistry 15*, 3627 (1976).

32. Ezra, F. S., Lee, C. H., Kondo, N. S., Danyluk, S. S., and Sarma, R. H., *Biochemistry 16*, 1977 (1977).
33. Cheng, D. M., and Sarma, R. H., *J. Am. Chem. Soc. 99*, 7333 (1977).
34. Dhingra, M. M., and Sarma, R. H., *Nature 272*, 798 (1978).
35. Cheng, D. M., Dhingra, M. M., and Sarma, R. H., *Nucleic Acids Research 5*, 4393 (1978).
36. Sarma, R. H., and Danyluk, S. S., *Int. Natl. J. Quant. Chem. QBS 4*, 269 (1977).
37. Lee, C. H., and Sarma, R. H., *J. Am. Chem. Soc. 97*, 1225 (1975).
38. Evans, F. E., and Sarma, R. H., *FEBS Lett 41*, 253 (1974).
39. Hart, P. A., and Davis, J. P., *J. Am. Chem. Soc. 93*, 753 (1971).
40. Imoto, T., Akasaka, K., and Hatano, H., *Chem. Letters 73* (1974).
41. Lemieux, R. U., Nagbhushan, T. L., and Paul, B., *Can. J. Chem. 50*, 773 (1972).
42. Delbaere, L. T. J., James, M. N. G., and Lemieux, R. U., *J. Am. Chem. Soc. 95*, 7866 (1973).
43. Lemieux, R. U., *Ann. N. Y. Acad. Sci. 222*, 915 (1973).
44. Dea, P., Kreishman, G. P., Schweizer, M. P., and Witkowski, J. T., *Proceedings of the Ist Int. Conf. on Stable Isotopes in Chemistry, Biology and Medicine.* Editor, Klein, P. D., p. 84 (1973).
45. Schweizer, M. P., and Kreishnan, G. P., *J. Magn. Reson. 9*, 334 (1973).
46. Sternhell, S., *Q. Rev. Chem. Soc. 23*, 236 (1969).
47. Sternhell, S., *Rev. Pure Appl. Chem. 14*, 15 (1964).
48. Barfield, M., and Chakrabarti, B., *Chem. Rev. 69*, 757 (1969).
49. Hall, L. D., and Manville, J. F., *Carbohydr. Res. 8*, 295 (1968).
50. Blackburn, B. J., Lapper, R. D., and Smith, I. C. P., *J. Am. Chem. Soc. 95*, 2873 (1973).
51. Hruska, F. E., Mak, A., Singh, H., and Shugar, D., *Can. J. Chem. 51*, 1099 (1973).
52. Sarma, R. H., Mynott, R. J., Hruska, F. E., and Wood, D. J., *Can. J. Chem. 51*, 1843 (1973).
53. Hall, L. D., and Malcolm, R. B., *Can. J. Chem. 50*, 2092 (1972).
54. Hall, L. D., and Malcolm, R. B., *Can. J. Chem. 50*, 2102 (1972).
55. Donaldson, B., and Hall, L. D., *Can. J. Chem. 50*, 2111 (1972).
56. Sarma, R. H., Mynott, R. J., Wood, D. J., and Hruska, F. E., *J. Am. Chem. Soc., 95*, 6457 (1973).
57. Grey, A. A., Smith, I. C. P., and Hruska, F. E., *J. Am. Chem. Soc. 93*, 1765 (1971).
58. Schleich, T., Blackburn, B. J., Lapper, R. D., and Smith, I. C. P., *Biochemistry 11*, 137 (1972).
59. Darnall, K. R., Townsend, L. B., and Robins, R. K., *Proc. Natl. Acad. Sci. U. S. A. 57*, 548 (1967).
60. Dhingra, M. M. and Sarma, R. H. (unpublished data).
61. Haneishi, T., Okazaki, T., Hata, T., Tamura, C., Nomura, M., Naito, A., Seki, I., and Arai, M., *J. Antibiotics 24*, 797 (1971).
62. Dea, P., Schweize, M. P., and Kreishman, G. P., *Biochemistry 13*, 1862 (1974).
63. Hruska, F. E., *Can. J. Chem. 49*, 2111 (1971).
64. Barfield, M., Spear, R. J., and Sternhell, S., *J. Am. Chem. Soc., 93*, 5322 (1971).
65. Barfield, M., McDonald, C. J., Peat, I. R., and Reynolds, W. F., *J. Am. Chem. Soc. 93*, 4195 (1971).
66. Evans, F. E., and Sarma, R. H., *Cancer Research 35*, 1458 (1975).
67. Pople, J. A., Schneider, W. G., and Bernstein, H. J., *High Resolution Nuclear Magnetic Resonance.* McGraw-Hill Book Co., Inc. (1959).
68. Emsley, J. W., Feeney, J., and Sutcliffe, L. H., *High Resolution Nuclear Magnetic Resonance Spectroscopy.* Pergamon Press, Oxford (1965).
69. James. T. L., *Nuclear Magnetic Resonance in Biochemistry*, Academic Press, New York (1975).
70. Bovey, F. A., *NMR Spectroscopy*, Academic Press, New York (1969).
71. Danyluk, S. S., and Hruska, F. E., *Biochemistry 7*, 1038 (1968).
72. Remin, M., and Shugar, D., *J. Am. Chem. Soc. 95*, 8146 (1973).
73. Remin, M., Darzynkiewicz, E., Dworak, A., and Shugar, D., *J. Am. Chem. Soc. 98*, 367 (1976).
74. Follman, H., and Gremels, G., *Eur. J. Biochem. 47*, 187 (1974).
75. Prestegard, J. H., and Chan, S. I., *J. Am. Chem. Soc. 91*, 2843 (1969).
76. Remin, M., Ekiel, I., and Shugar, D., *Eur. J. Biochem. 53*, 197 (1975).
77. Schweizer, M. P., Banta, E. B., Witkowski, J. T., and Robins, R. K., *J. Am. Chem. Soc. 95*, 3770 (1973).
78. Giessner-Prettre, C., and Pullman, B., *J. Theor. Biol. 65*, 171 (1977).
79. Giessner-Prettre, C., and Pullman, B., *J. Theor. Biol. 65*, 189 (1977).
80. Sarma, R. H., Dhingra, M. M., and Feldman, R. J. in *Stereodynamics of Molecular Systems*, Sarma, R. H. (Ed.), Pergamon Press, Inc., New York, 1979.
81. Giessner-Prettre, C., and Pullman, B., *Biochem. Biophys. Res. Commun. 70*, 578 (1976).

82. Dhingra, M. M., Sarma, R. H., Giessner-Prettre, C., and Pullman, B., *Biochemistry 17*, 5815 (1978).
83. Bovey, F. A. in *Stereodynamics of Molecular Systems*, Sarma, R. H. (Ed.) Pergamon Press, Inc., New York, 1979.
84. Bolton, P. H., and Bodenhausen, G., *J. Am. Chem. Soc. 101*, 1080 (1979).
85. Maudsley, A. A., and Ernst, R. R., *Chem. Phys. Lett. 50*, 368 (1977).
86. Maudsley, A. A., Muller, L., and Ernst, R. R., *J. Magn. Reson. 28*, 463 (1977).
87. Aue, W. P., Bartholdi, and Ernst, R. R., *J. Chem. Phys. 64*, 2229 (1976).
88. Muller, L., Kumar, A., and Ernst, R. R., *J. Chem. Phys. 63*, 5490 (1975).
89. Bodenhausen, G., Freeman, R., and Turner, D. L., *J. Chem. Phys. 65*, 839 (1976).
90. Aue, W. P., Karhan, J., and Ernst, R. R., *J. Chem. Phys. 64*, 4226 (1976).
91. Nagayama, K., Wuthrich, K., Bachmann, P., and Ernst, R. R., *Biochem. Biophys. Res. Commun. 78*, 99 (1977).
92. Nagayama, K., Wuthrich, K., Bachmann, P., and Ernst, R. R., *Natur Wissenschaften 64*, 581 (1977).
93. Bodenhausen, G., Freeman, R., Niedermeyer, R., and Turner, D. L., *J. Magn. Reson. 26*, 133 (1977).
94. Kaptein, R., Dijkstra, K., Muller, F., Vanschagen, C. G., and Visser, A. J. W. G., *J. Magn. Reson. 31*, 171 (1978).
95. Kapstein, R., in *Nuclear Magnetic Resonance Spectroscopy in Molecular Biology. The Jerusalem Symposium on Quantum Chemistry and Biochemistry*, XI, Pullman, B., (Ed.), p. 211 (1978).
96. Lauterbur, P. C. and House, W. V. Jr., Simon, M. H. Dias, M., Jacobson, M. J., Lai, C. M., Bendal, P. and Rudin, A. M. in *Stereodynamics of Molecular Systems*, Sarma, R. H. (Ed.), Pergamon Press, Inc., New York, 1979.
97. Danyluk, S. S. and Schwartz, H. M. in *Stereodynamics of Molecular Systems*, Sarma, R. H. (Ed.), Pergamon Press, Inc., New York, 1979.
98. Permanent Address: Tata Institute of Fundamental Research, Homi Bhabha Road, Bombay, India.

Nuclear Magnetic Resonance Spectroscopy Stereodynamics of Small Molecules

C. Hackett Bushweller
Department of Chemistry
University of Vermont
Burlington, Vermont 05405

and

Institute of Biomolecular Stereodynamics
State University of New York at Albany
Albany, New York 12222

Much research in molecular stereodynamics over the past twenty years or so has determined beyond any doubt that a myriad of molecular systems are capable of a variety of intramolecular conformational exchange processes. Some of these rate processes include rotation about bonds, pyramidal inversion, pseudorotation, ring reversal, and fluxional behavior in organometallic or inorganic complexes. Many of the rate processes cited above have energy barriers in the range of 4-25 kcal/mole thus making such conformational exchange processes detectable using variable temperature or "dynamic" nuclear magnetic resonance (DNMR) spectroscopy.[1][2]

Consider for example a proton which can exchange between two different environments in an equilibrated system undergoing fast molecular conformational exchange. In this kind of kinetic process, one is dealing with a system in rapid equilibrium and no irreversible chemical reactions are occurring, i.e., the processes involved are nondestructive. Let us also assume that there is no spin-spin coupling of this proton (spin $I = 1/2$) to any other magnetic nuclei. This situation can be illustrated very simply in equation 1. H_a represents the proton in one environment and H_b another different environment. The populations at each site ("a" or "b") are a function of

$$H_a \underset{k_b}{\overset{k_a}{\rightleftharpoons}} H_b \tag{1}$$

the equilibrium constant associated with equation 1 ($K_{eq} = [H_b]/[H_a]$). This system will obey simple first order kinetics and, at a fixed temperature, $K_a[H_a] = k_b[H_b]$ where k_a is the first order rate constant (sec^{-1}) for the forward reaction (conversion of H_a to H_b) and k_b is for the reverse reaction. Normally, k_a and k_b will increase with increasing temperature. The equilibrium constant will also change with temperature if H_a and H_b relate to species of *different* enthalpy and/or entropy. If H_a and H_b

relate to different environments in two *equivalent* species, then $K_{eq} = 1.00$ and there is no change in the populations at H_a and H_b with temperature. In addition, at a given temperature, k_a is equal to k_b under these specific conditions.

For a *static* molecular system (i.e., H_a is not exchanging with H_b) in which the populations of H_a and H_b are equal, the ¹H NMR spectrum consists of two different singlet resonances revealing protons in two different molecular environments. A hypothetical computer-generated example of such a situation is illustrated in the lower right hand corner of Figure 1. The spectrum is composed of two singlets of

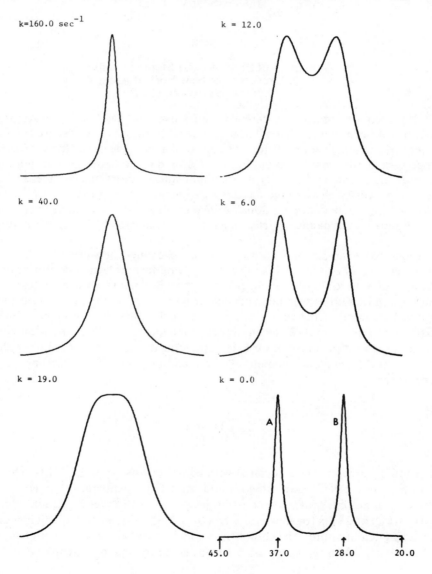

Figure 1. Two site exchange, $\nu_a = 37$ Hz, $\nu_b = 28$ Hz; population ratio 1 to 1; $T_2 = 30$.

equal area separated by 9.0 Hz. The ''k'' in the upper left of each spectrum is the value of the first order rate constant (k_a and k_b, eq 1) for exchange. Thus, the spectrum in the lower right of Figure 1 represents a situation in which the NMR spectrometer may be viewed as a camera with a very fast shutter. The energies of the spin state transitions at each resonance can be accurately defined and two sharp lines result. As the rate of exchange increases (k_a, k_b, eq 1) in proceeding vertically up the right hand column and then up the left hand column of Figure 1, significant changes in the NMR line shape occur. At relatively high rates (e.g., 160 sec^{-1}), the spectrum

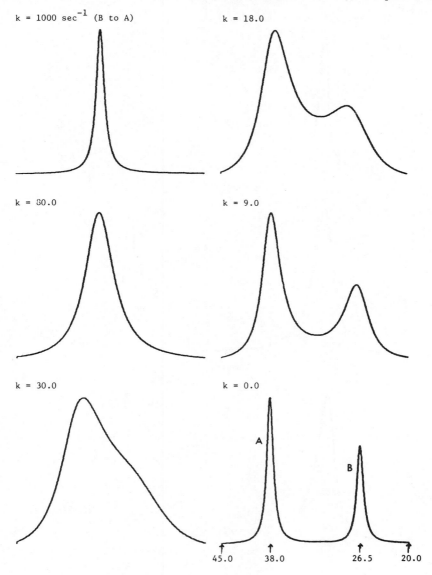

Figure 2. Two site exchange, $\nu_a = 38.5$ Hz, $\nu_b = 26.5$ Hz; population ratio 1.5 to 1 $T_2 = 30$.

consists of just one singlet located at a chemical shift which is a weighted time-average value of the values for the two singlets in the lower right hand spectrum. As the exchange rate increases, the NMR spectrometer becomes a camera with an increasingly slower shutter relative to the rate process in question. It is indeed clear from Figure 1 that over a certain range of rate constants (6 to 40 sec^{-1}), a changing value of the rate constant for exchange can have a significant differential effect on the NMR line shape.

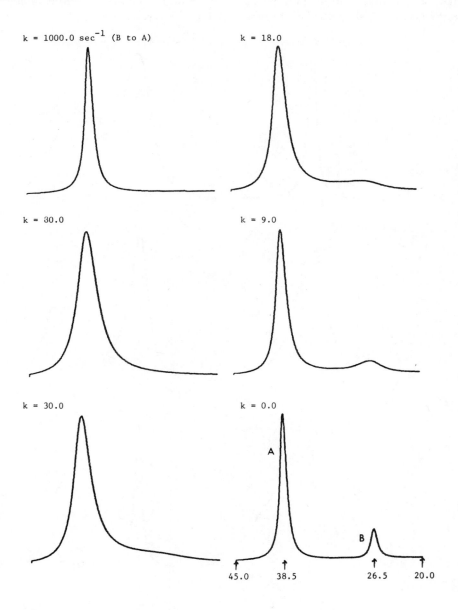

Figure 3. Two site exchange, $\nu_a = 38.5$ Hz, $\nu_b = 26.5$ Hz; population ratio 5 to 1; $T_2 = 30$.

One may encounter a situation in which the populations at the two sites are not equal. In Figures 2 and 3, examples are given for exchange between two unequally populated sites. The rate constants listed are those for disappearance from the less populous site. The rate constant for the reverse process can be computed from a simple kinetic expression for the system at equilibrium ($k_a[H_a] = k_b[H_b]$).

For most simple first order kinetic processes, the rate increases with increasing temperature. Thus, in order to observe experimentally the line broadening and coalescence phenomena illustrated in Figures 1-3, one needs to record NMR spectra as a function of temperature. Since the shape of the NMR spectrum will change significantly over a certain temperature range and that shape is determined essentially by the rate of a dynamic molecular process, the term dynamic nuclear magnetic resonance (DNMR) evolved.[1]

There exists an extensive literature on the theoretical treatment of static and dynamic NMR spectra.[1/3] It is not the purpose of this short article to review such approaches in depth but to provide a qualitative overview of the tecniques employed. An NMR spectrum is essentially a plot of intensity (g) on the vertical axis versus frequency (ν) or "chemical shift" on the horizontal axis. For a static NMR spectrum (no exchange effects), peaks or resonances will occur in the spectrum when the frequency of an externally applied radiofrequency signal is equal to the precessional frequency of pertinent magnetic nuclei in a strong magnetic field (H_o). From the Block phenomenological equations for nuclei of spin 1/2 can be derived an expression which deals with simulation of the DNMR behavior illustrated in Figures 1-3.[4] The result of that derivation is embodied in equation 2.[3]

$$g(\omega) = -\gamma H_1 M_o (1 + \tau/T_2) \frac{P + QR}{P^2 + R^2} \qquad (2)$$

$$P = \tau\{(1/T_2)^2 - [\tfrac{1}{2}(\omega_a + \omega_b) - \omega]^2 + \tfrac{1}{4}(\omega_a - \omega_b)^2\} + 1/T_2$$

$$Q = \tau[\tfrac{1}{2}(\omega_a + \omega_b) - \omega - \tfrac{1}{2}(p_a - p_b)(\omega_a - \omega_b)]$$

$$R = [\tfrac{1}{2}(\omega_a + \omega_b) - \omega](1 + 2\tau/T_2) + \tfrac{1}{2}(p_a - p_b)(\omega_a - \omega_b)$$

γ = magnetogyric ratio of the magnetic nucleus observed
H_1 = amplitude of observing radiofrequency signal
M_o = magnetization

τ_a = lifetime in state a
τ_b = lifetime in state b
$\tau_a = 1/k_a$ (see eq 1)
$\tau_b = 1/k_b$ (see eq 1)

$$\tau = \frac{\tau_a \tau_b}{\tau_a + \tau_b}$$

T_2 = transverse relaxation time
ω_a = chemical shift in radians/sec at site a
ω_b = chemical shift in radians/sec at site b
ω = chemical shift at any arbitrary point in the spectrum in radians/sec
$\omega = 2\pi\nu$ (ν is in units of cps or Hz)
p_a = fractional population of nuclei at site a
p_b = fractional population of nuclei at site b
$p_a + p_b = 1.00$

The magnetization M_o is a vector sum of an ensemble of nuclear magnetic moments. Under normal circumstances, the M_o vector is coincident and aligned with the applied magnetic field (H_o). For the practical purpose of using equation 2 to calculate DNMR spectra, the term $\gamma H_1 M_o$ may be used simply as a proportionality factor to adjust the height of the calculated spectrum. The term π in equation 2 can be related directly to the rate constants for exchange in equation 1. The transverse relaxation time (T_2) is a measure of the time it takes for an ensemble of magnetic nuclei precessing in phase about the H_o axis just after resonance to dephase. For practical purposes, T_2 may be estimated by equation 3. The term

$$T_2 \cong \frac{1}{\pi \Delta\nu_{1/2}} \qquad\qquad (3)$$

$\Delta\nu_{1/2}$ is the width-at-half-height (cps or Hz) of a singlet resonance *not* subject to dynamic exchange effects. Thus, if ω_a, ω_b, T_2, p_a, and p_b can be defined, one can proceed to use equation 2 to compute a DNMR spectrum as a function of τ. Again, τ is a direct measure of the rate of exchange between sites "a" and "b" (eq 1). Thus, for a certain range of values of τ, the DNMR spectrum will undergo dramatic changes (Figures 1-3). In order to generate a theoretical spectrum using equation 2, one sets the values of τ, ω_a, ω_b, p_a, p_b, T_2 and $\gamma H_1 M_o$ and then calculates $g(\omega)$ as a function of incrementally changing values of ω. A plot of these points produces a DNMR spectrum.

If one is able to observe *experimentally* the kinds of changes illustrated in Figures 1-3, then accurate theoretical simulations using equation 2 of such experimental spectra will give a series of rate constants for exchange as a function of temperature. It should be stated that the DNMR line shape is most sensitive to changes in τ in the regions of broadening and coalescence. Indeed, it is from these spectra that the most accurate values of the rate constants may be extracted. However, in performing such a DNMR line shape analysis, one must be cognizant of possible sources of error. The most important source of error in computing activation parameters or energy barriers using rate constants obtained from a DNMR line shape analysis is the measurement of temperature. In many commercial variable temperature NMR probes, the insert is not efficiently insulated and serious temperature gradients may exist along the sample tube. Much of this problem can be eliminated by constructing evacuated and silvered Dewars for probe inserts.[5] Critical NMR parameters such as chemical shifts and T_2 values may change with temperature. Any sound line shape analysis should include an attempt to assess the temperature dependence of pertinent chemical shifts involved in the DNMR spectral behavior. T_2 values vary with temperature and can become quite short giving broad lines especially at very low

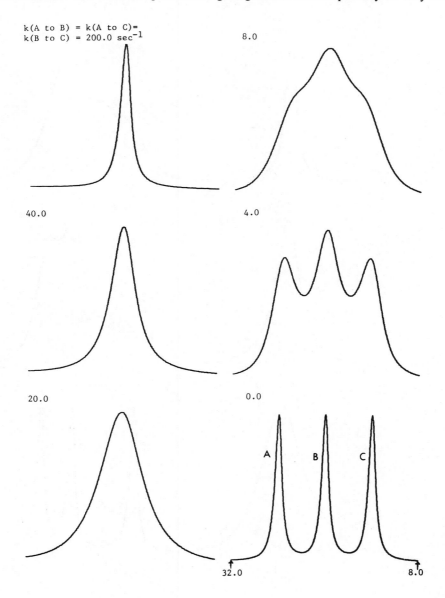

Figure 4. Three site exchange, $\nu_a = 26$ Hz, $\nu_b = 20$ Hz, $\nu_c = 14$ Hz; population ratio 1:1:1; $T_2 = 30$.

temperatures in highly viscous solutions. Thus, it is necessary to separate line-broadening due to T_2 effects from broadening due to the conformational exchange phenomenon.

Finally, if one is dealing with a dynamic process involving molecular species of different energies, it is necessary to determine any variations in pertinent equilibrium constants with temperature. The values of such population ratios or equilibrium

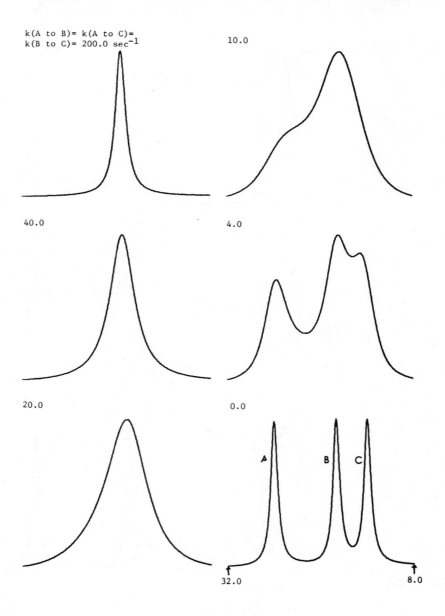

Figure 5. Three site exchange, $\nu_a = 26$ Hz, $\nu_b = 18$ Hz and $\nu_c = 14$ Hz population ratio 1:1:1; $T_2 = 30$.

constants can indeed affect DNMR line shapes (Figure 1-3). If one remains aware of the possible problems outlined above, DNMR line shape analysis can be an accurate means of determining kinetic parameters for systems in rapid equilibrium.

Many DNMR spectra can be substantially more complex than the two-site examples discussed above. Two more complicated examples involve equal rates of exchange among *three* equally populated but different sites in Figures 4 and 5. Figure 6 il-

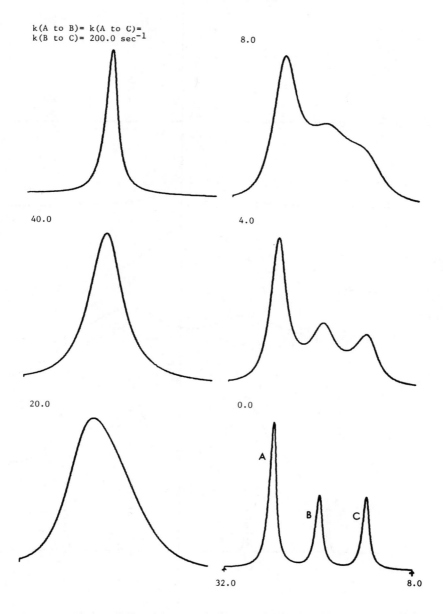

Figure 6. Three site exchange, $\nu_a = 26$ Hz, $\nu_b = 20$ Hz, and $\nu_c = 14$ Hz population ratio 2:1:1, $T_2 = 30$.

lustrates exchange among three sites two of which are of equal population but smaller than the third. Figure 7 illustrates equal rates of exchange among four equally populated sites. Equation 2 is not able to handle the cases illustrated in Figures 4-7. However, the theoretical treatment of such multi-site exchange has been developed[6] and computer programs are available to simulate such behavior.[7]

If a DNMR phenomenon involves spins which are coupled, the "classical"

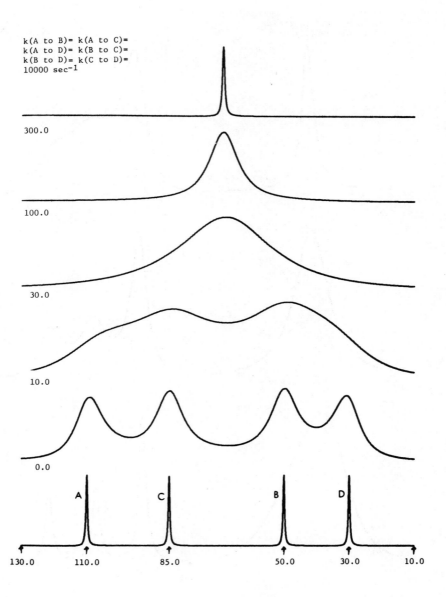

$k(A \text{ to } B)= k(A \text{ to } C)=$
$k(A \text{ to } D)= k(B \text{ to } C)=$
$k(B \text{ to } D)= k(C \text{ to } D)=$
10000 sec^{-1}

300.0

100.0

30.0

10.0

0.0

A   C   B   D

130.0   110.0   85.0   50.0   30.0   10.0

Figure 7. Four site exchange, $\nu_a = 110$ Hz, $\nu_b = 50$ Hz, $\nu_c = 85$ Hz and $\nu_d = 30$ Hz.

theoretical approaches outlined above (e.g., equation 2) are not relevant. One must treat each system of coupled spins as a single ensemble of spins and consider *all* of the NMR transitions resulting from spin-spin coupling. It is necessary to treat such systems quantum-mechanically and density-matrix theory is usually employed.[8] Excellent computer programs are available to simulate such DNMR behavior.[9] A hypothetical example of a simple mutual exchange of two protons which are coupled (J = 5.0 Hz) is illusdtrated in Figure 8. Another more complicated case is shown in Figure 9. In Figure 9, a proton giving a signal at 110.0 Hz is coupled to another at 50.0 Hz (J = 12 Hz). A proton at 85.0 Hz is coupled to another at 30.0 Hz (J = 6 Hz). All four sites are equally populated and exchange rates among all four sites are equal for a given spectrum.

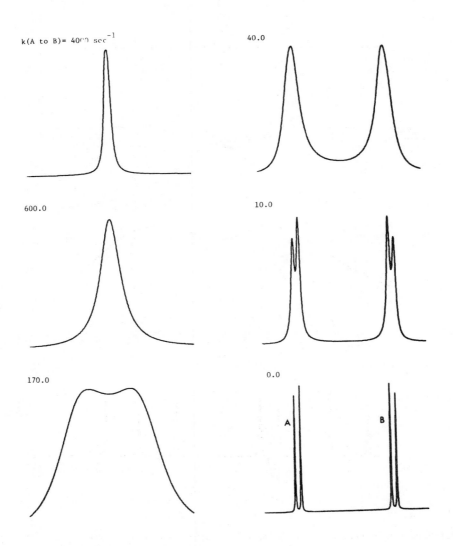

Figure 8. Calculated AB spectra, $\nu_a = 100$ Hz, and $\nu_b = 15$ Hz, $J_{AB} = 5.0$ Hz.

Once a series of experimental DNMR spectra have been fit theoretically thus giving rate constants as a function of temperature, one can use the familiar Eyring approach (equations 4-5) to calculate an enthalpy of activation (ΔH^{+}), entropy of activation (ΔS^{+}), or free energy of activation (ΔG^{+}) for the process observed. A least-squares analysis of a plot of ln (k/T) versus 1/T gives a straight line with a slope of $-\Delta H^{+}/R$.

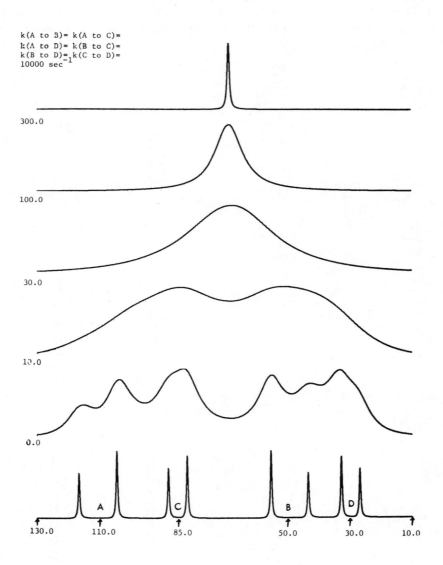

Figure 9. Four site exchange, $\nu_a = 110$ Hz, $\nu_b = 50$ Hz, $\nu_c = 85$ Hz and $\nu_d = 30$ Hz; $J_{AB} = 12$ Hz, $J_{CD} = 6$ Hz.

$$k = \frac{\varkappa KT}{h} \; e^{-\Delta G^{\pm}/RT} \tag{4}$$

$$k = \frac{\varkappa KT}{h} \; e^{(-\Delta H^{\pm}/RT + \Delta S^{\pm}/R)} \tag{5}$$

k = rate constant
$\varkappa$ = transmission coefficient
K = Boltzmann's constant
h = Planck's constant
T = absolute temp (°K)
R = ideal gas constant

Another approach to calculating "barrier" or activation energies would employ the Arrhenius equation (equation 6). A plot of ln k versus 1/T gives a straight line of slope

$$k = Ae^{-E_a/RT} \tag{6}$$

$-E_a/R$ (E_a = Arrhenius activation energy). The values of such barriers for a series of related compounds can give valuable insight into the nature of the rate process observed.

References and Footnotes

1. L. M. Jackman and F. A. Cotton (Editors), "Dynamic Nuclear Magnetic Resonance Spectroscopy,"Academic Press, Inc., New York, 1975.
2. W. G. Orville-Thomas (Editor), "Internal Rotation in Molecules," John Wiley and Sons, New York, 1974.
3. (a) J. A. Pople, W. G. Schneider, and H. J. Bernstein, "High Resolution Nuclear Magnetic Resonance," McGraw-Hill, New York, 1959 (b) A. Abragam, "The Principles of Nuclear Magnetism," Oxford Univ. Press, London, 1961.
4. F. Bloch, *Phys. Rev., 70,* 460 (1946).
5. F. R. Jensen, L. A. Smith, C. H. Bushweller, and B. H. Beck, *Rev. Sci. Instr., 43,* 894 (1972).
6. R. A. Sack, *Mol. Phys., 1,* 163 (1958).
7. M. Saunders, *Tetrahedron Lett.,* 1699 (1963).
8. G. Binsch, *Mol. Phys., 15,* 469 (1968).
9. G. Binsch and D. A. Kleier, "The Computation of Complex Exchange-Broadened NMR Spectra,"Program 140, Quantum Chemistry Program Exchange, Indiana University, Bloomington. For modifications to improve computational efficiency in this program, see: C. H. Bushweller, G. Bhat, L. J. Letendre, J. A. Brunelle, H. S. Bilofsky, H. Ruben, D. H. Templeton, and Z. Zalkin, *J. Am. Chem. Soc., 97,* 65 (1975).

Nuclear Magnetic Resonance Spectroscopy
Solution Dynamics of Polymer Chains

F. A. Bovey
Bell Laboratories
Murray Hill, New Jersey 07974

Introduction

In recent years, since the development of pulsed Fourier transform spectroscopy, many investigators have employed measurements of carbon-13 nuclear relaxation to the study of the dynamics of both small and large molecules in the liquid state. Such information can in turn be related to the geometry, solvation, and interaction of organic molecules.[1,2] The relaxation measurements with which we shall be concerned in the present discussion are the *spin lattice relaxation*, characterized by T_1, and the *nuclear Overhauser enhancement* or NOE, symbolized by η. Their interpretation for macromolecules leads to substantial new insights into polymer chain motions.

Mechanisms of Spin-Lattice Relaxation

The process of spin-lattice relaxation is the return of the carbon-13 nuclear spin populations from a disturbed state to the equilibrium or Boltzmann distribution. The "lattice" serves as a heat sink for the spins in this thermal equilibration, but in order to do so, there must be a link by which thermal energy may be exchanged. This link is provided by molecular motion. Each carbon-13 nucleus experiences the presence of local magnetic fields, originating principally from motions of the same molecule. There are several mechanisms by which such fields may arise. They may, for example, originate from the spinning of the molecules themselves (aside from the nuclear magnetic fields), the molecules being rotating charge systems. Rotational variations and the tumbling of the molecules modulate these magnetic fields. Such *spin-rotation* contributions increase with temperature but are generally negligible for macromolecules because their relatively slow motions do not generate appreciable magnetic moments in this way.

The magnetic shielding of nuclei is anisotropic—i.e., directional—and this also gives rise to fluctuating magnetic fields as the molecules tumble in solution, but such *chemical shift anisotropy* contributions to spin-lattice relaxation depend on the square of the laboratory magnetic field and are negligible in ordinary practice.

The only significant mechanism for the spin-lattice relaxation of carbon-13 nuclei in polymers is *dipole-dipole* interaction with neighboring magnetic nuclei, principally protons. The motions of these neighboring nuclei give rise to fluctuating magnetic fields and are characterized by a broad range of frequencies. To the extent that these motions have components at the resonant frequency of the carbon nuclei, they will induce spin-lattice relaxation. The distribution of motional frequencies is given by a *spectral density function*. In order to deduce the form of this function for a polymer chain, one must adopt a dynamic model. The simplest model views the macromolecule as a rigid sphere immersed in a viscous continuum and reoriented by small, random diffusive steps. For such a model, the motional frequency distribution will be described by a spectral density function of the form

$$J_n(\omega) = \frac{\tau_c}{1 + \omega_2^2 \tau_c^2} \tag{1}$$

The chain motion is thus described by a single correlation time τ_c. In Figure 1, this function is shown for three values of the correlation time: a long τ_c, corresponding to relatively slow motions, as for large molecules or low temperatures; a short τ_c, corresponding to rapid motions as for small molecules, flexible chains, or high temperatures; and an intermediate value. It will be noted that the *total* spectral density is constant and independent of τ_c, i.e., the areas under the curves are the same. For long correlation times, the component at the resonant frequency of the observed nucleus, ω_0, is weak; at short τ_c, the frequency spectrum is broad, and so no one component, in particular that at ω_0, can be very intense. At an intermediate value, the resonant component is at a maximum. Since the rate of spin lattice relaxation depends on the intensity of the component at ω_0, one expects it to be relatively small

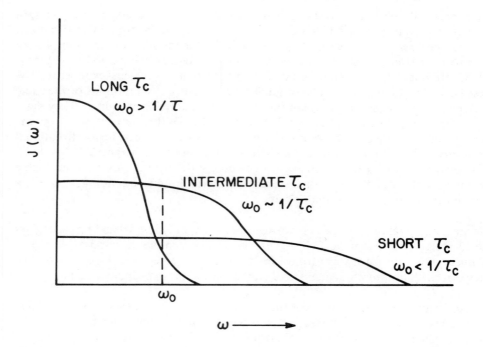

Figure 1. Spectral density functions (motional frequency) as a function of long, medium, and short values of the correlation time τ_c.

at very short and very long τ_c and to reach a maximum when $\tau_c \simeq 1/\omega_0$. For a particular system, T_1 will then pass through a minimum as the temperature is lowered and increase again at low temperature. For a ^{13}C nucleus relaxed by the fluctuating magnetic field arising from N equivalent protons at distance r_{C-H}, it is found that:[4]

$$\frac{1}{NT_1} = \frac{1}{10} \; \frac{\gamma_H^2 \gamma_C^2 \hbar^2}{r_{C-H}^6} \tag{2}$$

$$\times \{ J_0(\omega_H - \omega_C) + 2J_1(\omega_C) + 6J_2(\omega_H + \omega_C) \},$$

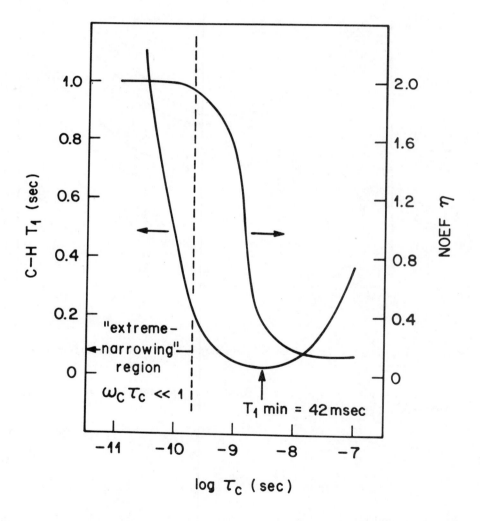

Figure 2. Semi-log plot of T_1 as a function of the correlation time τ_c at 2.35 Tesla (23.5 kilogauss) for ^{13}C relaxed by a single proton at a distance r of 0.109 nm, according to equation (2). (For discussion of the η function, see Sec. 4.)

where the J_n are of the form of equation (1) and prescribe the components of the motional frequency spectrum at $\omega_H - \omega_C$, ω_C, and $\omega_H + \omega_C$. Here, ω_H and ω_C are the resonant frequencies of proton and carbon-13, respectively; γ_H and γ_C are the corresponding magnetogyric ratios. A plot of equation (2) is shown in Figure 2 for a field H_0 of 2.35 Tesla (23.5 kilogauss). At higher values of H_0, the T_1 minimum increases and moves to smaller values of τ_c, as shown in Figure 3.

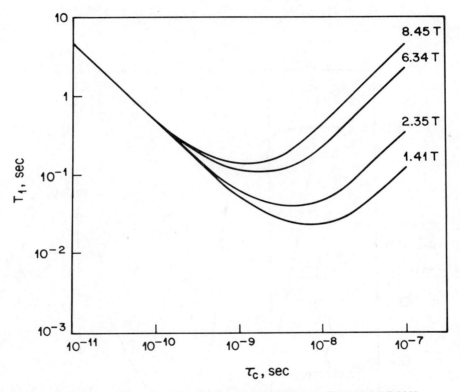

Figure 3. Log-log plots of T_1 vs. τ_c for ^{13}C relaxed by a proton (equation (2)) at varied field H_0.

For rapid motions, when $\tau_c\omega_0 < < 1$, equation (1) becomes $J_n(\omega) = \tau_c$, and equation (2) simplifies to:

$$\frac{1}{NT_1} = \tau_C \cdot \frac{\gamma_H^2\gamma_C^2h^2}{r_{C\text{-}H}^6} \qquad (3)$$

Under these conditions of "motional narrowing," T_1 is independent of H_0 (Figure 3). Because of the inverse 6th-power dependence of the rate of dipolar spin-lattice relaxation on the internuclear ^{13}C-^1H distance, it is commonly observed in both small molecules and macromolecules that only the protons directly bonded to the observed carbon contribute significantly to its T_1 value. Carbons without bonded protons, e.g., carbonyl and quaternary carbons, accordingly exhibit relatively long T_1 values. For aliphatic carbons (CH, CH_2, CH_3), with which we shall be mainly concerned, $r = 0.109$ nm and so:

$$\frac{1}{NT_1} = 2.03 \times 10^{10} \cdot \tau_C \tag{4a}$$

or

$$\tau_c = \frac{4.92 \times 10^{-11}}{NT_1} \tag{4b}$$

We shall employ the isotropic single-τ_c model in much of the subsequent discussion, but in the full form of equation (2) rather than in the motional narrowing limit implied by equations (3) and (4). Even so, this model is too simple although it is in many cases a useful approximation. Its shortcomings will be evident when we discuss the interpretation of the nuclear Overhauser effect (Sec. 4, *et seq.*).

Measurement of T_1

The most commonly employed method for the determination of carbon-13 spin-lattice relaxation times is that of *inversion-recovery*. In this method, an intense pulse of r.f. radiation at the resonant carbon frequency is applied for a time (a few tens of microsec.) just sufficient to turn the macroscopic moment through 180°, i.e., to invert the spin populations. This is shown in sequences (b) and (c) in Figure 4, and occurs non-selectively for all the lines of the spectrum. (Figure 4 actually represents the spin system not in the usual laboratory coordinate frame but in a frame rotating at the resonant carbon frequency ω_C; in this rotating frame H_1 and $x'y'$ components of the magnetization may be represented as static.) At the end of the 180° pulse, the magnetization begins to return to its equilibrium value (d). As we have seen (Sec. 1), the time constant for this process is T_1. To measure it, the magnetization is sampled at a time τ (not to be confused with τ_c) by applying a 90° pulse to turn it back into the $x'y'$ plane(e), where it now gives rise to a detectable signal, I_τ. After a delay time equal to about $5T_1$, the pulse sequence is applied again with a different value of τ, yielding another value of I_τ. The experiment is repeated to provide several values of I_τ and also the value of the equilibrium magnetization (long τ), I_0. In Figure 5, Fourier-transformed signals for the CH_2 and CH carbons of polyphenylthiirane (weight average molecular weight 241,000) in 20 percent solution in deuterochloroform at 55° are shown. As expected, the early signals are inverted; their intensity passes through zero and recovers to I_0. T_1 is obtained from the slope of a plot of $\ln[1-(I_\tau/T_0)]$ vs. τ, as shown in Figure 6. One may note from Figure 6 the important observation that T_1 for the CH_2 carbon is half that for the CH carbon, relaxation being proportional to the number of interacting protons, in accordance

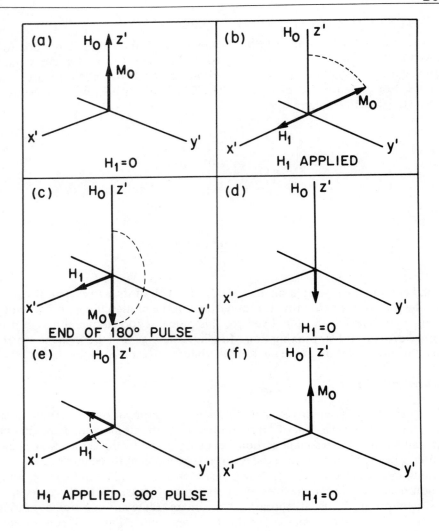

Figure 4. Schematic representation in the rotating frame of the 90°-τ-180° inversion-recovery method of T_1 measurement. In (a), the macroscopic magnetization M_0 is directed along the z-axis. In (b) and (c) a pulse of resonant energy H_1 inverts M_0. Its regrowth to the equilibrium value is sampled by applying a 90° pulse in (d) and (e), which turns it into the x'y' plane. After a time of several T_1, the magnetization again attains its equilibrium value (f).

with equation (2). This is important evidence for the conclusion that the dipole-dipole mechanism is dominant, being consistent with our expectation that only the directly bonded protons contribute observably to $1/T_1$. One may also conclude (employing equation (4)) that τ_c for this system is approximately 0.13 nanosec.

Nuclear Overhauser Enhancement

It is standard practice in carbon-13 spectroscopy to irradiate the protons with a randomly modulated r.f. field which eliminates the scalar couplings between the car-

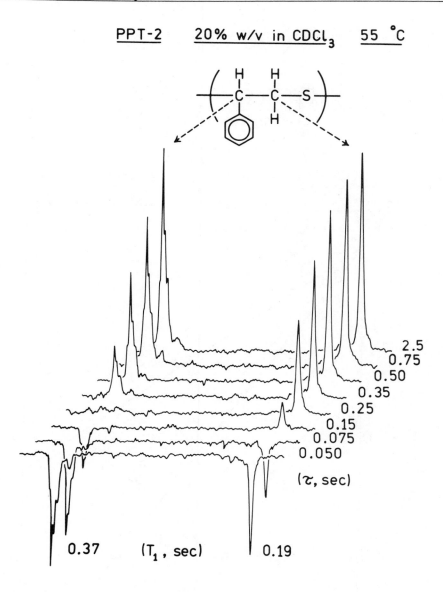

Figure 5. A "stack plot" of Fourier-transformed signals for α-CH and β-CH$_2$ carbons of polyphenylthiirane ($\overline{M}_w$ = 241,000) observed at 25 MHz and 55° as a 20 percent (w/v) solution in CDCl$_3$.

bons and the protons, irrespective of the chemical shifts of the latter.[5] This procedure not only collapses the carbon multiplets ($J_{C-H} \simeq$ 150-250 Hz for directly bonded protons) but, owing to cross-relaxation, that is, simultaneous spin flips of both carbons and protons, perturbs the carbon energy levels in such a way as to increase the intensities of their resonances by an amount η, termed the *nuclear Overhauser enhancement* or NOE. For a carbon-proton system under irradiation at the proton frequency, the ratio of the carbon intensity I_i compared to that with irradiation, I_0, is:

$$\frac{I_i}{I_0} = 1 + \eta = 1 + \frac{\gamma_H}{\gamma_C} \left\{ \frac{6J_2(\omega_H + \omega_C) - J_0(\omega_H - \omega_C)}{J_0(\omega_H - \omega_C) + 3J_1(\omega_C) + 6J_2(\omega_H + \omega_C)} \right\} \quad (5)$$

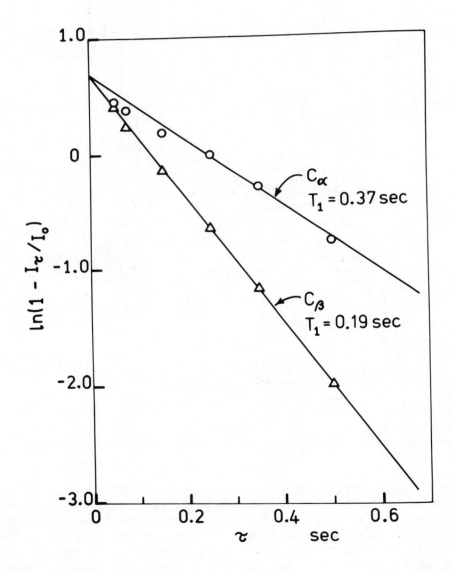

Figure 6. Plot of $\ln[1 - (I_\tau/I_0)]$ for inversion-recovery determination of T_1 for α-CH and β-CH$_2$ carbons of polyphenylthiirane, as in Figure 5.

In the motional narrowing limit η becomes independent of τ_c:

$$I_i/I_0 = 1 + \tfrac{1}{2}(\gamma_H/\gamma_C) = 2.988, \tag{6}$$

since $\gamma_H/\gamma_C = 3.976$. In Figure 2, η is plotted vs. τ_c according to equation (5). It will be seen that as molecular motion becomes slower, η decreases very sharply in the vicinity of the T_1 minimum, reaching a small but non-zero value.

The nuclear Overhauser enhancement may of course be obtained by simply measuring the carbon signal intensity with and without proton irradiation. A disadvantage of this procedure is that the carbon multiplicity returns when the decoupling is removed, and in the complex carbon spectra often encountered in polymers, it may be hard to measure with accuracy owing to low intensity and peak overlap. An alternative practice is to gate the decoupling field on only during data acquisition following each pulse; the scalar coupling is thus immediately abolished but the NOE, reappearing with a time constant of the order of T_1, does not become appreciable.

Carbon-13 Relaxation in Macromolecules

Polyethylene

Although polyethylene might seem to be one of the simplest of polymers, the structure of high pressure polyethylene is actually somewhat complex when considered in detail owing to the presence of branches. These can be examined and measured with a high degree of discrimination by carbon-13 spectroscopy.[6-13] In Figure 7 is shown a 25 MHz spectrum of a branched polyethylene, observed at 110° using a 44 percent (w/v) solution in 1,2,4-trichlorobenzene.[14] It has been reported[13/15] that under these conditions, polyethylene exhibits a full nuclear Overhauser enhancement. In this spectrum the very large peak at 30 ppm corresponds to the strongly predominant (ca. 80 percent) fraction of CH_2 groups which are four carbons or more removed from a branch point, branch end, or chain end. Carbons near branch or chain ends appear upfield, whereas α-carbons (see Figure 7 for notation employed), branch carbons, and certain others are less shielded than those of long methylene sequences. T_1 values, determined in 1,2,4-trichlorobenzene solution at 110°, are presented in column 2 of Table I, together with those of a small model analogue, n-$C_{44}H_{90}$. Correlation times are calculated from equation (4b).[16] It is evident that mobility is greatest near chain ends, particularly for methyl groups, which can rotate independently of the neighboring chain units, and is least at branch points, since here reorientation requires sweeping sizable chain elements through the solution. (No data is reported for the ethyl branches, as their resonances are too weak for meaningful measurements.) Similar findings have been reported by Axelson et al.[10] For n-$C_{44}H_{90}$, only the three carbons near the end give distinguishable resonances. The remaining 38 carbons give a single peak from which only a single T_1 value can be measured. The correlation time is about half that for the linear portions of polyethylene, so that a substantial contribution to this mobility is made by overall tumbling. (More detailed ^{13}C relaxation studies of paraffinic hydrocarbons have been reported by several groups.[17-19])

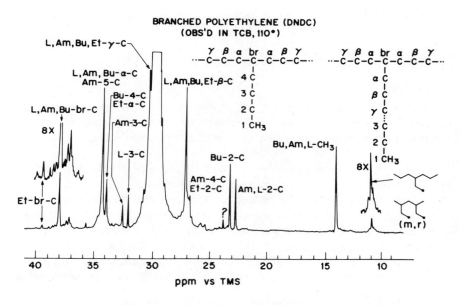

Figure 7. 25 MHz C-13 Spectrum of branched polyethylene, observed in 1,2,4-trichlorobenzene at 110°. (The ethyl branches exhibit a complex structure; the CH₃ resonances near 11 ppm and the branch resonance, 37-40 ppm, suggest they may in part occur in pairs.)

Table I

Carbon-13 T_1 Values and Correlation Times for Branched Polyethylene ($\overline{M}_w = 143,000$) Observed at 25 MHz for a 33 percent (w/v) Solution in 1,2,4-Trichlorobenzene at 110°. For Carbon Identifications, Refer to Figure 7. Corresponding Values for n-$C_{44}H_{90}$ Are Shown Under the Same Conditions, Except 20 Percent (w/v) Concentration.

Carbon	T_1, Sec.	τ_C Nanosec.
Polyethylene:		
CH₃	6.25	0.0025
Bu-2C	3.57	0.0066
Am,L-2C	5.45	0.0043
Am-3C	4.83	0.0049
L-3C	6.00	0.0041
Bu-4C	1.60	0.015
L,Am,Bu-br-C	1.24	0.040
L,Am,Bu-α-C	0.63	0.039
L,Am,Bu-β-C	0.85	0.029
$(CH_2)_n^*$	1.26	0.020
n-$C_{44}H_{90}$:		
CH₃	11.4	0.0025
2C	10.7	0.0043
3C	8.00	0.0039
$(CH_2)_n^*$	2.83	0.0083

*CH_2 groups 4 or more carbons removed from a branch point or chain end.

For polymer chains, such tumbling of the whole molecule becomes a negligible factor at quite low molecular weights.

Polysulfones

A chain dynamics problem of particular interest is presented by the *polysulfones*; C-13 relaxation measurements have contributed in an important way to the understanding of their behavior. Polysulfones are copolymers of vinyl monomers with sulfur dioxide, prepared with free radical initiators. All terminal olefins, beginning with ethylene and continuing on up to the higher olefins, copolymerize with sulfur dioxide and, with the exception of ethylene, give sulfone copolymers of strictly 1:1 alternating structure, since neither the chains ending in SO_2 nor those ending in olefin can add their own monomer:

$$\left(\begin{array}{c} O \\ \| \\ S-CH_2-CH \\ \| \quad \quad | \\ O \quad \quad R \end{array}\right)_n$$

Monomers which can add to their own radicals are capable of copolymerizing with sulfur dioxide to give products of variable composition. Among such monomers is *styrene*; we shall consider styrene—SO_2 copolymers a little later.

Some years ago, Bates, Ivin, and Williams[21] reported that measurements of the dielectric dispersion in solutions of alternating 1:1 copolymers of sulfur dioxide with hexene-1 and 2-methylpentene-1 showed no loss in the high frequency region, i.e., beyond 1 MHz. A high frequency loss region is to be expected for flexible polymers having electric dipole components perpendicular to the main chain direction.[22/23] Instead, a loss maximum was exhibited at low frequency and was found to be inversely proportional to the degree of polymerization raised to the power *ca.*. 1.5-2.0, being for example at about 25 kHz at 25 °C in benzene for a poly(hexene-1 sulfone) having a number average molecular weight of 210,000, as shown in Figure 8. From these observations, it was concluded[21] that overall tumbling is the only motion effective in relaxing the molecule in an oscillating electric field. This conclusion is quite surprising, since relatively "stiff" vinyl polymers such as polystyrene and poly(methyl methacrylate), with glass-transition temperatures of *ca.* 100°, comparable to those of the polysulfones (for example, 97° for poly-(butene-1 sulfone)), have correlation times for segmental motions in solution of the order of nanoseconds.[24/25] In recognition of this paradoxical behavior, three groups[26/27/28] have simultaneously and independently investigated a variety of olefin sulfone polymers by ^{13}C relaxation measurements.

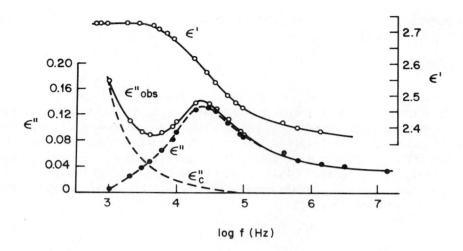

Figure 8. Dielectric data for poly(hexene-1 sulfone) of $\overline{Mn}$ = 210,000 in benzene at 25°. Full lines are the observed dielectric ϵ' and loss ϵ''_{obs}; dashed lines show the contribution to ϵ''_{obs} from dipolar loss ϵ'', and loss due to d.c. conductance ϵ''_c. (From Bates, *et al.*, ref. 21.)

Let us consider the results of poly(butene-1 sulfone)[26], R = CH₂CH₃ above. In Figure 9 is shown the 25 MHz ¹³C spectrum of a polymer of degree polymerization n equal to *ca.* 700, observed in deuterochloroform solution at 50°C. The resonances, particularly those of the backbone, are broadened by unresolved tacticity effects (the chains being probably very nearly random with respect to the relative handedness of the asymmetric methine carbons) and by the phenomenon of *dipolar broadening*, common for large molecules of limited flexibility. In Table II are presented the T_1 and nuclear Overhauser η value (in brackets) for each carbon at each of these temperatures. (It is not feasible to observe this system at temperatures

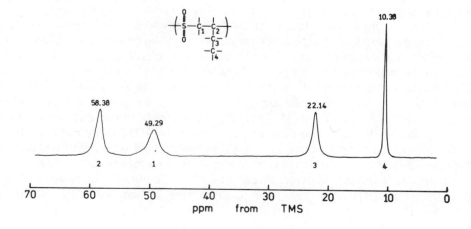

Figure 9. 25 MHz C-13 spectrum of poly(butene-1 sulfone) of degree of polymerization *ca.* 700, observed in CDCl₃ solution at 40° (R. E. Cais and F. A. Bovey, ref. 26).

Table II

Effect of Temperature and the ^{13}C T, and η Values for Poly(butene-1 sulfone) Observed as a 25 Percent (w/v/) Solution in CDCL$_3$ at 25 MHz

		T$_1$, millisec. [η]		
Temp	Backbone		Ethyl Side-Chain	
K	CH	CH$_2$	CH$_2$	CH$_3$
328	86	46	88	599
	[0.63]	[0.83]	[0.97]	[1.36]
313	90	47	78	549
	[0.46]	[0.74]	[0.85]	[1.32]
298	108	51	69	419
	[0.54]*		[0.70]	[1.30]

*Because of line broadening at this temperature, individual backbone multiplets were not satisfactorily resolved, particularly in the absence of proton irradiation.

as high as used for polyethylene in the earlier discussion because of an inherent tendency of these chains to decompose into the monomers.) It is particularly noteworthy that T$_1$ for the main-chain *decreases* with increasing temperature, indicating that we are on right-hand side of the T$_1$ minimum in Figure 2, whereas the side-chain carbons show the opposite trend and are on the left-hand or short-τ_c branch of the curve. This shows that when observing macromolecules it is important to measure T$_1$ values as a function of temperature, since otherwise their interpretation may be ambiguous. Equation (2) is double valued in τ_c at all points except the minimum. It is also very useful to measure T$_1$ at two or more values of the laboratory magnetic field; it is evident from Figure 3 that on the right of the minimum T$_1$ increases with the field strength, whereas on the left side, it is independent of it.

In Table III are presented the correlation times deduced from the T$_1$ and η values and assuming the isotropic single-τ_c model. It is very clear that we must not in this case assume the motional narrowing condition and use equation (4), since in addition to the T$_1$ trend pointed out above for the main-chain carbons, the values of η are all very far from the maximum value of 2. We further find an order-of-magniudtude discrepancy in τ_c depending upon whether it is deduced from T$_1$ or η.

Table III

Correlations Time τ_c in nanoseconds for Backbone and Side-Chain Motions in Poly(butene-1 sulfone), Calculated from the T$_1$ and η Values of Table II

Temp	Backbone:			Ethyl Side-Chain		
			CH$_2$		CH$_3$	
K	from NT$_1$	from η	from NT$_1$	from η	from NT$_1$	from η
328	22	2.6	0.29	1.9	0.027	1.3
313	23	3.1	0.33	2.2	0.030	1.3
298	27	3.4	0.38	2.7	0.039	1.3

For the side-chain CH_2, the correlation times calculated from T_1 differ by nearly two orders of magnitude from those for the backbone CH_2, a result which is not reasonable. The discrepancies between τ_c values from T_1 and η are large and in the opposite direction from what is observed for the main-chain. It is clear that the motional model employed here is not adequate, and that a more elaborate one employing a multiplicity of correlation times is necessary. We shall address this question in the discussion of another polymer system in the next section. Qualitatively, we may conclude that the polysulfone chain is a fairly flexible one, with segmental rotation times of the order of nanoseconds, as would be expected.

This being the case, we must then ask: why are these rapid motions dielectrically inactive, so that in loss measurements the chains act like completely stiff "monoliths," the only observable motion of which is overall tumbling? A possible type of segmenal motion which could account for this is shown in Figure 10. This is

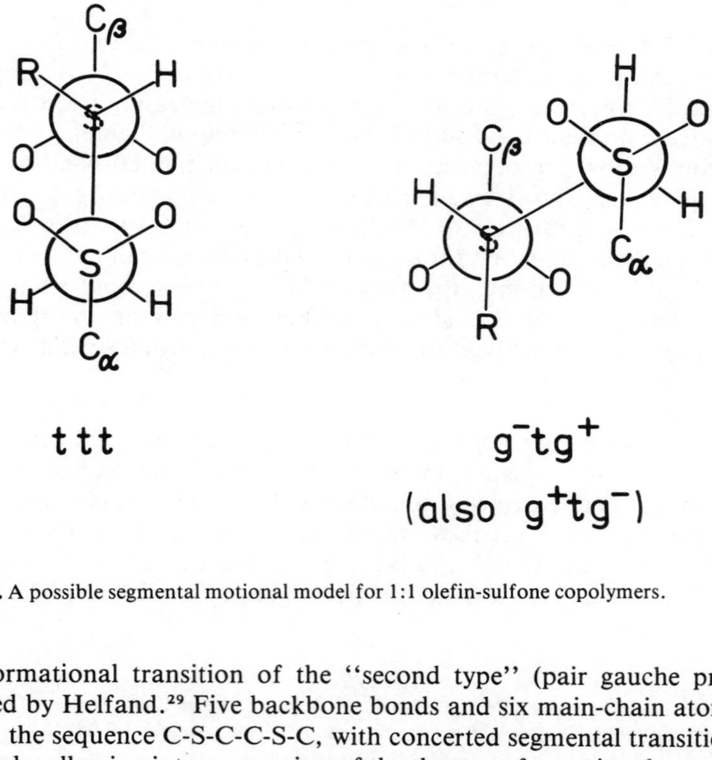

$$t\ t\ t \qquad\qquad g^-tg^+$$

$$(also\ g^+tg^-)$$

Figure 10. A possible segmental motional model for 1:1 olefin-sulfone copolymers.

a conformational transition of the "second type" (pair gauche production), as classified by Helfand.[29] Five backbone bonds and six main-chain atoms are involved, i.e., the sequence C-S-C-C-S-C, with concerted segmental transitions about two C-S bonds, allowing interconversion of the three conformational states ttt, g^+tg^- and g^-tg^+. The backbone C-C bond always remains trans. A principal driving force would be the unusually large interaction energy between sulfone dipoles, the dipole moment of the sulfone group being about 4.5 Debye. The sulfone dipoles are viewed as remaining antiparallel and therefore cancelling each other. They are thus not reoriented with respect to each other by chain segmental motions which rapidly reorient the C-H bond vectors.

The segment mobility of styrene-SO$_2$ copolymers in which the styrene:SO$_2$ ratio is greater than 1 is more than an order of magnitude greater than in the 1:1 butene-SO$_2$ copolymers.[26] The T$_1$ values for both CH and CH$_2$ carbons *increase* with temperature rather than decreasing, showing that we are now on the short-τ_c branch of the T$_1$-τ_c relationship. Copolymers with styrene: SO$_2$ ratios of 2:1 and 1.5:1 show similar relatively short correlation times, about the same as those in polystyrene itself.[30] It is not to be expected that a pendent phenyl group would be more permissive than a pendant ethyl group in allowing segmental reorientation, and we must therefore conclude that the presence of relatively mobile styrene-styrene sequences in the chain permits this greater flexibility. The restricted styrene-SO$_2$-styrene sequences simply ride along with this more rapid motion.

The styrene-SO$_2$ copolymers are of interest in that they allow a test of the motional model proposed above for the poly(butene-1 sulfone). If the sulfone dipoles are moved farther apart on the average by additional vinyl monomer "spacers," the interaction energy between them would be expected to be substantially reduced, and consequently they might not be able to maintain an antiparallel conformation. Such copolymers would therefore be expected to show a normal dielectric loss in the high frequency region. This has been very recently found to be the case by Stockmayer and coworkers.[31] Their results on a 2:1 styrene: SO$_2$ copolymer of Cais and Bovey are shown in Figure 11, which exhibits the dielectric loss for a 1.03 percent solution

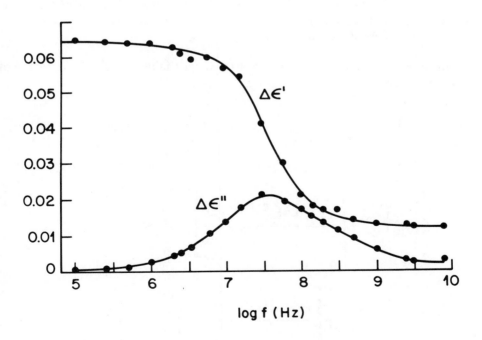

Figure 11. Dielectric data for a *ca.* 2:1 styrene: SO$_2$copolymer, observed as a 1.03% solution in dioxane at 20°. The increment in dielectric constant (over solvent) is represented by $\Delta\epsilon'$ and corresponding loss increment by $\Delta\epsilon''$. (Stockmayer, *et al.*, ref. 31).

of this polymer in dioxane at 25°. The loss peak falls at 35 MHz, corresponding (according to Stockmayer *et al.*) to a segmental reorientation time of *ca.* 2 nanosec., compared to an NMR value of *ca.* 0.6 nanosec. under slightly different conditions. Such agreement can be regarded as satisfactory.

More Realistic Models: Polybutene-1

We have pointed out that much of the carbon-13 relaxation data on macromolecular systems have been interpreted using the isotropic, single-correlation time model. The shortcomings of this model can be conspicuous, as witness the interpretation of the polysulfone data in the previous section, and were realized some time ago. For example, Schaefer[32/33] reported in 1973 that the assumption of a distribution of correlation times was required to explain the T_1 and nuclear Overhauser enhancement observations on both bulk and dissolved polymers. Heatley and Begum[25] found that both the so-called log-χ^2 distribution proposed by Schaefer and the Cole-Cole distribution[34/35] used in the interpretation of dielectric relaxation could be fitted to the T_1 and NOE data for solutions for poly(methyl methacrylate), polystyrene, and poly(propylene oxide). A generally somewhat more satisfactory and physically appealing model is that of Monnerie and coworkers,[36] which pictures the chain reorientation as occurring by localized conformational jumps of chain segments on a tetrahedral lattice ("diamond lattice"), involving three- and four-bond motions without conformational bias. It derives from the defect diffusion model of Glarum[39] and Hunt and Powles,[40] and involves two correlation times: τ_D characterizing local segmental motions and τ_0 characterizing larger-scale molecular tumbling, but not of the whole molecular chain, for we have seen that such whole-chain tumbling is not important for segmented polymers of high molecular weight.

For this model, the spectral density term for main-chain motions (the analogue of equation (1)) is:

$$J_n(\omega) = \frac{\tau_0 \tau_D (\tau_0 - \tau_D)}{(\tau_0 - \tau_D)^2 + \omega^2 \tau_0^2} \times \tag{7}$$

$$\left\{ \left(\frac{\tau_0}{2\tau_D}\right)^{1/2} \times \left[\frac{(1 + \omega^2 \tau_0^2)^{1/2} + 1}{1 + \omega^2 \tau_0^2}\right] + \left[\frac{\tau_0}{2\tau_D}\right]^{1/2} \times \right.$$

$$\left. \frac{\omega \tau_0 \tau_D}{(\tau_0 - \tau_D)} \left[\frac{(1 + \omega^2 \tau_0^2)^{1/2} - 1}{1 + \omega^2 \tau_0^2}\right]^{1/2} - 1 \right\}$$

Table IV

Carbon-13 T_1 Values and Nuclear Overhauser Enhancements
for Polybutene-1 Observed at Two Magnetic Field Strengths
in Pentachloroethane Solution (0.1 g./cm³).

T_1, millisec.

Temp. °C	130	100	85	40	0	-15
				Field = 84600 Gauss		
α-CH	520	--	230	200	230	--
β-CH₂	290	--	160	120	160	--
side-chain CH₂	410	--	190	140	150	--
CH₃	2370	--	1240	770	580	--
				Field = 21000 Gauss		
α-CH	450	310	230	110	60	60
β-CH₂	260	190	150	60	35	27
side-chain CH₂	340	240	180	70	55	29
CH₃	--	--	--	590	--	--

η

Temp. °C	130	100	85	40	20	0	-10
				Field = 84600 Gauss			
α-CH	1.5	1.3	1.3	0.6	0.4	0.3	0.2
β-CH₂	1.6	1.4	1.1	0.6	0.4	0.2	0.2
side-chain CH₂	1.3	1.3	1.1	0.7	0.5	0.4	0.3
CH₃	1.3	1.3	1.3	0.9	0.6	0.6	0.5
				Field = 21000 Gauss			
α-CH	2.6	--	2.8	2.8	--	--	--
β-CH₂	2.7	--	2.6	2.5	--	--	--
side-chain CH₂	--	--	--	--	--	--	--
CH₃	--	--	--	--	--	--	--

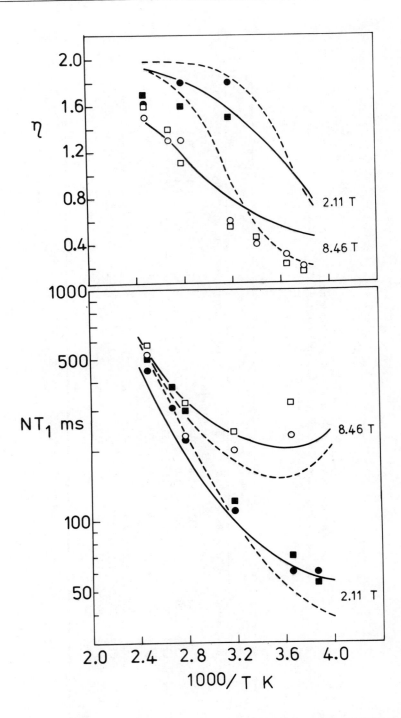

Figure 12. Comparison of the relaxation curves computed from Model I (dashed curves; see text) and Model II (solid curves) with the observed data for the chain backbone of polybutene-1: Circles, α carbons; squares, β carbon; open points, data obtained at 8.46 Tesla; closed points, data obtained at 2.11 Tesla.

Both this function and the continuous distribution functions have the effect of broadening and raising the T_1 minimum and also of broadening the η transition. In addition, as the assumed τ_D/τ_0 ratio in the Monnerie function is increased, the minimum in the T_1-τ_D curves moves markedly to longer τ_D.

In Table IV are presented the T_1 and nuclear Overhauser effect data for a polybutene-1 of weight average molecular weight 69500 in pentachloroethane solution (10 percent w/v) at temperatures from -15° to 130°C and at two field strengths, 21,100 gauss (ω_C = 22.62 MHz) and 84600 gauss (ω_C = 90.52 MHz).[41] (The T_1 values are considered reliable to within ± 5 percent, then η values to no better than ± 10 percent, particularly at the lower temperatures.) The data were interpreted using both the single correlation time model (Model I) and the Monnerie model (Model II). Fitting to Model I was not very satisfactory, but for comparison purposes two reasonable assumptions were made: (a) Model I was considered to be valid at the highest temperature (130°) so that τ_c could be obtained from NT_1; (b) it was assumed that τ_c has an Arrhenius-type dependence on temperature with a value of apparent activation energy of 20 kJ-mol^{-1}, a typical potential barrier value for a polymer chain. Then τ_c could be calculated for lower temperatures and NT_1 and η predicted for this model.

The correlation times calculated for single-τ_c (Model I) and the two-τ_c (Model II) schemes as a function of temperature are shown in Table V, and the calculated η and NT_1 are shown in Figure 12. Neither model fits the data within experimental error

Table V
Parameters for Motional Models to Describe the C-13
Relaxation of Polybutene-1 as a Function of Temperature

Temp., °C	Model I	Model II	
	τ_c, nanosec.	τ_D, sec.	τ_D/τ_0
130	0.10	0.056	0.10
100	0.16	0.17	0.20
85	0.21	0.21	0.25
40	0.56	0.54	0.33
20	0.93	1.2	0.50
0	1.7	2.1	0.70
-10	2.4	3.5	0.80
-15	2.8	4.7	0.90

over the entire temperature range, but Model II seems closer; two correlation times appear to be better than one. The NT_1 minima predicted by Model I are, as expected, too low; it fits the NT_1 data at higher temperatures but at these temperatures predicts values for η at 8.46 Tesla which are greater than observed. (On the other hand, Model I predicts the low temperature η values more satisfactorily at this field.)

It is probably significant that at high field and low temperatures the NT_1 values for the α- and β-carbons are not equal, as they should be if the dipole-dipole isotropic model is basically correct, regardless of assumptions concerning the properties of the correlation times. This may mean that the basic model as embodied in equation (2) is not quite accurate, and that the assumption of isotropic motion may be called into question. Similar behavior has been reported for poly(propylene oxide) in bulk[32/33] and in solution[42].

It can be seen from Table V that the ratio τ_D/τ_0 (calculated for optimum fitting to the experimental data) increases as the temperature decreases, indicating a larger activation energy for local segmental jumps than for the longer range reorientations, a conclusion also drawn for polystyrene and poly(propylene oxide)[34]. The values of the activation energies are 22 kJ-mole^{-1} for segmental jumps and 12 kJ-mole^{-1} for the larger scale motions. The first value approximates the main-chain torsional barrier and the second is nearly equal to the 13 kJ viscosity activation energy of the solvent, reasonable values and supportive of the proposed motional model but of course no proof of its correctness.

References and Footnotes

1. J. R. Lyerla, Jr. and G. C. Levy, *Topics in Carbon-13 NMR Spectroscopy*, Vol. 1, Chap. 3 (1974).
2. G. C. Levy, *Acc. Chem. Res. 6*, 161 (1973).
3. E. Breitmaier, K.-H. Spohn and S. Berger, *Angew. Chem. (Internat. Ed. In Engl.) 14*, 144 (1975).
4. D. Doddrell, V. Glushko and A. Allerhand, *J. Chem. Phys. 56*, 3683 (1972).
5. R. R. Ernst, *J. Chem. Phys. 45*, 3845 (1966).
6. D. E. Dorman, E. P. Otocka and F. A. Bovey, *Macromolecules, 5*, 574 (1972).
7. J. C. Randall, *J. Polym. Sci. Polym., Phys. Ed., 13*, 901 (1973).
8. M. E. A. Cudby and A. Bunn, *Polymer, 17*, 345 (1976).
9. F. A. Bovey, F. C. Schilling, F. L. McCrackin and H. L. Wagner, *Macromolecules, 9*, 76 (1976).
10. D. E. Axelson, L. Mandelkern and G. C. Levy, *Macromolecules, 10*, 557 (1977).
11. T. N. Bowmer and J. H. O'Donnell, *Polymer, 18*, 1032 (1977).
12. J. C. Randall, *J. Appl. Polym. Sci., 22*, 585 (1978).
13. D. E. Axelson, G. C. Levy and L. Mandelkern, *Macromolecules, 12*, 41 (1979).
14. F. A. Bovey, F. C. Schilling and W. H. Starnes, Jr., *Polymer Preprints*, in press.
15. Y. Inoue, A. Nishioka and R. Chujo, *Makromol. Chem., 168*, 163 (1973).
16. F. C. Schilling and F. A. Bovey, unpublished observations (1974).
17. Y. K. Levine, N. J. M. Birdsall, A. G. Lee and J. C. Metcalfe, *Biochemistry, 11*, 1416 (1972).
18. J. R. Lyerla, H. M. McIntyre and D. A. Torchia, *Macromolecules, 7*, 11 (1974).
19. R. C. Long, Jr., J. H. Goldstein and C. J. Carman, *Macromolecules, 11*, 574 (1978).
20. K. J. Ivin and J. B. Rose in *Advances in Macromolecular Chemistry*, Vol. 1, W. M. Pasika, Ed. (Academic Press, New York), 336. (1968)
21. T. W. Bates, K. J. Ivin, and G. Williams, *Trans. Faraday Soc., 63*, 1964 (1967).
22. W. H. Stockmayer, *Pure Appl. Chem., 15*, 539 (1967).
23. H. Block and A. M. North, *Adv. Mol. Relaxation Processes, 1*, 309 (1970).
24. J. Schaefer and D. F. S. Natusch, *Macromolecules, 5*, 416 (1972).
25. F. Heatley and A. Begum, *Polymer, 17*, 399 (1976).
26. R. E. Cais and F. A. Bovey, *Macromolecules, 10*, 757 (1977).
27. W. H. Stockmayer, A. A. Jones, and T. L. Treadwell, *Macromolecules, 10*, 762 (1977).
28. A. H. Fawcett, F. Heatley, K. J. Ivin, C. D. Stewart, and P. Watt, *Macromolecules, 10*, 765 (1977).
29. E. Helfand, *J. Chem. Phys., 54*, 4651 (1971).
30. J. Schaefer and D. F. S. Natusch, *Macromolecules, 5*, 416 (1972).
31. W. H. Stockmayer, K. Matsuo, J. A. Guest, and W. B. Westphal, *Macromol. Chem. 180*, 281 (1979).
32. J. Schaefer, *Macromolecules 6*, 882 (1973).
33. J. Schaefer, *Topics in Carbon-13 NMR Spectros., 1*, 150 (1974).
(1976).
34. K. S. Cole and R. H. Cole, *J. Chem. Phys. 9*, 341 (1941).
35. T. M. Connor, *Trans. Faraday Soc. 60*, 1574 (1964).

36. B. Valeur, J.-P. Jarry, F. Geny, and L. Monnerie, *J. Polym. Sci., Polym. Phys. Ed, 13*, 667, 675, 2251 (1975).
37. F. Heatley and M. K. Cox, *Polymer, 18*, 225 (1977).
38. F. Heatley, A. Begum, and M. K. Cox, *Polymer, 18*, 637 (1977).
39. S. H. Glarum, *J. Chem. Phys., 33*, 639 (1960).
40. B. I. Hunt and J. G. Powles, *Proc. Phys. Soc. (London), 88*, 513 (1966).
41. F.C. Schilling, R.E. Cais and F. A. Bovey, *Macromolecules, 11*, 325 (1978).
42. F. Heatley, *Polymer, 16*, 493 (1975).

Single Crystal Crystallography[1]

Nadrian C. Seeman
Institute of Biomolecular Stereodynamics
Center for Biological Macromolecules
and Department of Biological Sciences
State University of New York at Albany
Albany, New York 12222

Introduction

X-ray diffraction is the most powerful tool at our disposal for the elucidation of the three-dimensional structures of biological molecules. It has played a crucial role in the development of our understanding of the ways in which these molecules form discrete structures and the ways in which these structures interact with each other. There are two prominent types of X-ray diffraction procedures which have been used for this purpose: (1) single crystal analysis and (2) helical diffraction from oriented fibers and gels. While extensive single crystal analysis of peptides and proteins dates from the 1950's,[2,3] it is only within the last decade that investigators have been able to successfully determine the structures of oligonucleotides[4-17] and short macromolecular nucleic acids such as transfer RNA.[18-20] Prior to that time, all of our knowledge of nucleic acid structures larger than monomers relied on helical diffraction studies. Since the other chapters in this volume do not deal with helical diffraction extensively, this chapter is restricted to being an introduction to single crystal analysis. Those interested in the basics of helical diffraction are referred to the fine introduction written by Wilson,[21] and to the papers of Arnott[22] and his colleagues.

The result of a successful X-ray crystallographic structure determination is a statue of the substance being investigated, at the resolution to which the crystallographic data has been obtained. Distances, angles and particularly conformational parameters are therefore directly observable from the crystallographic study. It is important to realize that a structure derived by single crystal techniques is a minimally inferential depiction of the molecule; the bias introduced by the investigator is small, compared with those spectroscopic techniques which rely heavily upon a combination of indirect structural data with molecular model building. The crystallographic theory for relating the positions of atoms within the crystal to the intensities of diffracted rays is sufficiently well developed, that the quality of the crystals and the quality of the diffraction data are the limiting variables in the determination of nucleic acid structures. The following sections will be a description of that theory, emphasizing those parts of particular interest for non-crystallographers studying nucleic acid structure. For a more complete introduction to crystallography, the reader is referred to the excellent book by Glusker and Trueblood,[23] or the somewhat more advanced treatment of Stout and Jensen.[24] A good description of macromolecular crystallography may be found in Blundell and Johnson.[25]

A Crystal is a Periodic Array of Material

Those crystals that we will be discussing will be 3-dimensional arrays, but our illustrations will frequently be limited to two-dimensional examples for clarity. Figure 1a shows a simple motif, a set of dots. Figure 1b shows a lattice of points which establish a given periodicity in two dimensions. Figure 1c shows a periodic array, or crystal, of the structural motif shown in Figure 1a, with the periodicities of Figure 1b.

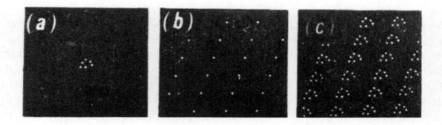

Figure 1. *Forming a crystal from a structural motif.* (a) A structural motif formed by a series of dots. (b) A repeating lattice. (c) A "crystal" of the dot pattern formed by placing the structure at each of the lattice points. This is analogous to a molecular crystal; each of the "molecules" is parallel to the others.

Figure 1c illustrates the importance of molecular crystals in the determination of molecular structure. A crystal is the only means which we have at present for holding a large number of molecules parallel to each other and allowing us to change their orientation (with respect to a laboratory reference frame) in unison. In solution, the orientation of molecules is random. Thus, any observation on a large number of molecules will yield a spherical average of their spatial properties. This is like trying to read the label on a phonograph record while it is playing. On the other hand, it should be remembered that crystal structures may be sensitive to small perturbations induced by the presence of the lattice. Furthermore, one conformation may crystallize under the conditions used to prepare the crystals, but others may exist in solution. Thus, crystallography should be used in conjunction with solution techniques in order to determine the complete range of structures available to the substance. The fundamental periodic unit of a crystal is called the *unit cell*. As shown in Figure 2, the three vectors which define the periodicity are termed **a**, **b**, and **c**. The angle between **b** and **c** is termed α, that between **a** and **c** is β, and the one between **a** and **b** is γ. Crystallographic coordinates are usually quoted as fractions of these fundamental repeat vectors. Note that there is no requirement for the repeat vectors to be equal in length, nor is there any requirement that they be perpendicular to each other. In the most general case, neither of these possibilities is fulfilled. However, the symmetry of a given unit cell's contents may impose such requirements.

The contents of a unit cell are frequently revealed to be a symmetric arrangement of

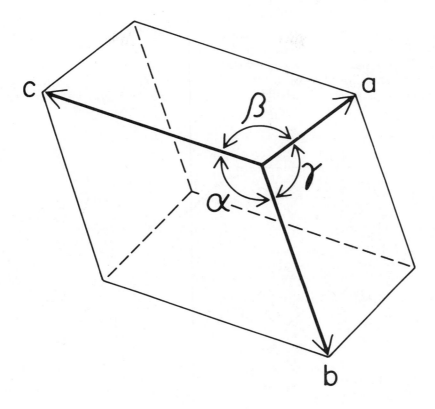

Figure 2. *The general unit cell showing the three vectors defining the periodicities in each of the three directions.* The three vectors are **a**, **b**, and **c**. The angle between **a** and **b** is γ, that between **a** and **c** is β and the one between **b** and **c** is α. Note that the lengths of **a**, **b**, and **c** are not equal, and that α, β and γ are neither equal to each other nor to 90°.

molecules which extends throughout the lattice. Only a subset of crystallographically allowed symmetries are available to molecules which contain asymmetric carbon (or other) atoms, such as proteins and nucleic acids. These are the proper rotation symmetries involving 2-fold, 3-fold, 4-fold and 6-fold axes. The rotational symmetry may be combined with a translation parallel to it to yield a screw axis. A complete listing of crystallographic symmetries may be found in the International Tables for Crystallography, Vol. 1.[26] If symmetry does exist within a unit cell, the unique portion is termed the *asymmetric unit*.

X-Rays are Light Waves of Very Short Wavelength

The X-rays used in structure determination have wavelengths of about 1 Å. The most popular radiation is copper K$_\alpha$, wavelength 1.5418 Å. Since we will be discussing the interaction of X-rays with matter, it is necessary that we understand the way in which the sinusoidally oscillating waves of which X-rays are composed interact with each other. A wave may be characterized for our purposes by its wavelength, λ, its amplitude, A, and its phase, ϕ, as shown in Figure 3. The wavelength is the distance between crests, the amplitude is just the height of the wave at the crest, and the phase is the distance of the crest from an arbitrary origin. It is convenient to describe waves in polar coordinates in the complex plane. A wave at any given instant is represented as a vector from the origin of the complex plane, with length A, cor-

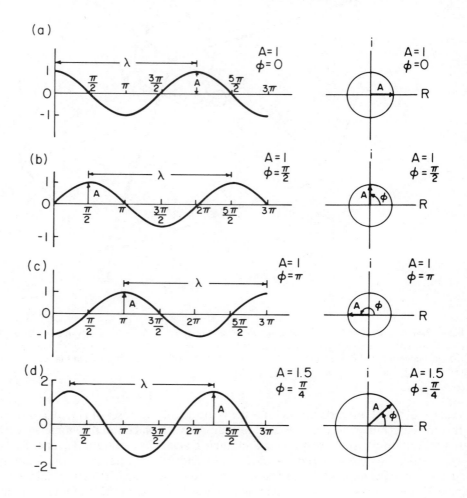

Figure 3. *Waves, amplitudes and phases.* The four parts of this figure illustrate waves both in their fully drawn form and by their representation in the complex plane. On the left of each part is a wave, whose amplitude is indicated as A, and whose phase, in radians is indicated by ϕ. The wavelength for each wave is the same, and is indicated by λ, the distance between the crests. The distance along the abscissa has been scaled in radians. On the right, the amplitudes are indicated by the length of the vector A, and the phase is indicated by the angle ϕ. The projection of each vector on the real axis of the complex plane is $A\cos\phi$, and the projection of each vector on the imaginary axis is $A\sin\phi$. The circles on the complex planes indicate the locus of the possible vectors corresponding to waves of the indicated amplitude. The actual wave illustrated is indicated by the vectors drawn in the complex plane. (a) A wave of unit amplitude and zero phase. (b) A wave of unit amplitude and phase of $\pi/2$, or 90°. (c) A wave of unit amplitude and phase of π, or 180°. (d) A wave of amplitude 1.5, and phase $\pi/4$, or 45°.

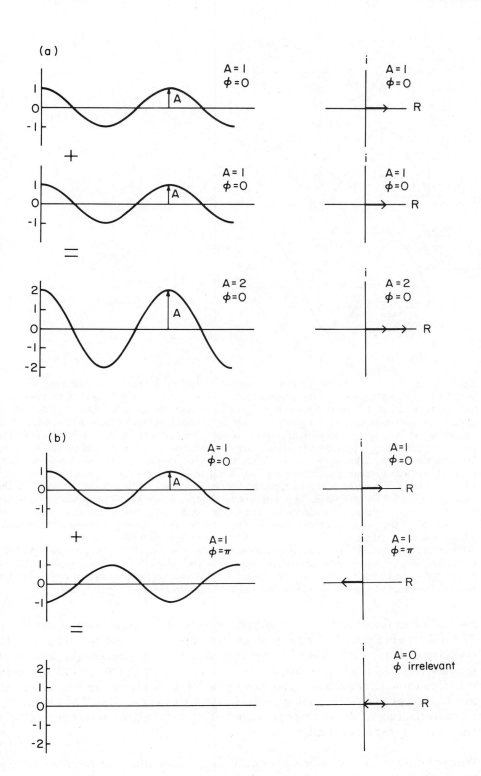

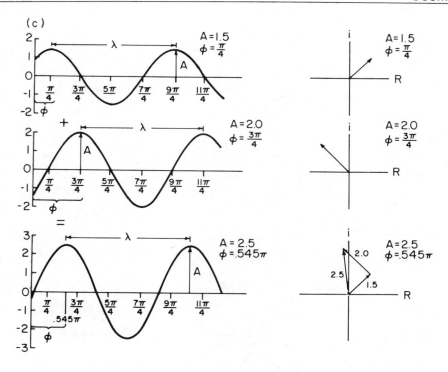

Figure 4. *Superposition of two waves.* The superpositions of waves are indicated in both wave and complex plane representations in this figure. Note that the waves represented on the left side of each figure are the same as those indicated by the vectors in the complex plane on the right side. Point by point addition of the two top waves in each part of the figure yields the wave at the bottom part. Similarly, addition of the two vectors in the top complex planes yields the complex plane representation of the wave which is the resultant. Note that the wavelengths of all waves are the same. Only amplitudes and phases are altered by the summation. (a) Addition of two unit amplitude waves of identical phase. The result is a wave of double amplitude with the same phase. Note that there is nothing special about the fact that the phase of 0 was chosen for this example. Whatever the phases of the two waves, so long as they are identical, the resultant wave will have that phase. (b) Addition of two unit amplitude waves of opposite phase. In this case, the two waves are exactly π radians (180°) out of phase, although their amplitudes are equal. The resultant wave is the complete cancellation of each wave by the other, resulting in a wave of 0 amplitude. (c) Addition of two arbitrary waves. In this case, the first wave is of amplitude 1.5, and phase $\pi/4$ radians (45°), while the second wave is of amplitude 2.0 and phase $3\pi/4$ radians (135°). The point by point addition of the two waves on the left yields the wave indicated at the bottom. This wave has amplitude 2.5 and phase 0.545 π radians (98.1°). Note that the addition of the two vectors representing the first two waves in the complex plane yields the vector which represents the resultant wave in the complex plane.

responding to the amplitude, and angular position ϕ, corresponding to the phase. This is shown in Figure 3. If we moved the wave, the vector would sweep out a circle, changing ϕ, but not A. Thus, the amplitude would not change, but the position of the crest from the origin would be altered. As it traversed a single period, the phase would go from 0 to 2π radians, and then repeat. Note that the projection on the real axis is $A\cos(\phi)$ and the projection on the imaginary axis is $A\sin(\phi)$. The wave is just the vectorial sum of these components, or $A(\cos(\phi) + i\sin(\phi))$, which may be written as $Ae^{i\phi}$, by the Euler relationship.

When two waves of the same wavelength are superimposed, the resultant wave is merely the point by point addition of the two waves. As shown in Figure 4, this can also be represented by the vectorial sum of their complex plane representations. The resultant wave will have real axis component $A_1(\cos(\phi_1)) + A_2(\cos(\phi_2))$ and imaginary axis component $A_1(\sin(\phi_1)) + A_2(\sin(\phi_2))$. This may also be written $A_1 e^{i\phi_1} + A_2 e^{i\phi_2}$.

The Scattering of X-rays by a Crystal Depends on The Superposition of Waves

Since no means has been found to focus X-rays, it is not possible to devise a lens which will give a direct picture of the contents of a crystal. Therefore we must make a more detailed analysis of the scattering process in order to interpret the pattern which is observed. Let us look at the general case, depicted in Figure 5. Our experimental setup consists of a source of X-rays at infinity, a scattering body in line with the source, and a detector at infinity. In order for scattering to be detected by the detector, it must be scattered through an angle at 2θ. For this experiment, we have thus fixed the orientation of the object with respect to the source, detector and an arbitrary origin, O within it. We have furthermore fixed the scattering angle at 2θ. The question we are asking is, what will be observed at the detector under these conditions?

Let us begin by considering the scattering from two points within the object, our origin, O and another point, M, removed from O by the vector **R**. We shall only be concerned with the elastic scattering experiment, so the wavelengths of our incident and scattered beams will be equal in this analysis. In Figure 5, **I** represents a unit vec-

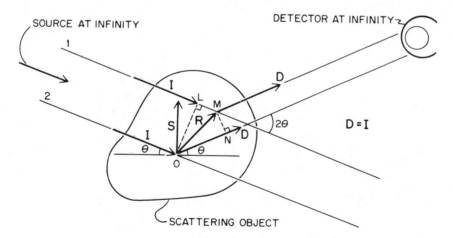

Figure 5. *Scattering of X-rays by an arbitrary object.* A single scattering experiment is depicted in this figure. The scattering object is being illuminated by a source which is at infinity, and we are considering scattering of X-rays by the object as detected by a device also at infinity. The deflection of the scattered beam being monitored from the incident beam is indicated by the angle 2θ. The 2θ angle is characteristic of the experimental setup. We consider the interference of two rays, one scattered from point O and the other scattered from point M. The vector from O to M is indicated by the vector **R**. **I** is a unit vector in the direction of the incident beam, while **D** is a unit vector in the direction of the scattered beam. The perpendicular from O, our origin, to incident ray 1 intersects that ray at point L, while the perpendicular from M to scattered ray 2 intersects that ray at point N. ON-LM is the path length difference between the two rays. **S** is the difference vector between **D** and **I**. **S** bisects the angle between **I** and **D**. Note that the perpendicular to **S** makes two equal angles θ with **I** and **D**.

tor in the direction of the incident beam, and **D** represents a unit vector in the direction of the scattered beam. Clearly, the waves detected at infinity will be the superposition of the waves scattered from the points O and M. Since they are of the same wavelength, the path-length difference between them will determine their relative phases. The difference between the distance traveled by the second wave compared with the first, Δ, is just the length of $\overline{NO}$ less the length $\overline{LM}$; otherwise, they travel the same distances:

$$\Delta = \overline{NO} - \overline{LM}$$

The length of $\overline{NO}$ is just $\mathbf{D} \cdot \mathbf{R}$ (i.e., the projection of **R** on **D**). Similarly, the length of $\overline{LM}$ is just $\mathbf{I} \cdot \mathbf{R}$ (i.e., the projection of **R** on **I**). Thus,

$$\Delta = \mathbf{D} \cdot \mathbf{R} - \mathbf{I} \cdot \mathbf{R} = (\mathbf{D}\text{-}\mathbf{I}) \cdot \mathbf{R} = \mathbf{S} \cdot \mathbf{R},$$

where **S** is the difference vector between **D** and **I**. Since the magnitudes of **D** and **I** are the same, **S** bisects the angle between them. If we construct the perpendicular to **S** through O, we can see that both **I** and **D** make an angle θ with this perpendicular. The scattering angle, as you will recall, is 2θ. The length of **S** is $2\sin\theta$, since both **I** and **D** are unit vectors (see Figure 6). If we wish to convert th path length difference to a phase difference in radians, we must multiply it by $2\pi/\lambda$. This makes our phase difference $2\pi/\lambda \, \mathbf{S} \cdot \mathbf{R}$.

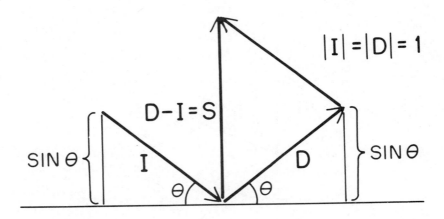

Figure 6. *Illustration of the magnitude of the* S *vector. The two components of the S vector are each sinθ, since D and I are both unit vectors. Hence the magnitude of S is 2sinθ.*

X-rays are scattered only by electrons.[27] The final result of a structure determination will therefore be the electron density map of the contents of unit cell, which will in turn be a statue of the molecules of which it is composed. Points of higher electron density will scatter X-rays more strongly than points of lower electron density. Thus, each of the diffracted waves originating at different points **R** will be weighted by a density term, $\varrho(\mathbf{R})$, and each scattered wave will be described by a term $\varrho(\mathbf{R}) \exp (2\pi i/\lambda \, \mathbf{S} \cdot \mathbf{R})$. For this orientation of the scattering object with respect to the source and detector, the scattered wave observable at the detector, G(S), will simply be the sum of the interfering waves throughout the object:

$$G(S) = \int_{object} \varrho(R) \exp [(2\pi i/\lambda)(S \cdot R)]dR$$

This G(S) term is just a complex number which characterizes the result of the scattering experiment described above. It is just the addition within the complex plane of all the scattering points within the object, weighted by the scattering density at those points and phased by their path length differences, as shown in Figure 7. In that figure, A_1, A_2 and A_3 are the amplitudes characteristic of scattering from three points in the object, and ϕ_1, ϕ_2, and ϕ_3 are the phases resulting from their path length differences from an arbitrary origin. The integral is the summation of all these vectors, of which we have shown three in the figure for simplicity. It is also clear from the figure that changing the origin would alter the phase, but not the amplitude of the resultant vector, since it would correspond to a rotation of the figure about the origin of the complex plane. The amplitude is determined by the relative distribution of scattering matter within the object.

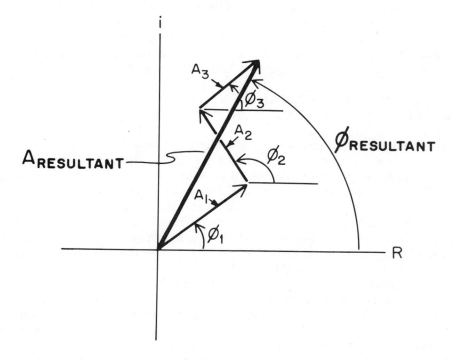

Figure 7. *Illustration in the complex plane of the summation of scattering from 3 points.* Each wave has an amplitude, A, indicated by the length of the vector, and a phase relative to an arbitrary origin, indicated by the phase angles ϕ. The resultant wave amplitude is indicated, with its resultant phase as well. This represents the contribution of each of 3 scattering points to the resultant wave observed at the detector. Clearly rotating the entire assemblage about the origin of the complex plane (changing the arbitrary origin to which phases are referred) would alter the resultant phase, but not the resultant amplitude. By adding more vectors, the scattering from each point in the body can be added into the summation to give a resultant for the entirely scattering object.

The Total Scattering From An Object Is Defined Within
A Sphere Called Reciprocal Space

If we left the object in Figure 5 fixed, and rotated the souce and detection apparatus about the origin, keeping 2θ fixed, we would get a new scattering situation. (In practice, the object is usually rotated, rather than the experimental apparatus.) S would be of the same length, but would be pointed in a new direction, corresponding to the new orientation of the source and detector. This new vector S would still bisect the angle between I and D, but S•R would be different, for each R vector considered. The complete set of vectors S, corresponding to the scattering angle, 2θ, and all possible orientations of the experimental apparatus with respect to the scattering object defines a spherical shell of radius corresponding to the magnitude of S, namely $2\sin\theta$. At each of the positions on this shell, the scattered wave, G(S) is defined.

The scattering angle, 2θ, can clearly be varied, in the range from 0 to 180°. Thus, S ($=2\sin\theta$) can range from 0 to 2. Thus, we have another shell of radius $2\sin\theta$ for each possible magnitude of S. This gives a solid sphere of radius 2 for the complete locus of the heads of vectors S. At each point within this sphere, the scattering function, G(S) may be observed. This sphere is called *reciprocal space*. It is important to remember that the G(S) function, characteristic of the scattering from the object, is tied to the object. If the object is rotated, G(S) rotates with it.

Scattering Factors for Individual Atoms
Characterize Molecular Scattering

It is very convenient to talk about the scattering arising from individual atoms, rather than from a continuum of scattering matter. For the resolutions with which we are dealing, we need not be concerned with the oriented aspects of atomic structure. Thus, the model used by crystallographers who deal with moderately large biological molecules (20 atoms or above) does not take into account the directionality of lone pairs or bonding electrons. The atom is treated as a spherically symmetric object whose oriented features have been eliminated in a spherical average. In deriving a scattering function f(S) (S now a scalar because of the spherical average), we assume that the electrons can be localized within atoms, that they scatter independently from each other, and that they scatter in the crystal just like they do when in isolated atoms. Representative scattering curves are shown in Figure 8. The only assumption which is not transparent to the treatment of data on nucleic acid fragments and tRNA is the independent scattering of electrons. In our treatment, we have assumed that the phase shift upon scattering (180°) is the same for all points in the scattering object. In certain cases, it is slightly different, resulting in *anomalous scattering*. This is a slight correction to the scattering from certain atoms (phosphorus is the only significant one amongst nucleic acids), resulting from some electrons scattering at a slightly different time from the others. Thus, the scattering factor f(S) is actually a sum of three terms,

$$f(S) = f_o(S) + f'(S) + if''(S),$$

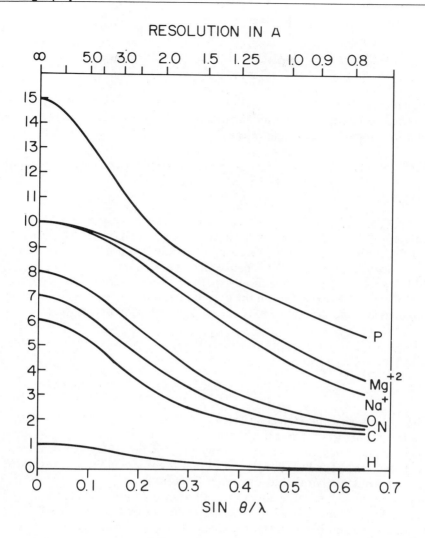

Figure 8. *Scattering curves for atoms found in nucleic acid structures.*[26] The scattering for each atom is indicated as a function of scattering angle, normalized for wavelength. Each curve is normalized for the number of electrons in the atom or ion. The resolution is indicated at the top of the figure. Note that the curves are not all the same shape, and that the contributions of all atoms decline with increasing scattering angle.

where f_o is the normal scattering factor and f' is a real (and usually negative) correction term, while f'' ($= .43$ electrons for P with CuK_α radiation) is an imaginary correction term. Use of anomalous dispersion is extremely important in determining the structures of both nucleic acid fragments[12] and tRNA.[18]

Scattering From A Crystal Samples The Complete Scattering Pattern

With this assumption of atomic scattering, we can now describe the scattering just by adding up the contributions from each of the atoms in the molecule or molecules which we are observing with our apparatus:

$$G(S) = \sum_{j=1}^{N \, atoms} f_j(S) \, \exp[(2\pi i/\lambda)(S \cdot R_j)]$$

To make a crystal of our molecule, all we have to do is place our molecules next to one another, separated by the fundamental periodicities **a**, **b**, and **c**, and add up the contributions for each of the unit cells in the crystals:

$$G(S) = \sum_{N_1=1}^{N_a} \sum_{N_2=1}^{N_b} \sum_{N_3=1}^{N_c} \sum_{j=1}^{N \, atom} f_j(S) \, \exp[(2\pi i/\lambda)S \cdot (R_{j+} N_1 a + N_2 b + N_3 c)]$$

where N_a, N_b and N_c are the number of unit cells in each of the **a**, **b**, and **c** directions, respectively. Thus, with a given value of the indices N_1, N_2, N_3 and j, we are accounting for the contributions to the scattering from the j'th atom in the N_1th unit cell in the **a** direction, N_2th unit cell in the **b** direction and N_3th unit cell in the **c** direction. It should be noted that we are seeing the spatial average of the crystal's content. If a mixture of structures exists within the crystal (e.g., two different solvent structures) the weighted average structure will be seen. This expression may be factored to give:

$$G(S) = (\sum_{j=1}^{N \, atom} f_j(S) \, \exp[(2\pi i/\lambda)(S \cdot R_j)]) \, (\sum_{N_1=1}^{N_a} \exp[(2\pi i/\lambda)(S \cdot N_1 a)]) \times$$
$$(\sum_{N_2=1}^{N_b} \exp[(2\pi i/\lambda)(S \cdot N_2 b)]) \, (\sum_{N_3=1}^{N_c} \exp[(2\pi i/\lambda)(S \cdot N_3 c)]).$$

The first term is just the molecular scattering term from an isolated molecule. The other terms are *fringe functions*, which are related to the periodicities of the lattice. Figure 9 demonstrates the effect of making a crystal out of a given structural motif in two dimensions. The boxes at the bottom of the figure are structures whose scattering has been demonstrated by optical diffraction techniques. In Figure 9a, we see the individual array and above it is its scattering function. This function is a continuum. In 9b, we have started to make a one dimensional crystal out of it, by placing it adjacent to another version of the same array. This results in a set of broad fringes across the transform. In 9d, an infinite (for our purposes) 1-dimensional crystal has been formed, and the fringes are very fine, although they have the same separation as in 9b. The function has a value of effectively 0 at all points off the fringe. By squinting at the figure, it may be readily seen that at the points where the fringes allow the function to come through, it looks the same as in 9a. Figures 9b, 9e and 9f show the same principles with respect to the addition of a second dimension of periodicity. The spacings of the fringes are inversely related to the periodicity. If the periodicity in Figure 9d is d, then the separation of the fringes will be λ/d. Thus, in Figure 9d, the central fringe runs through the origin of reciprocal space, and the first one over is removed from it by λ/d, the second one by $2\lambda/d$, the third by $3\lambda/d$ and so on. These separations may be rewritten as λ/d, $\lambda/(d/2)$, $\lambda/(d/3)$ ... $\lambda/(d/n)$, indicating that they contain information about the structure of resolutions d, d/2, d/3, ..., d/n, respectively. Thus,

$$\lambda/(d/n) = 2\sin\theta.$$

This expression may be rearranged to yield Bragg's law,

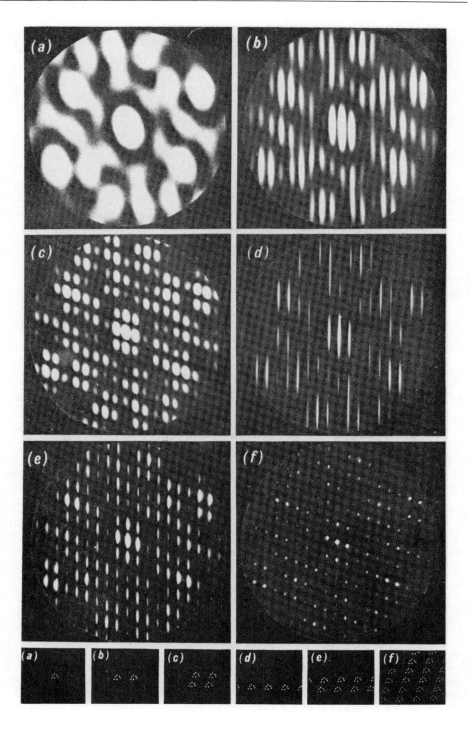

Figure 9. *The effect on scattering of incorporating a structural motif into a crystal.* The small blocks at the bottom of the figure are the objects whose scattering functions are shown in the large blocks with corresponding letters at the top. This is an emulation of the X-ray scattering case by using masks and optical diffraction tecniques.[46] (a) The structural motif and its scattering function. In this two dimensional exam-

$$n\lambda = 2d\sin\theta.$$

which is the Bragg condition for diffraction by a crystal.

The reader should examine Figure 9f in detail and realize that even in the case of the two-dimensional crystal (and of course in the three-dimensional case as well), the parts of the scattering function which do come through the fringe function have the same intensity as they do in 9a. Thus, by putting the molecule in a crystal, and fixing many parallel copies of it so that we can manipulate it macroscopically, we find that we can only *sample* its scattering function, rather than being able to observe it continuously. There is no scattering between the fringes for a crystal since the number of unit cells is so high. Because of the value of being able to deal with oriented molecules, this is a relatively small price to pay.

Scattering From A Crystal Defines The Reciprocal Lattice

This array of intersections of the fringe functions forms a lattice in its own right, a *reciprocal lattice*. The fundamental lattice repeats in the reciprocal lattice are called a*, b*, and c*, and the angles between them are α^*, β^*, and γ^*. Thus, $a^* \cdot a = \lambda$, $b^* \cdot b = \lambda$ and $c^* \cdot c = \lambda$. These reciprocal vectors are defined as

$$a^* = \lambda b \otimes c/V, \; b^* = \lambda c \otimes a/V \text{ and } c^* = \lambda a \otimes b/V,$$

where V is the volume of the unit cell. The different points in the reciprocal lattice (termed, naturally enough, reciprocal lattice points) are indexed from the origin of reciprocal space in terms of positive or negative whole numbers. These indices are called *Miller indices*, h,k and l. Thus, the point which is 3rd along a*, 5th in the opposite direction from b* and 4th along c* in the three dimensional lattice would have miller indices h = 3, k = -5, and l = 4. This represents a vector in reciprocal space with components 3a*, -5b*, 4c*.

Figure 10 shows the same structural motif as in Figure 9 (and Figure 1), but this time assembled into lattices of different periodicities. The reader should note that the intensities are the same at points with the same S value in reciprocal space, regardless

ple, reciprocal space is the contents of a circle, rather than a sphere. Each point within the circle corresponds to an S vector, from the origin at the center. The alteration in intensity corresponds to G(S), the scattering function characteristic of the structure. (b) A simple crystal has been constructed by juxtaposing two of the structures seen in (a). The effect on the scattering function is to overlay it with a fringe, caused by the interference between the two parallel structures. Because the crystal is very short, the fringe is very broad. Note that where the function is visible, however, it appears to have the same intensity as the scattering function seen in (a). (c) The crystallinity has been extended to a second dimension, resulting in a second set of fringes reflecting the newly added periodicity. Again the fringes are broad because of the limited nature of the crystal. (d) An infinite crystal of the structure in one dimension. The fringes are now narrow because of the infinite character of the crystal. (e) An infinite crystal in the horizontal direction but only two lattice rows in the other direction. Now the fringes are narrow in the horizontal direction, but are still broad in the other direction. (f) An infinite crystal in both directions. The fringes are sharp in both directions. Note that by squinting at the picture you can notice that the intensity distribution in (f) is the same, where sampled, as in (a). The complete optical masks for (d), (e), and (f) are not shown.

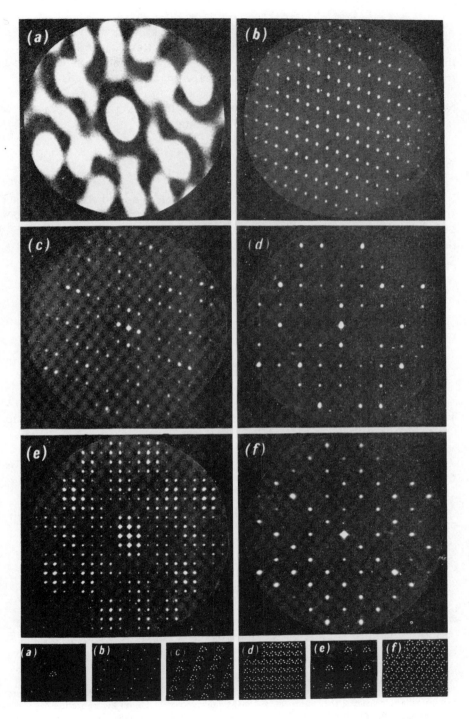

Figure 10. *The effect of changing the lattice on the scattering function.* The same protocol is used in this figure as in Figure 9. (a) The structural motif of Figure 9 (a) and its scattering function. (b) The lattice of Figure 9 (f) and its scattering function. Since there is only one point per unit cell, just the fringes are seen, due to the effect of the lattice. (c) The structural motif of (a) incorporated into the lattice of (b). This is the same pattern as seen in Figure 9 (f). (d) Incorporation of the structural motif into a lattice with a very

of the lattice which has been used. The other important fact to note is the demonstration of the reciprocal natures of the direct and reciprocal lattices. In 9d the unit cell is small, but the reciprocal lattice spacings are large; in 9e, the reverse is true: the unit cell is large, and the spacing in reciprocal space is small.

Now that we can index the points in our diffraction pattern, we no longer need to talk only about a G(S) function which is continuous,

$$G(S) = \sum_{j=1}^{N \, atoms} f_j(S) \exp [(2\pi i/\lambda)(S \cdot R_j)].$$

Rather, we can use the Miller indices to refer to a given point in reciprocal space and fractions of a unit cell edge to refer to a point in direct space. Thus, the components of R_j will be $x_j a$, $y_j b$, $z_j c$ and those of S will be ha^*, kb^*, lc^*. Thus, we may rewrite the scattering from the crystal at the points of interest, namely reciprocal lattice points, as

$$F(h,k,l) = \sum_{j=1}^{N \, atoms} f_j(S) \exp [(2\pi i/\lambda)(ha^* \cdot ax_j + kb^* \cdot by_j + lc^* \cdot cz_j)$$

Remembering that $a^* \cdot a = b^* \cdot b = c^* \cdot c = \lambda$, this expression may be simplified to be

$$F(h,k,l) = \sum_{j=1}^{N \, atoms} f_j(S) \exp [(2\pi i)(hx_j + ky_j + lz_j)].$$

The complex number F is called the *structure factor* associated with the reciprocal lattice point h,k,l.

The Scattering From A Crystal Corresponds To The Fourier Transform Of The Unit Cell Contents

At this point, we must ask the meaning of this array of structure factors, F(h,k,l). In order to do this, let us leave discretely scattering atoms for a moment. Following the treatment of Stout and Jensen,[24]

$$F(h,k,l) = \int_{volume \, of \, unit \, cell} \varrho(x,y,z) \exp [(2\pi i)(hx + ky + lz)]dxdydz,$$

where $\varrho(x,y,z)$ is the electron density at point x,y,z within the unit cell. Since ϱ is a periodic function, it must be possible, by the Fourier theorem, to expand it in a Fourier series,

small unit cell. In this case, because of the inverse relationship between direct lattice vectors and reciprocal lattice vectors, the reciprocal lattice spacings are very large. (e) Incorporation of the structural motif into a lattice with a very large unit cell. This results in a very fine sampling of reciprocal space. Note that since the periodicities in both (d) and (e) are orthogonal, the reciprocal lattices are also orthogonal. (f) A small unit cell with a skewed periodicity. Note that in (c), (d), (e), and (f) the intensities of the sampled points are the same as they are in (a). Only the positions available for sampling are different. The complete optical masks for (b), (c), (d), (e), and (f) are not shown.

$$\varrho(x,y,z) = \sum_{h'=-\infty}^{\infty} \sum_{k'=-\infty}^{\infty} \sum_{l'=-\infty}^{\infty} C_{h'k'l'} \exp\left[(2\pi i)(h'x + k'y + l'z)\right],$$

where the $C_{h',k',l'}$ are the Fourier coefficients of $\varrho(x,y,z)$. Plugging this into our expression for $F(h,k,l)$ we get:

$$F(h,k,l) = \int_{\substack{\text{volume of unit} \\ \text{cell}}} \sum_{h'=-\infty}^{\infty} \sum_{k'=-\infty}^{\infty} \sum_{l'=-\infty}^{\infty} C_{h'k'l'} \exp\left[(2\pi i)((h+h')x + (k+k')y + (l+l')z)dxdydz\right.$$

Since trigonometric functions are orthogonal, when we carry through the integration, all terms will be 0, unless $h' = -h$, $k' = -k$ and $l' = -l$. If this is the case, the integral will just equal the volume of the unit cell, so

$$F(hkl) = VC_{-h,-k,-l}$$

Thus

$$\varrho(x,y,z) = 1/V \sum_{h,k,l=-\infty}^{\infty} F_{(h,k,l)} \exp\left[(-2\pi i)(hx + ky + lz)\right].$$

Recalling our earlier equation,

$$F(h,k,l) = \int_{\text{volume of unit cell}} \varrho(x,y,z) \exp\left[(2\pi i)(hx + ky + lz)\right]dxdydz,$$

we can see that the F and ϱ functions are Fourier transforms of each other. This means that we can get from one to the other directly. Given the complete set of structure factors, F, we can reconstruct the electron density function ϱ, through some trivial arithmetic in the computer. Figure 11 shows how this is done for a few terms in a one dimensional case. In the multi-dimensional case, the waves are pointed along the reciprocal lattice vectors $H(=h,k,l)$, and the wavelength corresponds to the d-spacing ($d = 2\sin\theta/n\lambda$) characteristic of that vector H.

The Quality of the Structure Depends on the Extent of the Data

Since we are talking about a finite representation of our structure, we must ask what the effect of a limited data set will be. You will recall that information about finer and finer details of the structure is found in reciprocal lattice points further and further from the origin of reciprocal space. Let us see what happens when data from the outer portions of the Fourier transform of the structure is omitted. Figure 12a shows an image of a duck and Figure 12b shows it Fourier transform. By the direct or inverse Fourier transformation procedure (here done optically), either the transform or the picture of the duck may be produced, one from the other. Figure 12d shows the image of the duck when only the portion of the transform shown in 12c is used to

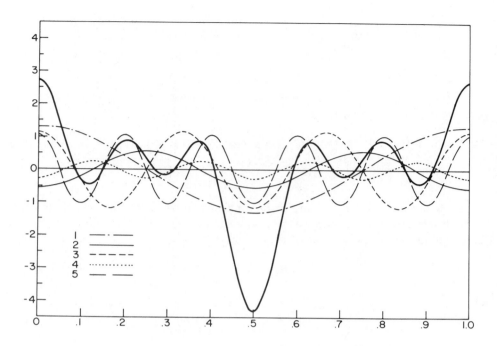

Figure 11. *A one-dimensional Fourier synthesis.* This drawing shows the summation (indicated by the dark line) of the first five terms of a Fourier synthesis. The structure corresponds to 6 unit-scattering atoms placed at fractional positions 0.05, 0.20, 0.35, 0.65, 0.80 and 0.95 in a unit cell 10Å long. The waves of periodicity 1,2,3,4 and 5 per unit cell are shown. Thus the resolution of this synthesis is 2Å. The wave of periodicity 1 has an amplitude of 1.311 and a phase of 0. The wave with periodicity 2 has an amplitude of 0.559 and a phase of π. The wave of periodicity 3 has an amplitude of 1.166 and a phase of 0. The wave of periodicity 4 has an amplitude of 0.256 and a phase of π. The wave of periodicity 5 has an amplitude of 1.071 and a phase of 0. It can be seen that those waves of phase 0 start at the origin of the unit cell on their crests, and those waves of phase π start at the origin of the unit cell at their minimum values. The presence of the mirror plane at 0.5 in this structure forces the phases to be only 0 or π in this example, although this is generally not true. A wave of 0 periodicity (just a constant value) of amplitude 6 has been omitted. Had it been included, all values in the resultant synthesis would be 6 greater than they are shown to be. Note that the atoms at 0.20, 0.35, 0.65 and 0.80 are fairly well resolved from each other, but those at 0.05 and 0.95 (which corresponds to -0.05) are not resolved, because they are separated by only 1Å, and this synthesis only includes terms of 2Å spacing or greater. The dip in the middle of the cell is indicative of series termination error. Addition of higher order terms (6,7,8...) would increase the resolution and make the image sharper, since their spacings are finer.

produce the image of the duck. It is clearly fuzzier than the image shown in 12a. This is because the portion of the transform which contains information about the fine details of the structure has been excluded. In like fashion, the fuzzy image in 12d would only produce a transform out to the indicated circle in 12c: There is no higher order (finer detail) information in the image, so its transform will be blank beyond the circle. A more extreme case is shown in Figures 12e and 12f. Here the image is only marginally recognizable.

Figure 13 shows the same thing, this time with respect to a molecular crystal. Figure 13a shows the diffraction pattern of the crystal simulation shown in 13b. On an

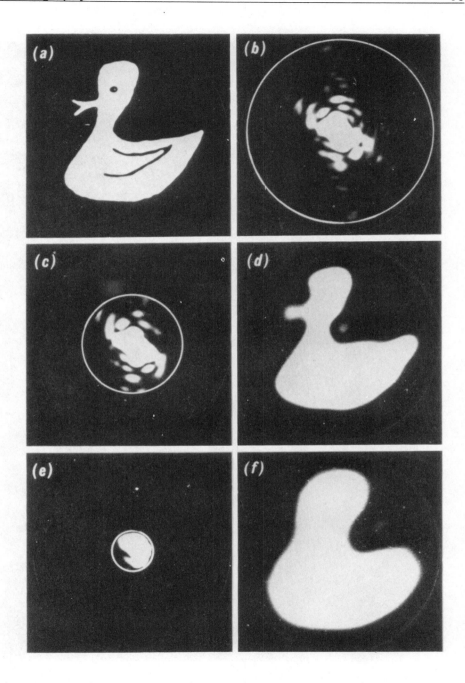

Figure 12. *Illustration of the effect of decreasing resolution on reconstructing an image from its transform.* (a) An image of a duck. (b) Transform of (a). (c) Part of the transform shown in (b) is covered up. (d) The reconstruction of the image shown in (a) using only that part of the transform shown in (c). Note that the fine details are lost, because those parts of the transform containing high resolution information have been suppressed. The transform of the fuzzy image shown in (d) would yield (c), since the outer parts of the transform can only arise if there are fine details, rather than fuzzy images. (e) and (f) are more extreme examples of (c) and (d) respectively.

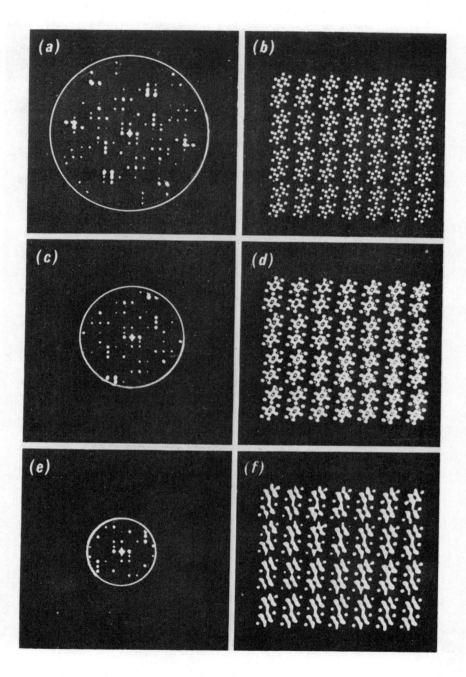

Figure 13. *Simulation of the effect of decreasing resolution on a molecular crystal.* (a) A diffraction pattern from the molecular crystal pattern shown in (b). The pattern shown in (b) has been obtained from the diffraction pattern shown in (a). This corresponds, on an atomic scale, to approximately 1Å resolution. In the pair (c) and (d), the resolution has been decreased to about 1.5Å resolution, and in the pair (e) and (f), the resolution has been further decreased to about 2.2Å resolution. Note the smeariness of the images reconstructed from the limited data sets. Note further, that if the *spatial average* of the images in (d) and (f) looked like any one of the reconstructed images (e.g., through slight misalignments), the patterns in (c) and (e), respectively, would result.

atomic scale, this would correspond to an approximate resolution of about 1Å. The atoms are clearly resolved from one another and the structure is readily apparent. With a more restricted dataset, shown in Figure 13c, the poorer quality image is seen in Figure 13d. This corresponds to approximately 1.5Å resolution—note that the individual atoms are no longer resolved from each other. The bottom pair in the figure, 13e and 13f represent a structure and its transform at about 2.2Å resolution. Fig. 14 is a picture of base pairs at 3Å resolution.

Why, one might ask, don't crystallographers always use data to 1Å resolution or above? Certainly, it is more work to get higher resolution data, since the number of data points increases with the cube of the reciprocal of the resolution. However, the answer lies not in professional slothfulness, but rather in dealing with the realities of nature. The data from a crystal are only so good as the crystal. The fine details of the structure, namely the data far from the origin of reciprocal space, may be lost through a number of circumstances. Thermal motion within the crystal is a prime cause of this problem for smaller structures. As the atoms in the molecule vibrate, they smear the time-averaged image that the crystallographer sees over the course of the scattering experiment. Besides that, crystal imperfection also contributes greatly to loss of data. If one unit cell is tilted slightly with respect to the next, data will be lost through the space averaging over all unit cells in the crystal. A typical crystal of an oligonucleotide will diffract to about 1Å resolution; tRNA crystals at best yield data to 2.5-2.2Å resolution, even when the temperature is lowered to reduce thermal motion. In order to interpret the electron density map from tRNA with a molecular model, it was necessary to understand the basic conformations which nucleic acids could assume from higher resolution studies of small fragments.[18] As is clear from Figure 14, a 3Å density map bears the same relationship to the underlying molecular structure that Figure 12f bears to a duck: If you know what can be there, the map is interpretable. It is therefore critical that crystallographers choose crystals which diffract to adequate resolution to answer the scientific questions in which they are interested, since a lot of time will be wasted if they do not.

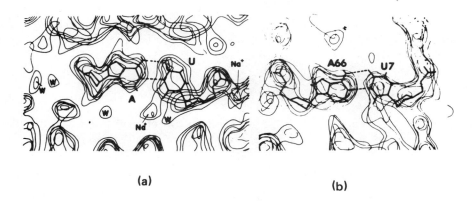

(a)   (b)

Figure 14. *Adenine-Uracil base pairs at 3Å resolution.* Base pairs are indicated by the molecular structures indicated. The electron density levels are indicated by the contouring. (a) A base pair from the structure of the dinucleoside phosphate, ApU.[9] This structure was determined to a resolution of 0.8Å, but only those terms to 3Å resolution have been used in this synthesis. (b) Base pair A66-U7 from the 3Å multiple isomorphous derivative map of Yeast tRNA[phe].[47]

The Phase Problem Complicates Crystallography Enormously

If the structure factors were directly observable through the X-ray diffraction experiment, the mere computation of structures would be all that was required of the crystallographer. Unfortunately, this is not the case. The phases of the complex structure factors are lost in the course of the diffraction experiment. The intensity of the diffracted ray is measurable, but this is proportional (when corrected for geometric and physical factors affecting the experiment) to the square of the amplitude of the structure factor: $I \propto F_{hkl} \cdot F^{*}_{hkl} = F_{hkl}\, e^{i\phi} \cdot F_{hkl}\, e^{-i\phi} = |F_{hkl}|^{2}$. The phase information is totally lost. This loss is termed the phase problem of crystallography. Without the phases, the structure is not directly knowable, and it must be solved by numerical and experimental techniques.

From the discovery of X-ray diffraction before World War I, until the middle 1930's, crystallographers were in the position of having to guess the structure, and then checking their guess against the diffracted intensities. This situation changed in the 1930's when Patterson asked if there were any meaning in the Fourier transform of the intensity distribution itself. Let this function be called $P(u,v,w)$:

$$P(u,v,w) = \sum_{hkl} |F_{hkl}|^{2} \exp\left[(-2\pi i)(hu + kv + lw)\right]$$

Now,

$$|F_{hkl}|^{2} = F_{hkl}F^{*}_{hkl},$$

and

$$F^{*}_{hkl} = \int \varrho(x,y,z) \exp\left[(-2\pi i)(hx + hy + lz)\right]dxdydz.$$

Thus,

$$P(u,v,w) = \sum_{hkl} F_{hkl} \int \varrho(x,y,z) \exp\left[(-2\pi i)(h(x+u) + k(y+v) + + l(z+w))\right]dxdydz$$

Reversing the integration and the summation yield

$$P(u,v,w) = \int \varrho(x,y,z) \left[\sum_{h,k,l} F_{hkl} \exp\left(-2\pi i(h(x+u) + k(y+v) + l(z+w))\right)\right]dxdydz$$

Ignoring scale factors, the term in brackets is simply $\varrho(x+u, y+v, z+w)$, thus yielding

$$P(u,v,w) = \int \varrho(x,y,z)\varrho(x+u, y+v, z+w)dxdydz.$$

We must now ask what this expression means. The Patterson function, as this synthesis is known, is a map of all the vectors in the structure. If we ignore the integral

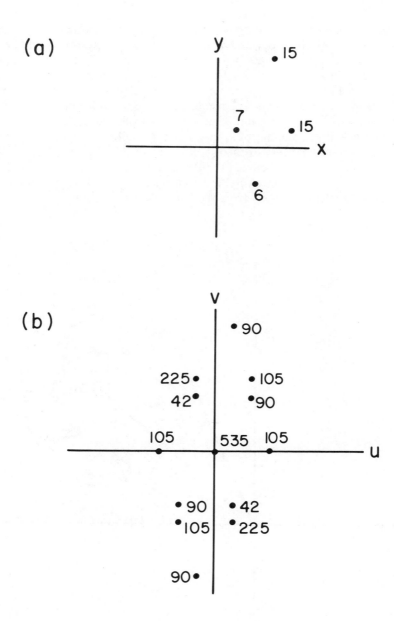

Figure 15. *The Patterson function of a simple distribution.* In (a) is shown a simple distribution of points, and in (b) its Patterson function is shown. Note that each point in the Patterson function corresponds to a vector between two points in the point structure, and is weighted by the product of the weights assigned to the points in (a). Note further, that the self-vectors pile up at the origin, and that for every vector from one point to another, the negative vector, from the second point to the first, also exists. This implies that both the structure and its enantiomorph may be derived from the same Patterson function. Note also that the same Patterson map could be constructed by placing each point of the structure on the origin of the Patterson function and redrawing the structure weighted by the weight of the point at the origin.

in the above equation, for a moment, and consider atomic point scatterers, we can see that if two atoms are separated by the vector (u,v,w) then there will be a peak in the Patterson function at (u,v,w) whose height will be proportional to the product of their electron densities. That is illustrated for a simple 4-point structure in Figure 15. The integral merely means that if there is more than one such pair of atoms in the structure those other products will also be added to the value of the Patterson function at u,v,w, as well. The number of vectors in the Patterson function increases with the square of the number of atoms, N, in the cell. N (self-vectors) of these will be concentrated in the origin of the function, however, leaving N(N-1) + 1 total vectors. Determining a structure from the Patterson function is not trivial, but at least this interpretation of the intensity distribution gave crystallographers a means of determining a structure directly from the X-ray diffraction data.

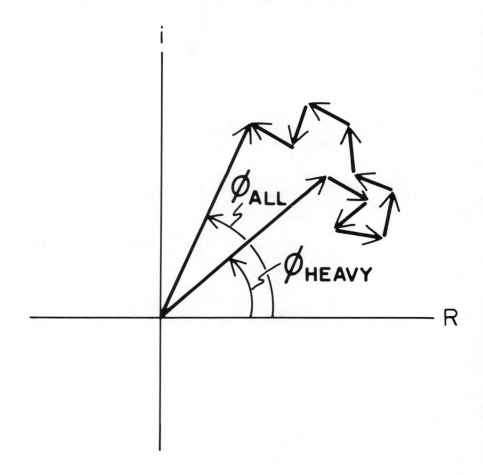

Figure 16. *The domination of the phase of a structure factor by the phase of a heavy atom.* The contribution of a heavy atom to the complete structure factor is illustrated in the complex plane. If the distribution of the contributions of the other lighter atoms, which have smaller scattering factors, is random as indicated, the phase of the heavy atom alone will be a reasonable approximation to the phase of structure factor.

It should be obvious from the expression for the Patterson function that if the structure contains a pair of atoms which are markedly denser (which correlates fairly well with higher atomic number) than the others, the vector between them should stand out in the map. Quite frequently, these may be the same atom in two different molecules related by symmetry. This has been useful in the phasing of a number of dinucleoside phosphate structures which contained derivatized bases or heavy cations.[14/15/28] Once one has identified the relative locations of the heavy atoms, it is possible to calculate a Fourier synthesis using the observed structure amplitudes and the phases derived merely from the identification of the sites of the heavy atoms. As shown in Figure 16, this will frequently be a good approximation to the actual phase of the structure factor because the phase is likely to be dominated by the heavy atom with the other atoms contributing randomly. When new atoms found in this synthesis have been added to the model, another Fourier synthesis is calculated, including the contributions of these atoms, and the procedure is iterated. This procedure is called Fourier refinement. Such a synthesis will frequently reveal much of the rest of the structure. Although the quality of the structure will not be as good as in the absence of the heavy atom, this technique will frequently solve the phase problem.

Short Nucleic Acid Fragments May Be Solved Directly From the Intensity Data

Since nucleic acids come with their own somewhat heavy atom, phosphorus, it is not necessary to derivatize them in order to solve their structures. Although phosphorus-phosphorus vectors will not be markedly visible in the Patterson function, due to the unusually heavy amount of overlap characteristic of Patterson functions (i.e., the summation over a lot of light atom-light atom vectors being greater than the product of two heavy atom vectors), it is possible to locate their position through the use of special techniques, such as resolution difference[4/9] and anomalous dispersion[12] procedures. Although the contribution of the phosphorus atom is not usually a dominant part of the structure factor, it is not too hard for the experienced investigator to determine the structure starting with an initial synthesis calculated with phosphorus phases. Generally several more cycles of Fourier refinement are necessary to deduce the rest of the structure in the absence of a very heavy atom. Such a procedure is akin to Patterson superposition procedures, for the details of which the reader is referred to references 29-31.

Another phasing technique which has found some use in the solution of oligonucleotide structures is termed *direct methods*. These procedures are based primarily on the non-negativity of the electron density distribution. If one invokes this constraint, strengthened by the localization of the electron density in a relatively small number of discrete scattering centers (atoms), it is possible to derive a very powerful formula relating the phases of the structure factors, the *tangent formula*.[32]

$$\tan(\phi_H) = \frac{\Sigma_{H'} |E_{H-H'}||E_{H'}|\sin(\phi_{H-H'} + \phi_{H'})}{\Sigma_{H'} |E_{H-H'}||E_{H'}|\cos(\phi_{H-H'} + \phi_{H'})}$$

In this formula, ϕ_H is the phase of the reflection $H(=h,k,l)$, $\phi_{H-H'}$ and $\phi_{H'}$ phases of

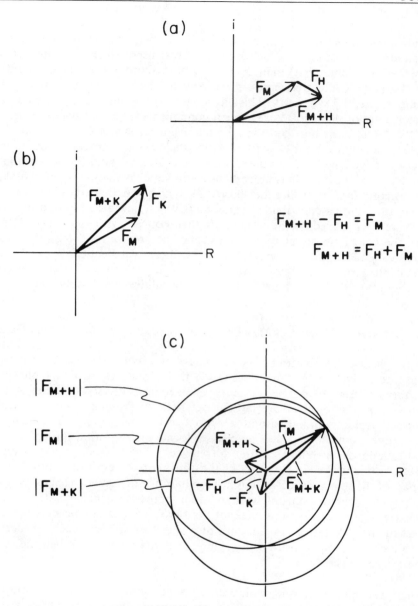

Figure 17. *The principle of multiple isomorphous replacement.* (a) The representation of the resultant structure factor when a macromolecule is derivatized. F_M represents the structure factor of the macromolecule, whose amplitude is known, but whose phase is sought. F_H represents the contribution of a heavy atom derivative. F_{M+H} is the resultant structure factor for the derivatized macromolecule. (b) The same structure factor, F_M, but this time combined with a different derivative, whose contribution F_K results in the structure factor F_{M+K} for the derivative. (c) Construction indicating how the phase of F_M can be derived from knowledge of the *amplitudes* of F_M, F_{M+H}, and F_{M+K} which are directly observable and the *amplitudes* and *phases* of F_H and F_K, which may be obtained by solving the derivative structure. A circle of radius $|F_M|$ can be constructed about the origin of the complex plane. A circle of radius $|F_{M+H}|$ can be constructed with its center at the point $(-F_H)$. This will intersect the F_M circle at two points, one of which corresponds to the correct phase. The ambiguity may be broken by constructing a second circle of radius $|F_{M+K}|$ with its center at the point $(-F_K)$. As can be seen from the figure, the intersection points obey the vector addition laws for the summation of structure factors.

two reflections (h-h′, k-k′, l-l′ and h′,k′,l′) whose indices vectorially add up to **H**. $E_{H-H'}$ and $E_{H'}$ are normalized structure amplitudes related to the ordinary structure amplitudes of $F_{H-H'}$ and $F_{H'}$. If one can assign or guess a few initial phases, it is possible to iteratively bootstrap oneself to the complete structure using this formula and associated probability relationships in the same way that one can bootstrap oneself to the complete structure by knowing the location of the phosphorus atoms.

Macromolecules Must Be Derivatized To Be Solved

Neither direct methods nor Patterson methods are directly applicable to the *de novo* solution of macromolecular structures. This is primarily due to the dependence of these procedures on the ability to resolve individual atoms. The theoretical basis of the tangent formula and the practical application of most Patterson procedures are greatly weakened if individual atoms are not resolvable.[33] Since macromolecular crystals rarely diffract beyond 2Å resolution, an alternative phasing means is necessary. This procedure, known as isomorphous replacement, had its first macromolecular application to protein crystals,[36] but it is quite general, and was used for the solution of Yeast tRNA^{phe}.[18/19/37]

Macromolecular crystals are different from crystals of small molecules, in that they tend to be about 50-70 percent solvent, where small molecule crystals usually range from 0 percent to 20 percent solvent. Because of the large amount of solvent in macromolecular crystals, much of it is in the liquid state, rather than tightly bound to the molecule. This enables the investigator to diffuse heavy atoms into the crystal. If these bind in a few places, without perturbing the original lattice, then one has prepared an *isomorphous derivative* of the macromolecule. How is this useful? If we look at Figure 17a, we see a triangle in the complex plane formed from the vectors corresponding to the structure factor for the native macromolecule, F_M, the structure factor for the derivitized macromolecule, F_{M+H} and the contribution to the derivatized macromolecules from the heavy atom derivative alone, F_H. Clearly, the following vector equations are valid:

$$F_{M+H} = F_M + F_H$$

$$F_M = F_{M+H} - F_H$$

If we draw a circle of radius $|F_M|$ about the origin of the complex plane, and draw another circle of radius $|F_{M+H}|$ about the point $(-F_H)$, they will intersect at two sites, both of which satisfy the above equations (Figure 17c). Both intersection points on the $|F_M|$ circle are possible phases for F_M. The ambiguity can be broken by using a second derivative, whose coefficients are F_{M+K} (Figure 17b) and following the same procedure (Figure 17c). This procedure must be repeated for all coefficients F_M. In practice, several derivatives are usually used, and the phases must be extensively refined before they can be used.[38] You will note that we have been using the amplitudes of the native macromolecules and the derivatized macromolecule (F_M and F_{M+H} or F_{M+K}, respectively), but we have been using the amplitude and *phase* of the derivative itself (F_H or F_K). One might reasonably ask where this phase and amplitude come from. What must be done is to solve the structure of the derivative by itself. One must use difference coefficients between the native and derivatized macromolecule, and solve this difference structure, with somewhat shaky amplitudes, in the same way that high resolution small molecule structures are solved.

Another technique used in conjunction with the multiple isomorphous replacement procedure described above involves the use of anomalous dispersion. Let us consider the two structure factors F_{hkl} and the one on the opposite side of reciprocal space from it, $F_{-h,-k,-l}$.

$$F_{hkl} = \sum_{j=1}^{N\,atom} f_j \exp[2\pi i (hx_j + ky_j + lz_j)]$$

$$F_{-h,-k,-l} = \sum_{j=1}^{N\,atom} f_j \exp[2\pi i((-h)x_j + (-k)y_j + (-l)z_j)] = \sum_{j=1}^{N\,atom} f_j \exp[-2\pi i(hx_j + ky_j + lz_j)].$$

These structure factors are clearly equal in amplitude, but of opposite phase. This is known as Friedel's Law, and the two reflections are called a Friedel pair. Friedel's Law only holds in the case of real scatterers, however, and breaks down in the case of complex, or anomalous, scattering, as shown in Figure 18. If one has an isomorphous derivative which is an anomalous scatterer for the wavelength of radiation used, one may then treat the data from Friedel pairs like the data from two separate derivatives, as shown in Figure 19.

It should be pointed out that phases derived from isomorphous replacement with or without anomalous dispersion are dependent on the small differences observed between large numbers. As a result, they are likely to be off by an average of 35° in the best cases, and more typically 50° or worse. With the map derived from these phases, at the best resolution obtainable, it is the crystallographer's job to fit a molecular model to the electron density; this is much like fitting the duck in Figure 12a to the density in Figure 12f. This is the point at which there is an interpretive aspect to macromolecular crystallography. Although the solution of the Patterson function or phasing via direct methods may require large amounts of chemical and crystallographic intuition, the crystallographer using these techniques at high resolution can quickly determine whether or not the structure is correct. This can be done by comparing structure factors calculated from the model, F_c, with those experimentally determined, F_o. A useful index in this respect is known as the R-factor,

$$R = \frac{\sum ||F_o| - |F_c||}{\sum |F_o|}$$

For a high resolution structure, before refinement, an R-value of .2 to .3 is indicative of a probably correct structure, with 0.59 being the expected result for a random array of atoms.[39] Correct macromolecular structures usually give R-values of 0.4 to 0.5, before refinement. Thus, the macromolecular crystallographer has

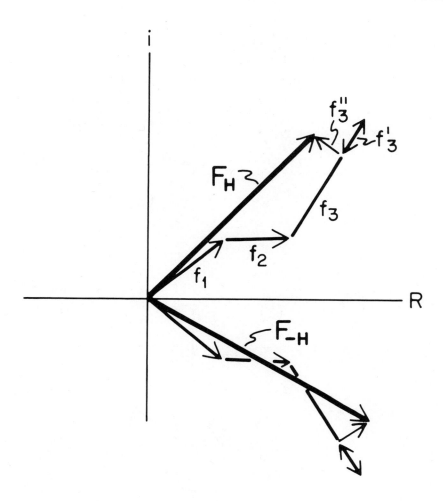

Figure 18. *Anomalous dispersion in the complex plane; violation of Fridel's Law.* This drawing indicates the contributions of three atoms to the structure factors F_H and F_{-H} which are on opposite sides of the origin of reciprocal space from each other. The first 2 atoms are both real scatterers, so that they contribute equally to the scattering of each reflection, but on opposite sides of the real axis; thus, the amplitudes of the resultant waves will be equal, but their phases will be opposite. The third scatterer is an anomalous scatterer which has a complex scattering factor. The f', or real portion of the anomalous scattering is taken to be negative here. The f'', or imaginary, portion of the anomalous scattering is indicated perpendicular to the real part. It can be seen that the resultant amplitudes for F_H and F_{-H} will not be equal, and their phases will also not be negatives of each other, since the imaginary component does not obey the mirror symmetry about the real axis.

much less feedback from the data about the correctness of the interpretation. Therefore, much more reliance must be placed on data from sequence, chemical modification and genetic studies, as well as high resolution crystallography of oligomers in order to make a meaningful interpretation of the map.

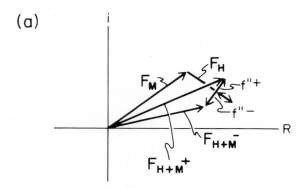

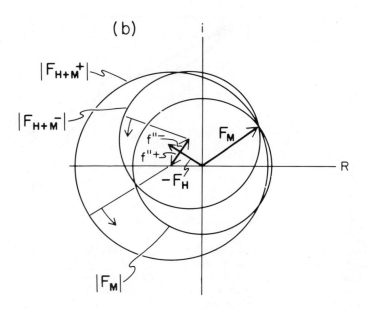

Figure 19. *Use of anomalous dispersion data in phasing.* (a) Vector diagram indicating the vector relationships between the structure factor for a macromolecule, F_M and the structure factors for an anomalously scattering heavy atom derivative, F_H, combining to give Friedel paired reflections F_{H+M}^+ and F_{H+M}^-. *The reader should note that the portion of the diagram corresponding to F_{H+M}^- has been reflected through the real axis to clarify the relationships.* The imaginary components, f'', are indicated. (b) In a similar manner to Figure 17(c), three circles have been constructed. One of radius $|F_M|$, as before, has its center at the origin of the complex plane. The complex scatterer, F_H, is drawn, and again, the sign of the vector is reversed to give $(-F_H)$. The imaginary and real portions of the anomalous scattering are then taken into account, giving the two indicated points on either side of the $-F_H$ vector. One point corresponds to the scattering contribution for F_H to F_{H+M}^+, and the other corresponds to the scattering contribution for F_H to F_{H+M}^-. Circles of these radii are constructed about these two points, respectively, and they can be seen to intersect at the phase of F_M on the F_M circle.

Refinement Is The Acid Test For A Structure

Once the crystallographer derives a model for the structure from the diffraction data, it must be refined so that the best fit to the data is obtained. The most common techniques for refinement involve least squares procedures, although others exist. The reader is referred to the treatment by Stout and Jensen[24] for the details and derivations of these methods. With small molecules, one initially refines positional coordinates for each atom, as well as an isotropic thermal parameter. During this phase of the refinement, one usually calculates difference Fourier syntheses,

$$\Delta\varrho(x,y,z) = \sum_{hkl} (|F_o| - |F_c|)\, e^{i\phi_c} \exp[-2\pi i(hx + ky + lz)],$$

where ϕ_c is the calculated model phase associated with $|F_c|$.

These syntheses reveal small errors in the structure and are very useful in filling in the solvent region of the unit cell if this has not been done previously. In oligonucleotide crystals, there is frequently a large amount of solvent, which must be interpreted correctly. It is frequently statistically disordered (namely, it might not be the same in all unit cells), and great care must be taken to insure that any solvent molecules added to the model are readily interpretable in a chemical sense: the solvent molecules in all ordered and disordered sites must form reasonable hydrogen bonded or non-bonded contacts with their putative neighbors. Once the solvent is filled in and isotropic convergence has been obtained (usually with an R-factor between 0.10 and 0.20), anisotropic temperature factors may be employed. The anisotropic thermal motion model for atomic motion increases the number of parameters per atom from 4 to 9, and thereby decreases the overdetermination of the number of data points to parameters varied, by more than a factor of two. Agreement between data and model will automatically improve by doing this, and care must be taken to see that the thermal motion is meaningful and that the improvement in agreement is greater than that expected just by increasing the number of parameters varied.[40] Figure 20 illustrates the anisotropic thermal motion derived in the dinucleoside phosphate ApU. At the end of the refinement, the R-factor will typically be about .1 or less.

In large high-resolution structures, such as oligonucleotides, the uninterpretablity of the solvent region combines with the poor scattering from hydrogen atoms to make these atoms very hard to find in the electron density map. As can be seen from Figure 8, hydrogen atoms will be very small contributors to the diffraction pattern, and their density in the Fourier synthesis will suffer from errors in the data and in other parts of the structure. The positions of some hydrogens are generable from the stereochemistry of the atoms to which they are bonded. Accurate positioning of hydrogen atoms is best accomplished by performing a neutron scattering experiment. Unfortunately, nobody has grown a crystal of an oligomeric nucleic acid fragment large enough to do this yet.

The refinement of macromolecules is more difficult, because the resolution is poorer and the overdetermination of observations to parameters is much less. The refine-

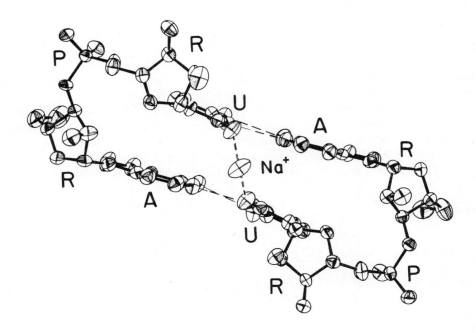

Figure 20. *Anisotropic thermal motion of the atoms in a miniature double helix.* This is an ORTEP plot[48] of the molecular structure of ApU,[9] indicating the anisotropic thermal motion of the individual atoms in the bimolecular complex. This is a view from the minor groove of the double helix. The principal components of the thermal motion are parallel to the major axes of the ellipsoids depicted, and the relative r.m.s. amplitudes of vibration in these directions are proportional to the relative lengths of the major axes. The sodium ion indicated at the center of the figure was found coordinated to the two uracil O2 atoms as indicated in the minor groove. The motion of the ion can be seen to be almost perpendicular to the liganding direction.

ment process also has much more of a combined "solution and refinement" character than with smaller molecules studied at higher resolution, since a larger number of solvent molecules are found in the course of the refinement. Furthermore, portions of the structure may have to be slightly reinterpreted as the refinement proceeds. In order to insure the chemical integrity of the structure, bond distances, bond angles and conjugated ring planarities are restrained to values determined by high resolution crystallography of smaller systems.[41] R-values between .1 and .20 have been reported for macromolecules nearing the end of their refinement. The R-value for Yeast tRNA[phe] is now 0.19.[42] The R-value by itself is not meaningful, however, and may be dramatically decreased if attention is not paid to retaining good stereochemistry throughout the course of the refinement. For refinement, in general, and macromolecular refinement in particular, the chemical reasonableness of the model is the *sine qua non*, with R values being of secondary importance.

Concluding Remarks

I have tried to present the reader with a brief introduction to X-ray crystallography, particularly that of nucleic acid fragments and macromolecules. Space has not permitted discussion of experimental matters, such as the generation and properties of X-rays, and the collection of diffraction intensity data. The reader is referred to the books by Stout and Jensen,[24] Arndt and Willis,[43] and Arndt and Wonacott[44] for a discussion of these matters. The growth of suitable crystals (a major stumbling block to many nucleic acid studies) is not yet systematized, and the reader is referred to original papers in the area and the review by McPherson.[45] Many areas of crystallography have only been briefly mentioned, in particular symmetry, structure solution procedures and refinement. I feel that the material presented above will allow the reader to read the structural papers in this volume critically, thereby removing some of the mystique from this field.

Nucleic acid crystallography is an exciting field which is still in its infancy. A relatively small number of oligonucleotide structures have been determined, and very few tRNA molecules have been solved. Because of the great importance of understanding the structural aspects of nucleic acids, I am sure that we will see many more structures in the near future. With those structures will come a greater appreciation for the molecular-structural aspects of life.

Acknowledgements

Figures 1, 9, 10, 12 and 13 are reprinted from G. Harburn, C. A. Taylor and T. R. Welberry: *Atlas of Optical Transforms*. Copyright © 1975 by G. Bell and Sons, Lts. Used by permission of the publishers, Cornell University Press.

This work was aided by a Basil O'Connor Starter Research Grant from The National Foundation-March of Dimes, by University Award Number 7421 from the Research Foundation of the State University of New York and by Grant GM26467 from the National Institute of Health.

References and Footnotes

1. Notation: The amplitudes of complex numbers and the lengths of vectors are represented by vertical bars, e.g., |F| is the amplitude of the complex number F and |D| is the length of the vector **D**. Vectors are represented by bold face type, e.g., **D**. The length of a line segment between two points A and B is indicated by $\overline{AB}$. The complex conjugate of a complex number F will be denoted F*. Structure factors will be denoted F_{hkl} when referring to the indices of interest, or will be denoted F_M or F_{M+H} when referring to Fourier coefficients from different crystals. Sometimes the triple indices (h,k,l) will be abbreviated as the reciprocal lattice vector **H**.
2. Kendrew, J. C., Bodo, G., Dintzis, H. M., Parrish, R. E., Wyckoff, H. and Phillips, D. C., *Nature 181*, 662 (1958).
3. Corey, R. B. and Pauling, L., *Rc. Ist. Lomb. Sci. Lett. 89*, 10 (1955).
4. Seeman, N. C., Sussman, J. L., Berman, H. M. and Kim, S. H. *Nature New Biology 233*, 90 (1971).
5. Sussman, J. L., Seeman, N. C., Kim, S. H. and Berman, H. M., *J. Mol. Biol. 66*, 403 (1972).
6. Rubin, J., Brennan, T. and Sundaralingam, M., *Biochemistry 11*, 361 (1972).
7. Camerman, N., Fawcett, J. K., and Camerman, A., *Sceienc 182*, 1142 (1973).
8. Rosenberg, J. M., Seeman, N. C., Kim, J. J. P., Suddath, F. L., Nicholas, H. B. and Rich, A., *Nature 243*, 150 (1973).
9. Seeman, N. C., Rosenberg, J. M., Suddath, F. L., Kim, J. J. P., and Rich, A., *J. Mol. Biol. 104*, 109 (1976).
10. Day, R. O., Seeman, N. C., Rosenberg, J. M., and Rich, A., *Proc. Nat. Acad. Sci. (USA) 70*, 849 (1973).

11. Rosenberg, J. M., Seeman, N. C., Day, R. O., and Rich, A., *J. Mol. Biol. 104*, 145 (1976).
12. Seeman, N. C., Day, R. O., and Rich, A., *Nature, 253*, 324 (1975).
13. Suck, D., Manor, P., Germain, G., Schwalbe, C. H., Weimann, G., and Saenger, W., *Nature New Biology 246*, 161 (1973).
14. Tsai, C. C., Jain, S. C., and Sobell, H. M., *Proc. Nat. Acad. Sci. (USA) 72*, 2626 (1975).
15. Sakore, T. D., Jain, S. C., Tsai, C. C., and Sobell, H. M., *Proc. Nat. Acad. Sci. (USA) 74*, 188 (1977).
16. Neidle, S., Achari, A., Taylor, G. L., Berman, H. M., Carrell, H. L., Glusker, J. P., and Stallings, W. C., *Nature 269*, 304 (1977).
17. Viswamitra, M. A., Kennard, O., Jones, P. G., Sheldrick, G. M., Salisbury, S., Falvello, L., and Shakked, Z., *Nature 273*, 687 (1978).
18. Kim, S. H. Suddath, F. L., Quigley, C., McPherson, A., Sussman, J. L., Wang, A. H. J., Seeman, N. C., and Rich, A., *Science 185*, 435 (1974).
19. Robertus, J. D., Ladner, J. E., Finch, J. T., Rhodes, D., Brown, R. S., Clark, B. F. C., and Klug, A., *Nature 250*, 546 (1974).
20. Schevitz, R. W., Podjarny, A. D., Krishnamachari, N., Hughes, J. J., Sigler, P. B., and Sussman, J. L., *Nature 278*, 188 (1979).
21. Wilson, H. R., *Diffraction of X-rays by Proteins, Nucleic Acids and Viruses*, London, Edward Arnold Publishers, Ltd. (1966).
22. Arnott, S. and Hukins, D. W. L., *J. Mol. Biol. 81*, 93 (1973) and references therein.
23. Glusker, J. P. and Trueblood, K. N., *Crystal Structure Analysis: A Primer*, New York, Oxford University Press (1972).
24. Stout, G. M. and Jensen, L. M., *X-ray Structure Determination: A Practical Guide*, New York, The Macmillan Company (1968).
25. Blundell, T. L. and Johnson, L. N., *Protein Crystallography*, New York, Academic Press (1976).
26. International Tables for X-ray Crystallography, Vol. 1-3, Birmingham, England, Kynoch Press (1969); Vol. 4 (1974).
27. The scattering is inversely proportional to the mass of the scattering particles, thus making electrons the only constituent of molecules to give a noticeable signal.
28. Wang, A. H. J., Nathans, J., van der Marel, G., Van Boom, J. H., and Rich, A., *Nature 276*, 471 (1978).
29. Buerger, M. J., *Vector Space*, New York, Wiley (1959).
30. Seeman, N. C., in *Crystallographic Computing*, F. R. Ahmed, ed, Copenhagen, Munksgaard (1970), pp. 87-89.
31. Seeman, N. C., Ph.D. Thesis, University of Pittsburgh (1970).
32. Karle, J. and Karle, I. L., *Acta Cryst. 21*, 849 (1966).
33. One important exception to this statement involves Patterson search procedures. If a crystal is expected to contain a well characterized electron density distribution, such as an adenine ring or a RNA double helix, it is possible to search the observed Patterson function for this distribution of vectors. This technique has been applied to the crystal structure of *E. coli* tRNA^{Metf} (34). While its applicability is very powerful, and it is not dependent on atomic resolution, it is not a *de novo* crystallographic phasing technique, since one must know what electron density distribution is being sought. For more information on this technique, see Rossmann.[35]
34. Woo, N. H., and Rich, A., Priv. comm. (1979).
35. Rossmann, M. G., *The Molecular Replacement Method*, New York, Gordon and Breach (1972).
36. Green, D. W., Ingram, V. M., and Perutz, M. F., *Proc. Roy. Soc. A 225*, 287 (1954).
37. Kim, S. H., Quigley, G. J., Suddath, F. L., McPherson, A., Sneden, D., Kim, J. J., Weinzierl, J., and Rich, A., *Science 179*, 285 (1973).
38. Blow, D. M. and Crick, F. H. C., *Acta Cryst. 12*, 794 (1959).
39. Wilson, A. J. C., *Acta Cryst. 3*, 347 (1950).
40. Hamilton, W. C., *Statistics in Physical Science*, New York, Ronald Press (1964).
41. Konnert, J. J., *Acta Cryst. A32*, 614 (1976).
42. Kim, S. H., Priv. comm. (1979).
43. Arndt, U. W. and Willis, B. T. M. *Single Crystal Diffractometry*, Cambridge University Press, Cambridge (1966).
44. Arndt, U. W. and Wonacott, A. S., *The Rotation Method in Crystallography*, Amsterdam, North-Holland Publishing Co. (1977).

45. McPherson, A., *Methods of Biochemical Analysis, 23*, 249 (1976).
46. Taylor, C. A. and Lipson, H., *Optical Transforms*, Ithaca, New York, Cornell University Press (1964).
47. Quigley, G. J., Wang, A. H. J., Seeman, N. C., Suddath, F. L., Rich, A., Sussman, J. L., and Kim, S. H., *Proc. Nat. Acad. Sci. (USA) 72*, 4866 (1975).
48. Johnson, C. K., *ORTEP*, ORNL-3794, Oak Ridge National Laboratory, Oak Ridge, Tennessee (1965).

Part II
Small Molecules and Ions

The Rotation-Inversion Dichotomy in Trialkylamines Diethylmethylamine and Triethylamine

C. Hackett Bushweller
Department of Chemistry
University of Vermont
Burlington, Vermont 05405

and

Bernard J. Laurenzi, John G. Brennan,
Martin J. Goldberg and Richard P. Marcantonio
Institute of Biomolecular Stereodynamics
Department of Chemistry
State University of New York at Albany
Albany, New York 12222

Two fundamentally important rate processes are associated with the nitrogen atom of an amine. One of these processes is pyramidal *inversion* at nitrogen.[1] The other process is *rotation* about bonds to nitrogen.[2] In assessing the stereodynamics of tertiary amines (i.e., no protons bonded to nitrogen), one must be cognizant of both rate processes. Indeed, it is becoming increasingly evident that there is no clear-cut dichotomy between such inversion and rotation processes in many amines and, in myriad cases, the two processes are coupled.

In a simple amine such as methylamine, microwave spectroscopy reveals a barrier to 3-fold rotation about the C-N bond of 2.0 kcal/mole (Figure 1).[3] In trimethylamine

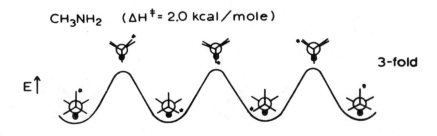

Figure 1

the C-N rotational barrier increases to 4.4 kcal/mole due presumably to increasing steric effects.[4] In a system such as nitromethane (sp² hybridized nitrogen), the potential surface for rotation about the C-N bond possesses 6-fold symmetry (Figure 2) and the barrier to rotation is very low (0.006 kcal/mole).[5]

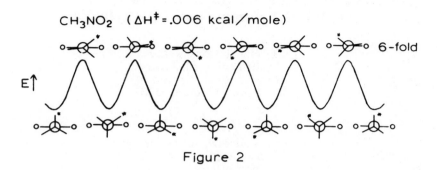

Figure 2

In contrast, the inversion process in a simple amine may be represented as a double-well potential surface (Figure 3). In Figure 3, "r" represents the distance between a plane defined by the three "X" groups and the nitrogen atom. The transition state for nitrogen inversion is usually assumed to be that geometry in which nitrogen is sp² hybridized (r=0; Figure 3). Substituent effects on the rate of inversion can be dramatic with barriers to inversion ranging from greater than 25 kcal/mole in N-haloaziridines[1] to 6 kcal/mole in N-tert-butyl-N,N-diethylamine.[6] The origins of these significant substituent effects can be traced to steric, electronic and angle strain effects.[1] Increasing steric crowding about the nitrogen of a tertiary amine will

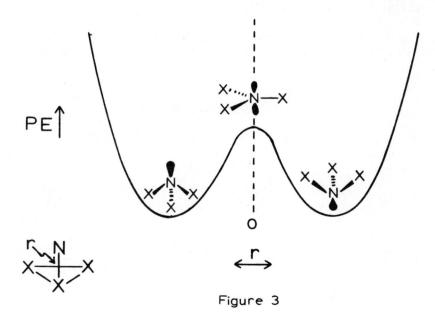

Figure 3

render the nitrogen less and less pyramidal (i.e., more sp² hybridized and closer to the transition state) thus lowering the barrier to nitrogen inversion. Increasing electronegativity of substituents bonded to nitrogen increases the s-character of the lone-pair electrons (i.e., stabilizes the pyramidal sp³ hybridized nitrogen) and *increases* the barrier to inversion. Incorporation of the nitrogen atom into a small ring such as an arziridine ($<CNC = 60°$) introduces severe angle strain into the transition state for inversion which of course prefers CNC bond angles of 120° and the barrier to inversion increases.

In a series of recent papers concerning the stereodynamics of a series of N-tert-butyl-N,N-dialkylamines[6] and N-tert-butyl-N-haloamines,[7] we presented the results of ¹H DNMR studies and semi-empirical molecular orbital calculations revealing that the tert-butyl rotation and nitrogen inversion processes occur in concert.

An example of the kind of ¹H DNMR results obtained for the tert-butyl-dialkylamines[6] is illustrated in Figure 4. The center column of Figure 4

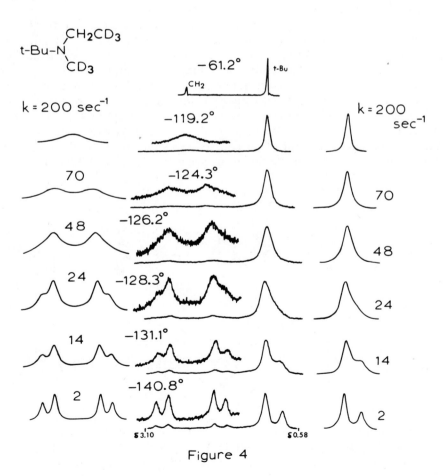

Figure 4

displays the *experimental* 60 MHz ^{1}H DNMR spectra of amine 1. The CH_2 region of the spectrum is also shown at increased amplitude. The left hand column includes theoretical simulations of the CH_2

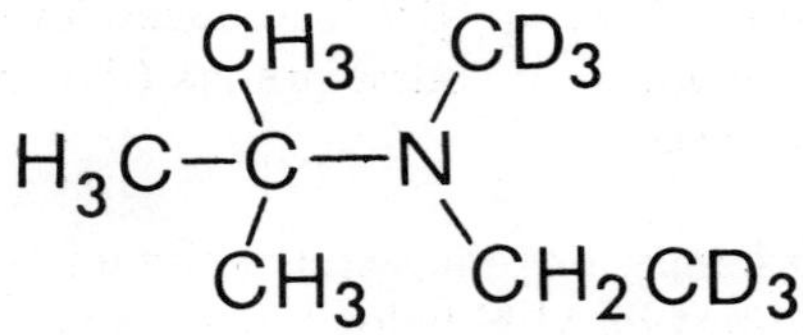

$$
\begin{array}{ccc}
 & CH_3 & CD_3 \\
 & \backslash & / \\
H_3C-C&-&N \\
 & / & \backslash \\
 & CH_3 & CH_2CD_3
\end{array}
$$

1

signal using a simple AB to BA spin exchange model. The right hand column illustrates theoretical simulations of the tert-butyl resonance as a simple two-site exchange model (no spin-spin coupling). The changes in the ^{1}H DNMR spectrum of the CH_2 protons can be attributed to slowing *nitrogen inversion* in 1.[6] However, the intriguing observation for 1 (Figure 4) is that the rate of conformational exchange of the tert-butyl methyl groups is *equal* to the rate of inversion. Experimental results such as this and semi-empirical molecular orbital calculations led us to postulate an optimized energy surface for conformational exchange in 1 which involves tert-butyl rotation occurring in concert with nitrogen inversion.[6,7] The implication in this model as applied to 1 is that the barrier to *isolated* tert-butyl rotation with no concomitant nitrogen inversion is higher than the barrier associated with the coupled process. A generalized example of the coupled inversion and C-N rotation process is shown in Equation 1. Simple rotation is illustrated in Equation 2. Since both simple

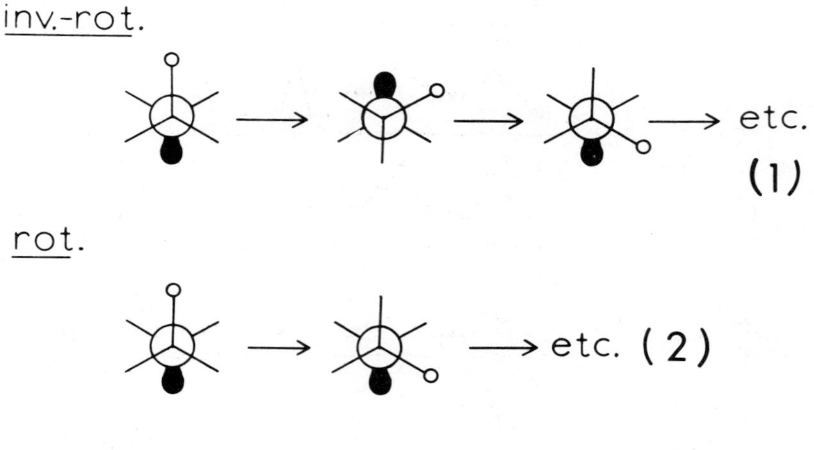

inv.-rot.

$$\tag{1}$$

rot.

$$\text{etc. } \tag{2}$$

isolated rotation (Eq. 2) and the coupled process (Eq. 1) can lead to the same net change in environments for the tert-butyl methyl groups in 1, the DNMR spectra are sensitive only to *lower* barrier coupled process and isolated tert-butyl rotation is invisible to the DNMR technique.

Indeed, a more complete picture of the optimized energy surface for conformational equilibration in tert-butyldimethylamine is shown in Figure 5.[6] The reader will note that not only is a coupled inversion- rotation process possible but also a lower barrier "wagging" process resulting from a preferred geometry in which the tert- butyl group is slightly skewed about the C-N bond.

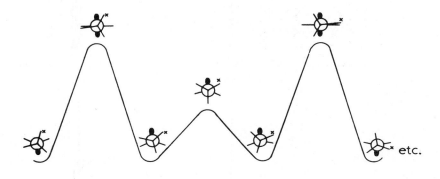

etc.

Figure 5

As a result of these studies, we postulated a dynamical model for the rotation- inversion dichotomy in all acyclic tertiary amines. Succinctly stated, it says that if the barrier to nitrogen inversion is greater than that for isolated rotation about C-N bonds, then the preferred route for conformational equilibration about the C-N bond is simple isolated rotation. However, if the barrier to inversion at nitrogen is lower than that for isolated C-N rotation, the lowest barrier process for geometric equilibration about the C-N bond involves *concomitant* C-N bond rotation and nitrogen inversion.[7]

Having completed our studies of the rather encumbered tert-butyl-dialkylamines, and postulating the dynamical model above, it was not clear to us that ^{1}H DNMR studies of *simpler* trialkylamines would bear fruit regarding inversion and rotation processes.[8] However, from data to be presented below, it will become clear that the combination of a high field NMR system and very low temperatures has great potential for yielding much valuable information about these two important rate processes in very simple trialkylamines.

Examination of the ^{1}H DNMR spectrum (270 MHz) of diethylmethylamine (3 percent v/v in $CBrF_3$) at 200°K shows a typical A_2X_3 spectrum ($\delta_A 2.32$, $\delta_X 1.06$, $^3J_{AX} = 7.0$ Hz) for the ethyl groups and a singlet ($\delta 2.10$) for the N-methyl protons (Figure 6). At lower temperatures, the CH_2 resonance undergoes broadening and separation into the AB portion of an ABX_3 spectrum ($\delta_A 2.50$, $\delta_B 2.13$, $\delta_X 1.06$, $^2J_{AB} = -13.8$ Hz, $^3J_{AX} = {}^3J_{BX} = 7.0$ Hz) shown at 140°K in Figure 6. The N-methyl singlet overlaps with the B resonance of the ABX_3 spectrum. It is important to note that the N-methyl and C-methyl resonances are independent of temperature from 200° to

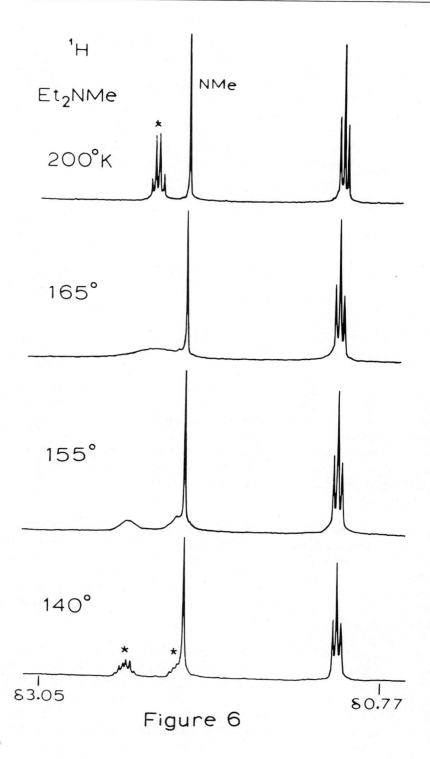

Figure 6

140°K (Figure 6). The DNMR coalescence phenomenon associated with the CH_2 resonance in Figure 6 is best rationalized in terms of slowing *nitrogen inversion* in diethylmethylamine. The nitrogen inversion process involves not just inversion at nitrogen but also rotation about the C-N bonds. The phenomenon is illustrated schematically in Equation 3 (R = Me). Starting from 2 (Eq. 3), inversion to a planar

R = Me, Et

nitrogen would give 3 and completion of the inversion process gives 4. Conformer 4 is of course an unstable eclipsed geometry and could decay to one of a number of lower energy forms (e.g., 5) by a 60° clockwise *rotation* of the front ethyl group. The same net conformational change may be achieved by a *concerted* inversion to planar nitrogen and a 30° clockwise *rotation* of the front ethyl group (2 to 6 Eq. 3) and a continuation of this process to 7 which is identical to 5. The point to be made here is that nitrogen inversion will always be accompanied by some rotation about C-N bonds and the process might better be called *inversion-rotation* rather than simple inversion.

A qualitative summary of all the possible stereodynamical processes in diethylmethylamine is given in Equation 4. The processes which involve *direct* conversions of 8 to 9, 9 to 10, 10 to 11, 11 to 12, 12 to 13, 13 to 8 and all associated reverse reactions are inversion-rotation (inv.-rot., Eq. 4) processes. The *direct* interconversions 8 to 12, 12 to 10, 10 to 8, 13 to 9, 9 to 11, 11 to 13 and corresponding reverse reactions are simple isolated rotations (rot., Eq. 4) about the N-CH_2 bond with *no* inversion. If all the inversion-rotation processes in Equation 4 become *slow* on the DNMR time scale but all N-CH_2 rotations remain *rapid*, the methylene protons H and H* in equation 4 will be nonequivalent ("diastereotopic") by virtue of the asymmetric environment presented on the nitrogen atom. While it is true that the environments of H and H* are averaged respectively via rotation, the respective time-averaged environments of H and H* are indeed different and H and H* are thus nonequivalent. It should be noted, however, that simple rotation is sufficient to time- average the C-methyl and N-methyl environments respectively even under conditions of static inversion (Eq. 4). Thus, if inversion-rotation is slow on the DNMR time scale but N-CH_2 rotation is fast, the ethyl spectrum should indeed be an ABX_3

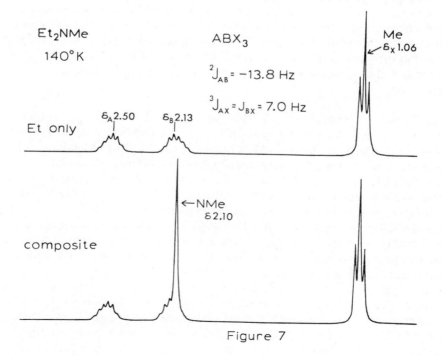

$$R = Me, Et$$

type spin system as observed at 140°K (Figure 6). A dissection of the spectrum at 140°K is illustrated in Figure 7. The bottom spectrum is a theoretical simulation[9] of

Figure 7

the complete spectrum at 140°K (Figure 6). The top spectrum of Figure 7 is a theoretical simulation of the ethyl spectrum *only* clearly showing the ABX_3 spin system. The experimental DNMR spectra shown in Figure 6 can be simulated accurately using an ABX_3 to BAX_3 exchange model with the N-methyl singlet superimposed.[9] These DNMR line shape analyses give the rate of inversion-rotation in diethylmethylamine as a function of temperature and thus provide activation parameters for the inversion-rotation process *($\Delta H^{\neq} = 8.7 \pm 0.5$ kcal/mole, $\Delta S^{\neq} = 7 \pm 4$ eu, $\Delta G^{\neq} = 7.6 \pm 0.2$ at 158°K).*

Since the spectrum of diethylmethylamine at 140°K in Figure 6 reflects conditions of slow inversion-rotation, any further changes in the ¹H DNMR spectrum at *lower* temperatures must be ascribed to slowing *rotational* processes most likely C-N rotation. Indeed, as shown in Figure 8, more changes in the ¹H DNMR spectrum (270 MHz) of diethylmethylamine do occur at lower temperatures. It is evident at 99°K (Figure 8) that the N-methyl resonance has separated into two singlets of substantially different area (see asterisks in spectrum at 99°K), the C-methyl resonance is split into at least two signals of unequal area and the CH_2 resonances have separated further into a series of peaks. An accurate theoretical simulation of the experimental spectrum at 99°K (Figure 8) is shown as the "composite" spectrum in Figure 9. The composite spectrum can be dissected into two spectra one of which corresponds presumably to *one* type of minor conformer (see top spectrum of Figure 9) and the other spectrum to a major species (see middle spectrum of Figure 9). The minor spectrum can be simulated as an ADY_3 spectrum ($\delta_A 2.93$, $\delta_D 2.19$, $\delta_Y 1.07$, $^2J_{AD} = -10$ Hz, $^3J_{AY} = {}^3J_{DY} = 7$ Hz; $T_2 = 0.028$ sec), a B_2Z_3 spectrum ($\delta_B 2.42$, $\delta_Z 0.89$, $^3J_{BZ} = 7$ Hz), and an N-methyl singlet ($\delta 2.24$, see asterisk). The ratios of the total areas of the ADY_3 to B_2Z_3 to N-methyl spectra are 5:5:3 suggesting strongly that this spectrum corresponds to *one* type of conformer for diethylmethylamine. The spectrum of the major conformer (Figure 9) is simulated as two CEX_3 spin systems ($\delta_C 2.37$, $\delta_E 2.04$, $\delta_X 1.14$, $^2J_{CE} = -10$ Hz, $^3J_{CX} = {}^3J_{EX} = 7$ Hz) and an N-methyl singlet ($\delta 2.12$). Such a spectrum indicates that the two different methylene protons of one ethyl group have equivalent counterparts on the other ethyl group. The area ratio of the CEX_3 spin system to the N-mehyl signal is 10:3 again suggesting that this spectrum corresponds to *one* type of conformer for diethylmethylamine.

Two important points need to be made at this juncture. The N-methyl peaks for diethylmethylamine are accurately simulated as broad *singlets* ($T_2 = 0.028$ sec), i.e., all protons on a given N-methyl group are equivalent. The C-methyl resonances are accurately simulated assuming all three methyl protons on a given C-methyl group are *equivalent*. This indicates that even at 99°K individual methyl group rotation is *fast* on the ¹H DNMR time scale. *Thus, the ¹H DNMR spectra from 116° to 99°K in Figure 8 are best rationalized in terms of slowing rotation about the N-CH₂ bonds.*

While the spin systems associated with the major and minor conformers of diethylmethylamine are reasonably well defined (Figure 9), a clear picture of the *actual* geometries present is not immediately evident. In order to obtain more insight into preferred molecular geometry, we executed a careful search for energy minima as a function of geometry using the MINDO/3 semi-emperical molecular orbital ap-

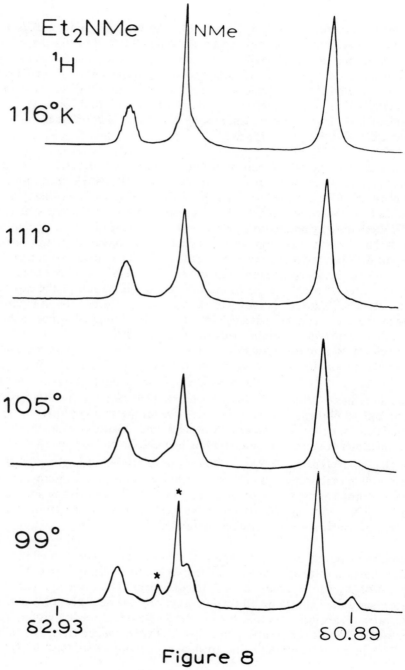

Figure 8

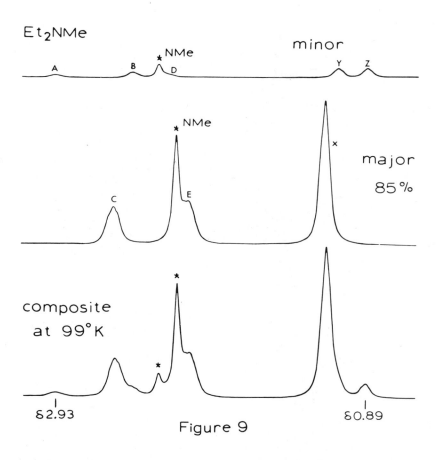

Figure 9

proach.[10] The most stable geometry as determined by MINDO/3 is 14. The dihedral angle between an axis through the nitrogen lone pair and the C1C2 bond is 10° and that between the nitrogen lone pair and the C3-C4 bond is 50° (methyl pointing back). However, there exists a broad potential minimum associated with "wagging" processes about the N-Cl and N-C3 bonds. Variations in the lone pair/C1C2 or lone pair/C3C4 dihedral angles can be as much as 40° with no significant change in energy. The net effect of these wagging processes is that the methylene protons marked with an asterisk in 14 can experience the same time-averaged environment approximately *trans* to the nitrogen lone pair. The other two methylene protons would also experience time-averaged environments but would be approximately *gauche* to the nitrogen lone pair. Finally, the wagging process would render the two C-methyl groups equivalent and result in one resonance as observed for the major conformer (Figure 9). Thus, 14 would give an ¹H NMR spectrum consistent with that for the dominant species illustrated in Figure 9. The only other potential minimum that we could find using the MINDO/3 approach is 15 which is calculated to be 1.5 kcal/mole higher in energy than 14. In 15, the lone pair/ C1C2 dihedral angle is 180° and the lone pair/C3C4 angle is 58° (methyl pointing down and back). It is the anisotropy of the nitrogen lone pair which will determine in large part the chemical shift difference between the two methylene protons on C1 and C3.[11] Since

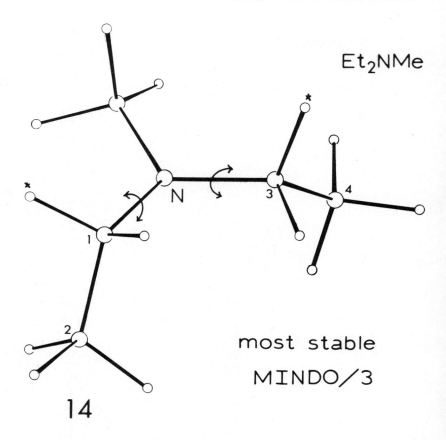

Et$_2$NMe

most stable

MINDO/3

14

the two methylene protons on C1 are both gauche to the nitrogen lone pair, they could easily have very similar chemical shifts giving rise to the B$_2$Z$_3$ spectrum of the minor isomer in Figure 9. The two methylene protons on C3 in 15 are in quite different orientations with respect to the lone pair and could easily give rise to the ADY$_3$ spectrum of the minor conformer (Figure 9).

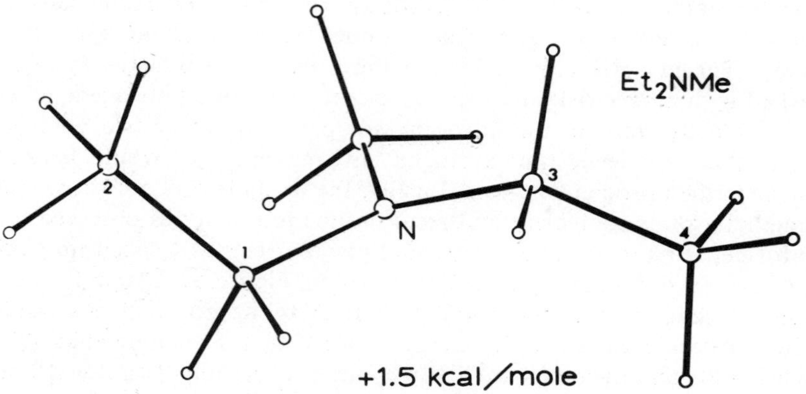

Et$_2$NMe

+1.5 kcal/mole

15

At 99°K, the ratio of the major to the minor conformer is 85:15 ($\Delta G°$ = -0.34 kcal/mole). It is noteworthy that the MINDO/3 method calculates a 1.5 kcal/mole energy difference between 14 and 15.

The ^{1}H DNMR spectra in Figure 8 can be simulated using a tentative DNMR model involving ECX_3 to B_2Z_3 to DAY_3 nuclear spin exchanges. The preliminary activation parameters derived from this analysis are ΔH^+ = 5.5 ±0.4 kcal/mole, ΔS^+ = 4 ± 4 eu, and ΔG^+ = 5.2 ± 0.2 kcal/mole at 108°K. It should be pointed out that these activation parameters relate to a *multiple* rotation process between species such as 14 and 15 and do not necessarily relate to rotation about just one N-CH$_2$ bond at a time. They may reflect a concerted double rotation.

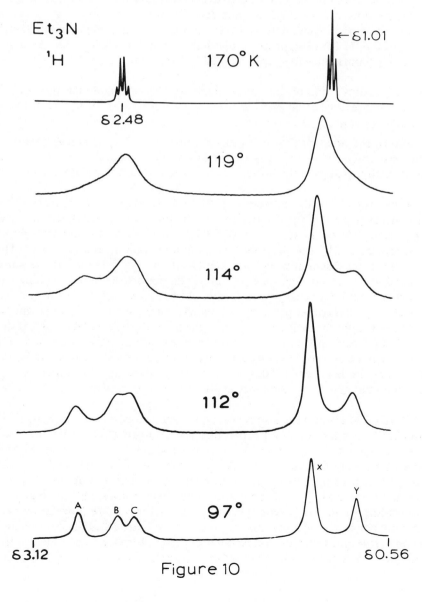

Figure 10

Examination of the ^{1}H DNMR spectrum (270 MHz) of triethylamine (3 percent v/v in CBRF$_3$) at 170°K shows a typical A$_2$X$_3$ spectrum (δ_A2.48, δ_X1.01, $^3J_{AX}$ = 6.7 Hz) in Figure 10. At temperatures below 170°K, the spectrum unerdergoes a clear-cut coalescence phenomenon and is separated into several peaks at 97°K (Figure 10). In rationalizing the DNMR spectral changes for triethylamine in Figure 10, it is useful to peruse Equation 4 (R = Et). Using arguments put forth previously for diethylmethylamine, it is evident from Equation 4 that if inversion-rotation is slow on the DNMR time scale for triethylamine but N-CH$_2$ rotation is very rapid, no change in the ^{1}H DNMR spectrum would be expected. Simple N-CH$_2$ rotation is indeed sufficient to swap the environments respectively of both the methylene protons and the methyl groups. Thus, any changes in the ^{1}H DNMR spectrum of triethylamine would require that *both* inversion-rotation *and* simple isolated N-CH$_2$ rotation be slow on the DNMR time scale. Since it is likely that the barrier to inversion-rotation in triethylamine is very similar to that in diethylmethylamine, the changes in the ^{1}H DNMR spectrum of triethylamine shown in Figure 10 may be attributed with confidence to simple isolated N-CH$_2$ rotation.

A theoretical simulation of the complete spectrum of triethylamine at 97°K is consistent with the presence of an A$_2$Y$_3$ subspectrum (δ_A2.79, δ_Y0.79, $^3J_{AY}$ = 7.0 Hz) and a BCX$_3$ subspectrum (δ_B2.52, δ_C2.37, δ_X1.13, $^2J_{BC}$ = - 11.0 Hz, $^3J_{BX}$ = $^3J_{CX}$ = 6.5 Hz). The ratio of the areas of the BCX$_3$ to A$_2$Y$_3$ spectra is 2:1. This indicates that there are at least three different types of methylene protons and at least two different types of methyl groups in the preferred conformation of triethylamine.

Such a spectrum is consistent with a MINDO/3 calculation that conformer 16 is the lowest energy geometry for triethylamine. The lone pair/C1C2 dihedral angle is 0°. The lone pair/C5C6 dihedral angle is 54°. The lone pair/C3C4 dihedral angle is 52°. The asterisked methylene protons are essentially trans to the lone pair. The C-H bonds on C1 are essentially eclipsing the N-C3 and N-C5 bonds. If the anisotropy associated with the nitrogen lone pair plays the major role in determining chemical shifts, then the two asterisked protons would have very similar or identical upfield chemical shifts. The other protons on C3 and C5 are gauche to the lone pair (16) and would have similar or identical downfield chemical shifts. Similar chemical shifts for the methyl protons in C4 and C6 could then lead to the BCX$_3$ spectrum of double weight which is observed. In contrast, the two protons on C1 (16) would be expected to have very similar chemical shifts and the very different environment at C2 could lead to the smaller methyl resonance at δ0.79 and the observed A$_2$Y$_3$ subspectrum.

The MINDO/3 calculations also reveal two other energy minima associated with the "propeller" geometry (17) and conformer 18. Conformer 17 is calculated to be 2.2 kcal/mole higher in energy than 16 and 18 is 2.5 kcal/mole higher than 16. In 17 and 18, the atoms marked with an asterisk are trans to the nitrogen lone pair. Clearly, 17 would give a spectrum at slow rotation inconsistent with the observed spectrum at 99°K. In 17, there are just *two* types of methylene protons and *one* type of methyl group leading to an ABX$_3$ type spectrum *only*. However, 18 possesses a symmetry completely consistent with the spectrum at 99°K (Figure 10). If one is to believe that semi-empirical molecular orbital calculations give at least an accurate *relative* scale of energies, 18 is to be ruled out as an important contributor to the spectrum at 99°K.

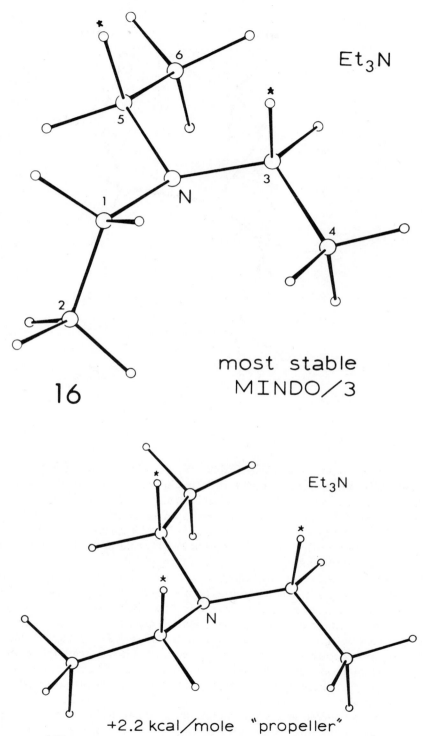

Et$_3$N

most stable
MINDO/3

16

Et$_3$N

+2.2 kcal/mole "propeller"

17

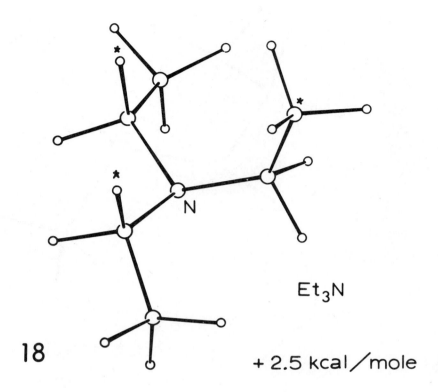

Et₃N

18   + 2.5 kcal/mole

From these studies, it is apparent that triethylamine strongly prefers the conformer 16 and it is that geometry which is reflected in the spectrum at 99°K (Figure 10). Obviously, there are two other equivalent forms of 16 resulting from rotation about pertinent N-CH₂ bonds. There also exists an enantiomer of 16 and two equivalent forms. The rate process reflected in the ¹H DNMR spectra of triethylamine (Figure 10) would then involve equilibration via N-CH₂ rotation only among these various conformers of the same symmetry (see 16). In order to proceed from one stable conformer to another, rotation about at least two N-CH₂ bonds must occur. In fact, the ¹H DNMR spectra of triethylamine (Figure 10) can be simulated using a dynamical model involving random exchange among A_2Y_3, BCX_3 and CBX_3 spin systems. The derived activation parameters for an apparent multiple rotation process are $\Delta H^+ = 5.0 \pm 0.3$ kcal/mole, $\Delta S^+ = -4 \pm 4$ eu, and $\Delta G^+ = 5.5 \pm 0.2$ kcal/mole at 114°K. While some problems remain to be resolved for the two systems above, these data indicate beyond any doubt that the combination of high field DNMR capability and the capacity to reach very low temperatures possesses great potential for studying fundamentally important rate processes in simple molecules. Such experimental DNMR data coupled with semi- empirical molecular orbital or molecular mechanics calculations should provide significant insight into the preferred stereodynamics of such simple molecular systems.

References and Footnotes

1. For reviews, see: Rauk, A., Allen, L. C., and Mislow, K., *Angew. Chem., Inst. Ed. Engl., 9*, 400 (1970); Lehn, J. M., *Fortschr. Chem. Forsch., 15*, 311 (1970); Lambert, J. B., *Top. Stereochem., 6*, 19 (1971).

2. Orville-Thomas, W. J., (Editor), "Internal Rotation in Molecules," John Wiley and Sons, London, 1979.

3. Nishikawa, T., Itoh, T., and Shimoda, K., *J. Chem. Phys., 23*, 1735 (1955).

4. Lide, D. R., Jr., and Mann, D. E., *J. Chem. Phys., 28*, 572 (1958).

5. Tannenbaum, E., Meyers, R. J., and Gwinn, W. D., *J. Chem. Phys. 25*, 42 (1956).

6. Bushweller, C. H., Anderson, W. G., Stevenson, P. E., Burkey, D. L., and O'Neil, J. W., *J. Am. Chem. Soc., 96*, 3892 (1974).

7. Bushweller, C. H., Anderson, W. G., Stevenson, P. E., and O'Neil, J. W., *J. Am. Chem. Soc., 97*, 4338 (1975).

8. Bushweller, C. H., Wang, C. Y., Reny, J., and Lourandos, M. Z., *J. Am. Chem. Soc., 99*, 3938 (1977).

9. The computer program used is a substantially revised local version of DNMR3 written by D. A. Kleier and G. Binsch (Program 165, Quantum Chemistry Program Exchange, Indiana University). Our local modifications are described in detail in C. H. Bushweller, G. Ghat, L. J. Lentendre, J. A. Brunelle, H. S. Bilofsky, H. Ruben, D. H. Templeton, and A. Zalkin, *J. Am. Chem. Soc., 97*, 65 (1975).

10. Dewar, M. J. S., and Ford, G. P., *J. Am. Chem. Soc., 101*, 783 (1979) and references therein.

11. Bushweller, C. H., Lourandos, M. Z., and Brunelle, J. A., *J. Am. Chem. Soc. 96*, 1591 (1974).

Pyramidal Carbanions?

John B. Grutzner, P. R. Peoples and T. Trainor
Department of Chemistry
Purdue University
West Lafayette, Indiana 47907

There are three limiting electronic states of tri-coordinate carbon found in reactive organic intermediates, carbocations (6ϵ), radicals (7ϵ) and carbanions (8ϵ). Of these, carbanions are the most widely employed in organic processes, but the least understood structurally. The extensive pioneering work of Cram[1] and Walborsky[2] has shown that base catalyzed hydrogen exchange at a chiral center occurs more rapidly than molecular inversion.

$$k_1 > k_2$$

These results provide the basis for the common text-book view that carbanions are pyramidal species whose rate of atomic inversion is slow with respect to their reaction rates. These concepts have been challenged by recent molecular orbital calculations which have calculated the inversion barrier in methyl anion to be of the order of 2 kcal/mole. In this article, the experimental and theoretical evidence for pyramidal carbanions is examined. The conclusion will be reached that the available evidence supports a pyramidal carbanion, but that the barrier to inversion in simple ions is unknown. The best estimates of the barrier predict a value comparable to or less than the 5.8 kcal/mole found in ammonia. Thus if it were not for counter-ion and solvent effects, a carbanion in solution would react as though it was effectively planar.

Much mechanistic, structural and theoretical work has been devoted to this challenging question. (For reviews, see references 3-16.) A sampling of old and new evidence which may be cited to support the hypothesis of pyramidal anions are: the solid state studies of the alkali methyls by Weiss[17] using X-ray and IR spectroscopy; extensive investigation of organolithiums by Stucky;[18] dynamic nmr studies pioneered by Fraenkel[19] and Roberts;[20] extensive mechanistic work interrelating reactant and product stereochemistries via intermediate carbanions[21-25] and a range of theoretical studies.[26-36] Despite this extensive work, the direct measurement of the inversion barrier in a simple carbanion remains a major experimental challenge. Our attempts to obtain a value for the barrier are described at the end of this paper.

Direct Evidence for Pyramidal Methyl Anion

The most direct evidence for pyramidal methyl anion is the x-ray study of methyl lithium by Weiss[17] in which the hydrogens were located. The one unsolved question is the best limiting structure to characterize the $(MeLi)_4$ unit—a covalent unit Me_4Li_4 in which carbon is tetra coordinate and consequently pyramidal or a closest packed ionic group in which the anion is free to assume its most stable geometry. Our personal prejudice agrees with Streitwieser's interpretation,[37] that the structure lies close to the ionic limit. All carbon lithium distances are the same within experimental error and Weiss's later reports[17] of the x-ray structure of MeK, MeRb and MeCs show monomeric closest packed units which, if extrapolated to lithium, would give the MeLi tetrameric structure. However, Stucky's results,[18] which established a preferred orientation for lithium coordinated by π-type anions (e.g., benzyl) and Lewis bases, show that covalent overlap effects cannot be neglected entirely. Weiss has also reported[17] two strong C-H stretching bands at 2810 and 2745 cm^{-1} in the IR spectrum of the isolated methyl groups of solid MeK. This observation is significant because the symmetrical C-H stretch should not be detectable in a planar species based on the symmetry requirements for IR active bands.

Perhaps the most striking data concerning the structure of methyl anion is its gas phase photo-electron spectrum (PES) obtained by Ellison, Engelking and Lineberger[38] (Figure 1). In particular, the observation of the intense and extended vibrational series requires a substantial geometry change in converting methyl anion to methyl radical. This should be contrasted with the lack of an intense vibrational series in the PES of methyl radical as it is excited to methyl cation.[39/40] The spacing in the vibrational series agrees with that of the symmetrical out-of-plane bending vibration of the methyl radical. The direction of major structural change on excitation is defined by this spacing. Methyl cation is planar and methyl radical is planar.[39-43] The conclusion that methyl anion is pyramidal appears inescapable. A more tentative piece of evidence was the detection of a weak band on the high binding energy edge of the spectrum. It was assigned to electron ejection from vibrationally excited methyl anion. If this assignment can be confirmed, the symmetrical out-of-plane bend has a frequency of about 460 cm^{-1} implying an anion inversion barrier[38] of 1 kcal/mole.

With this focus on structure, the most important aspect of this work has been neglected so far—the methyl anion exists in the gas phase. The number of anions of saturated hydrocarbons obtained in the gas phase has been severely limited.[44-47] The best recent estimates[36/44] suggested that the electron affinity of gas phase alkyl radicals is essentially zero. In fact, the most commonly agreed result from theoretical calculations is that the electron affinity of radicals is negative (i.e., the

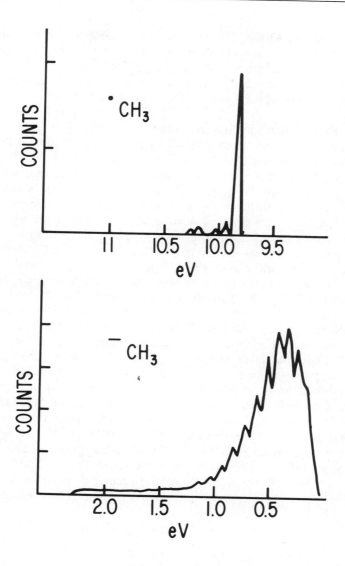

Figure 1. The photoelectron spectra of methyl radical (top)[39] and methyl anion (bottom).[38]

anion electron is unbound).[28-35] The value of 1.8 ± 0.7 kcal/mole measured by Ellison, Engelking and Lineberger[38] now shows that simple gas phase carbanions can exist for extended periods. In principle then, IR and microwave studies could be used to establish the structure and inversion barrier in gas phase ions. The conversion of this principle to practice will be challenging.

The Inversion Barrier in Methyl Anion

In principle, the experimental determination of the inversion barrier in methyl anion is exactly analogous to the classical study of inversion in ammonia[48/49] by detection of doubling in the IR spectrum. Figure 2 shows the calculated[34] energy surface for methyl anion as adapted by Lineberger and coworkers.[38/50] It shows the expected double minimum surface. In addition to tunneling, passage over the barrier is possible by excitation of the symmetrical out-of-plane bending mode (ν_2). Thus an inversion barrier could be estimated if the magnitude of ν_2 could be predicted.

With the following admonition from Bertrand Russell firmly in mind, an attempt will now be made to estimate the inversion barrier in methyl anion:—"Logic is the art of going wrong with confidence." As part of his classic series of papers on molecular shape, Walsh examined the factors influencing the planar/pyramidal preference of AX_3 molecules[51] (Figure 3). Extensive tests over many years have confirmed the accuracy and utility of these qualitative ideas.[52] Examination of individual geometric contributions in more advanced calculations (see ref. 39, for example) further support the Walsh analysis. The dominant feature in controlling geometry is the number of valence electrons. Thus in the series CH_3^+ (6 electrons) planar[39/40]; $CH_3^{\bullet}$ (7 electrons) planar; CH_3^- (8 electrons) pyramidal; addition of electrons to the a_2'' (lone pair) orbital enhances pyramidalization. Similarly the isoelectronic compounds, CH_3^{-38}, NH_3^{49} and H_3O^{+53} are all pyramidal with the lone pair orbital filled. These qualitative observations may be placed on a more quantitative basis by considering the ν_2 "umbrella" mode for a series of AH_3 species[26]—see Table I. The data for the first electronically excited states of methyl radical and ammonia have also been included as electron promotion from the lone pair orbital should lead to structures comparable to those obtained by complete electron loss by ionization. It should be noted that the vibrational frequencies in the pyramidal species are approximately twice those for planar species because of the imposed inversion barrier. The data in Table I may be extrapolated to predict a value in the region 400-1000 cm^{-1} for ν_2 in methyl anion, with a most probably value around 750 cm^{-1}. An experimental upper limit may be placed on this range because Andrews[55] has reported ν_2 as 1158 cm^{-1} for monomeric methyl lithium in liquid Ar together with confirmatory isotopic shifts. Coulombic attraction between cation and anion should selectively lower the energy minimum of the pyramidal form and consequent-

Table I
Out-of-Plane Symmetrical Bending Mode Frequencies
(ν_2 cm^{-1}) for AH_3 Molecules[a]

Electrons			
6	*$^+CH_3$*		
	1380 (39)		
	[1360] (41)		
7	*$^{\bullet}CH_3$*	*$^{\bullet+}NH_3$*	
	610 (43)	950 (49)	
		[878] (49)	
8	$^-CH_3$	NH_3	$^+OH_3$
	?	1060 (48)	1125 (54)

[a]Species in italic are planar. Numbers in parentheses are references and numbers in square brackets are for excited state species (see text).

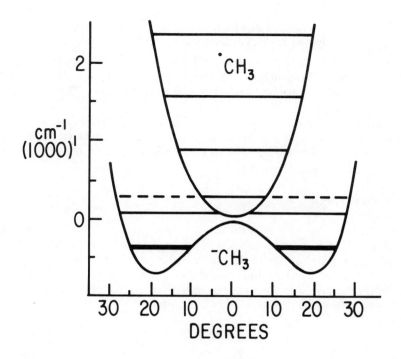

Figure 2. The potential energy surface for methyl anion[34/38] as a function of out-of-plane bending angle.

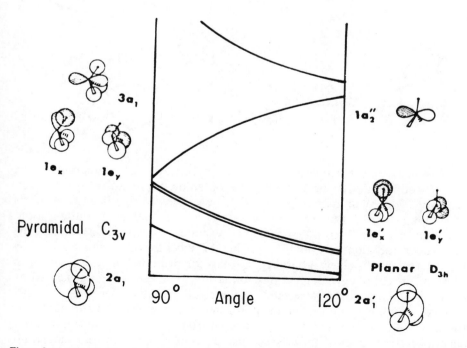

Figure 3. The Walsh diagram for AX_3 molecules showing orbital energy as a function of H-C-H angle.

ly increase both the inversion barrier and the ν_2 frequency in the ion pair. If it is now assumed that the geometry of the methyl anion is not significantly different from that of ammonia (cf. MeLi X-ray structure), the inversion barrier for free methyl anion is estimated to be 2-6 kcal/mole.[26] This experimentally based estimate agrees with both the best theoretical estimates and the value estimated from PES.

The experimental implications of this result could almost be termed revolutionary. Most dramatically it would mean that the many examples of retention of chirality in reactions are a reflection of the chiral environment maintained by the counter-ion and solvent and not of carbanion chirality!

Carbanion Inversion in Solution

In the previous section, the idea was introduced that simple carbanions may be pyramidal species but with inversion barriers which are so low as to be inconsequential for chemical reactions. Can this postulate be supported by solution results? Recent work by Stille[56] and Shechter[57] has shown that anion inversion can be a facile process provided the carbanion is generated by loss of nitrogen from an intermediate diazo anion.

The conclusions are clear, either anion inversion occurs at a rate comparable to the normal diffusion limit for protonation,[58/62] or anion protonation by a proton source, more acidic by at least 20 pK units, has been slowed significantly. The implicit assumption here is that protonation occurs with retention of configuration. The difference between the results for anions generated by nitrogen loss and by the traditional deprotonation methods[4-7] is striking. Whether the difference is caused by nitrogen inhibition of solvent approach slowing the protonation rate, or nitrogen disruption of the chiral environment of counter-ion, base and solvent generated by the chiral anion, remains to be answered. The intriguing results reported by Kessler[63] for ion recombination,   may offer an approach to this problem. An alternative explanation for the uniqueness of anions generated from nitrogen precursors in terms of radical processes appears unlikely because of Stille's failure to trap radical intermediates[46] and the reluctance of DMSO to act as a hydrogen atom source.[64]

Whatever the origin of the observed rate differential, between anion inversion and protonation, the conclusion is inescapable—carbanion inversion in solution is a rapid process.

Measurement of Anion Inversion Rates by Carbon-13 NMR

In this section, the results of our attempts to obtain a quantitative measurement of an anion inversion barrier are given. The current state of our research was summarized earlier.[65] The system chosen for study was the 7-phenyl-norbornyl anion. This bicyclic molecule possesses two orthogonal symmetry planes which bisect the center of interest and so is ideal for detection of an inversion process. The two possible invertomers are energetically equivalent, but their reduced symmetry is spectroscopically observable. Alternative processes, such as rotation, which complicate the interpretation in acyclic and monocyclic systems are readily differentiated by alternative symmetry criteria.

While it would be desirable to obtain the results for the unsubstituted norbornyl anion, consideration of both the method of generation and anion basicity suggested that a phenyl substituent would be required for stability. Clearly the phenyl substituent will favor anion planarity. In the case of nitrogen inversion in aziridines,[66] replacement of a methyl group by phenyl lowered the barrier by 8 kcal/mole. This lowering is partially compensated by the effect of angle strain.[12] The barrier to inversion in 7-methyl-aza-norbornene[67] is 6 kcal/mole higher than that in trimethylamine. Similarly, angle strain permitted the freezing out of inversion in the methyl oxiranium ion but not in other cyclic oxygen salts of larger ring size.[68] Thus the 7-phenylnorbornyl anion is a compromise between π conjugation favoring the planar form and angle strain favoring the pyramidal form. The barrier obtained for this ion should provide a first approximation to the barrier in a simple anion.

The 7-phenylnorbornyl anion was prepared by sodium/potassium alloy cleavage of 7-phenyl-7-methoxy-norbornane in tetrahydrofuran (THF) or 1,2 dimethoxyethane (DME) at -10 to 0°C for 1 hour under vacuum line conditions.[69]

It was quickly discovered that the anion was strongly basic and was converted to 7-phenylnorbornane by deprotonation of the solvent on standing in DME for 30 minutes at room temperature. (For related ether solent deprotonations, see Reference 70.) In all subsequent experiments, the anion was manipulated and stored at temperatures below 0°C. It is well established[71-75] that the choice of counter-ion and solvent can greatly influence inversion barriers. Accordingly, the anion has been prepared with lithium (LiCl metathesis[76]) potassium (standard cleavage) and cesium (Cs/Na/K cleavage[77]) counterions in both THF and DME. One of the intriguing features of the Na/K alloy cleavage reaction is the fact that the carbanion is associated with a potassium cation and sodium methoxide is precipitated. Similarly, the cesium salt is obtained from the cesium alloy cleavage.

Carbon-13 nmr is the tool of choice for sensitive differentiation of molecules on the basis of symmetry. For example, the two carbon bridges in the hydrocarbon, 7-phenylnorbornane are differentiated by 3 ppm by the remote phenyl substituent (Figure 4). Furthermore, the selective line broadening associated with the phenomenon of dynamic nmr[78] provides a convenient method for determination of the inversion barrier. Three possible situations can be differentiated. If anion inversion is slow on the nmr time scale—rate (k_I) less than 16 sec^{-1} in the present example assuming the chemical shift difference in the hydrocarbon is retained in the anion—two distinct resonances (C2 and C5) will be detected. If the inversion rate is comparable to the nmr time scale, 16 sec^{-1} < k_I < 1600 sec^{-1}, line broadening corresponding to 2-site exchange may be used to determine the rate. If inversion is fast, k_I > 1600 sec^{-1} or, in fact, the anion is planar, then C_2 and C_s will absorb as a single sharp resonance.

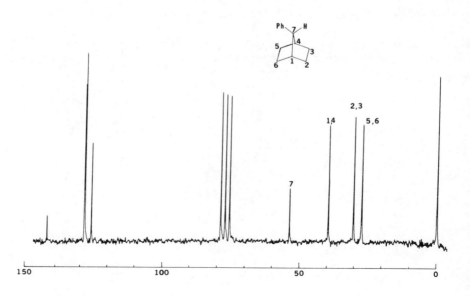

Figure 4. The 20 MHz ^{13}C spectrum of 7-phenylnorbornane in CDCl$_3$. Note the separation between C2 and C5.

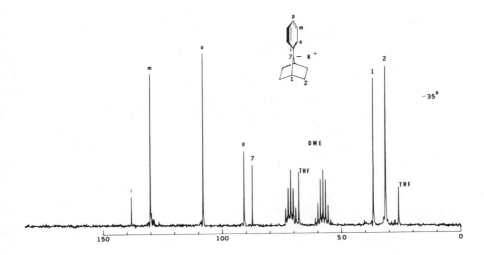

Figure 5. The 20 MHz ^{13}C MHz spectrum of potassium 7-phenylnorbornyl anion in DME-D10 at -35°. The anion was initially prepared in THF solvent and solvent removal was incomplete.

The experimental 20 MHz ^{13}C spectrum of the potassium salt in DME-D10 at -35° is shown in Figure 5. The spectrum shows that clean anion spectra are obtained and that the anion is effectively planar on the nmr time scale at this temperature. Peak assignments are shown in the figure. Change of counter-ion or solvent causes small changes in chemical shift (~1 ppm), but no other discernable effect. On further cooling to the -90° to 100° range, the peak assigned to C2, C5 shows the onset of selective broadening. The broadening is clearly selective as peaks due to minor amounts of ether precursor and hydrocarbon product show no additional broadening (Figure 6). While it is very tempting to ascribe the selective broadening to the onset of detectable pyramidal inversion, a degree of caution is necessary.

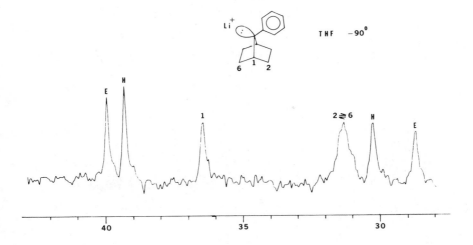

Figure 6. An expanded view of the aliphatic region of the 20 MHz ^{13}C spectrum of lithium 7-phenylnorbornyl anion in THF at -90°. Peaks labelled H and E arise from 7-phenylnorbornane and 7-phenyl-7-methoxynorbornane respectively.

While the anion peaks are broadened selectively and the C-2 resonance is broader than C-1, the aromatic carbons also show selective broadening. This totally unexpected observation is unexplained at this time. The broadening appears to be unrelated to the choice of counter-ion. The most plausible explanation is that the anion is associated in solution and that either chemical exchange, or more likely, restricted motion at low temperature is responsible for the broadening. Further experimental work is in progress in an attempt to clarify the interpretation. If the broadening is attributed solely to pyramidal inversion, an upper limit of 8.5 kcal/mole is established for the 7-phenylnorbornyl anion. This upper limit then provides additional experimental support for the notion that a carbanion in solution is a configurationally labile entity.

Carbanion Inversion in the Gas Phase

The recent development of negative ion, mass analyzed ion kinetic energy spectroscopy (MIKES)[79-81] by Cooks and coworkers has opened up the possibility of detecting anion inversion in the gas phase. If an anion decomposition could be found which was sensitive to anion stereochemistry, then measurement of ion lifetimes could be used to define gas phase anion inversion rates. The anti-periplanar requirement for anion and leaving group in the Grob fragmentation[82/83] would be one possible example. A reaction which may provide the necessary stereochemical selectivity is the retro-Diels Alder reaction observed in the 7-norbornenyl anion in

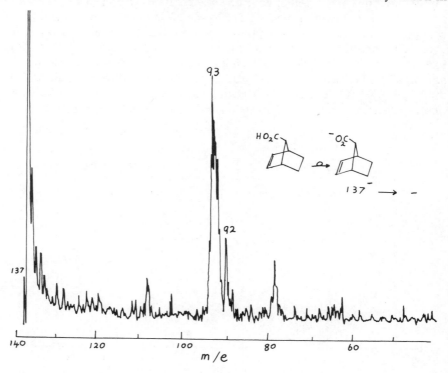

Figure 7. The negative ion MIKES spectrum of syn-7-norbornyl carboxylate. N_2 was the collision gas. Ion 137⁻ was monitored.

solution.[84] The precursors for the reaction, syn and anti-7-norbornenyl carboxylic acids have recently been prepared and separated in our laboratories.[85] Chemical ionization in the gas phase can generate the corresponding carboxylate ions which have been detected by the MIKES technique. The ions are then subjected to collision induced dissociation and electron loss. A typical spectrum of the negative ions derived from the starting carboxylate are shown in Figure 7. The subsequent fragmentation of the daughter ion at mass 93 is shown in Figure 8. A summary of the ionic species detected are shown below for both syn and anti compounds together with their saturated analog 7-norbornyl carboxylic acid. It should be clearly stated that while individual structures for many of the ions are shown, their precise identity is unknown (apart from their mass of course). Thus the structures are no more than convenient speculation at present.

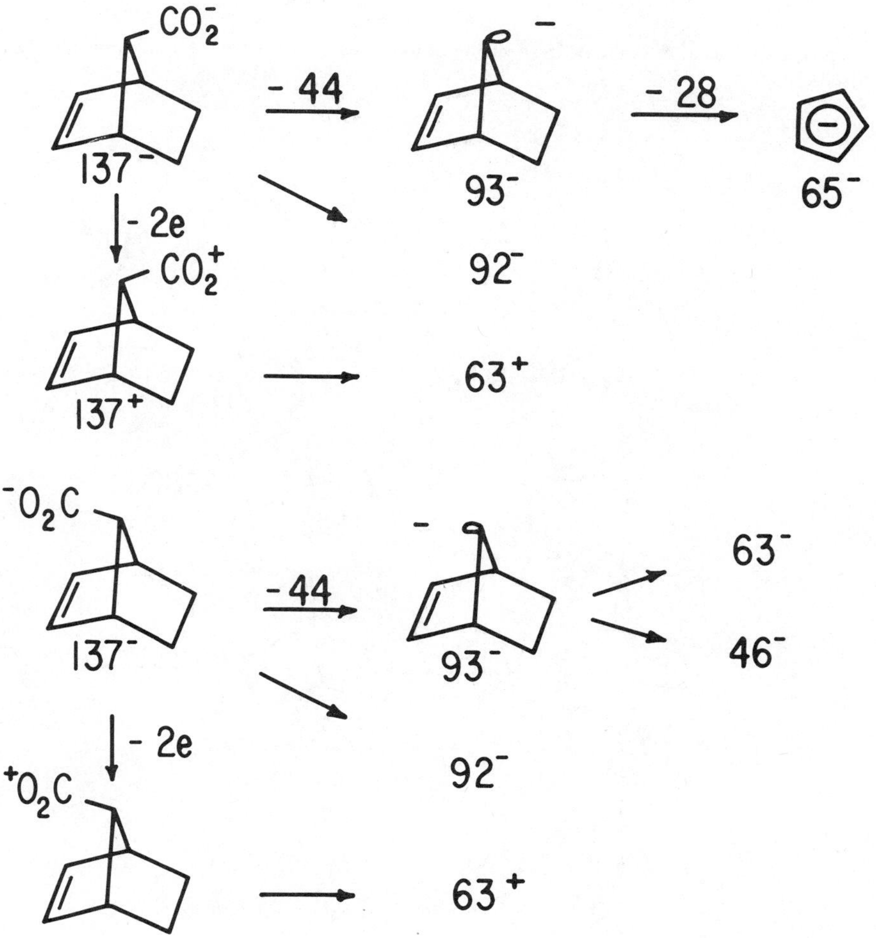

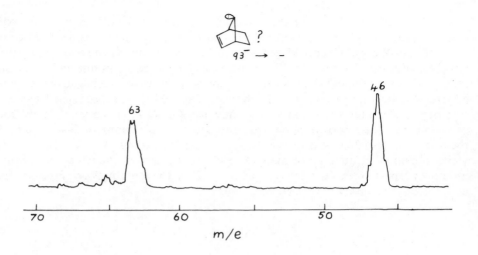

Figure 8. The negative ion MIKES spectrum of ion 93 derived from syn-7-norbornyl carboxylate after N_2 collision.

The most important observation is that these anions can exist in the gas phase. This is significant for two reasons—the electron affinity of the corresponding radical must be positive (or there is a barrier to electron loss) and the collision is sufficiently energetic to permit decarboxylation—a process which is endothermic by at least 50 kcal/mole. The question of the barrier to anion inversion in the gas phase could be answered if the retro-Diels-Alder reaction could be induced to occur stereospecifically from the syn and anti norbornenyl anions. The specific sequence required is the conversion of the ion of mass 93 to an ion of mass 65. The results to date have been variable (65 is detected in some spectra) and regrettably no reliable data pertinent to anion inversion has been obtained. The collision induced loss of 2 electrons and subsequent cationic fragmentation provide an independent check on the identity of the ions. The identity of the peak at mass 46 derived from mass 93 is perplexing. If this is an ion which contains only carbon and hydrogen, as seems likely, the only possibility for a mass 46 peak is the formation of a doubly negatively charged ion! Clearly, this data needs to be carefully checked with appropriately labelled compounds and this research is in progress at present.

Conclusion

Data from photo-electron spectroscopy, vibrational spectroscopy, anion protonation in solution and low temperature carbon-13 nmr concur with the theoretical estimates that carbanion inversion is a low barrier process in simple ions. Thus, simple anions should be effectively planar in all but diffusion controlled reactions.

Acknowledgements

The generous support of our research by the Petroleum Research Fund administered by the American Chemical Society is acknowledged with gratitude. Without the willing collaboration, technical skill and support of Professor R. G. Cooks and Mr. Kirk Harmon, the gas phase study would have been impossible. A number of carbon-13 nmr spectra were obtained at 37 MHz using the Purdue University Biological Magnetic Resonance Laboratory Nicolet 150 spectrometer purchased and operated with funds provided by the National Institutes of Health RR01077.

References and Footnotes

1. Almy, J., Hoffman, D. H., Chu, K. C., and Cram, D. J., *J. Am. Chem. Soc.*, **95**, 1185 (1973) and references therein.
2. Periasamy, M. P. and Walborsky, H. M., *J. Am. Chem. Soc.*, **99**, 2631 (1977) and references cited therein.
3. Le Noble, W. J., "Highlights of Organic Chemistry," Dekker, 1974, p. 821.
4. Buncel, E., "Carbanions: Mechanistic and Isotopic Aspects," Elsevier, 1975, Ch. 2.
5. Cram, D. J., "Fundamentals of Carbanion Chemistry," Academic Press, New York, 1965.
6. Dart, E. C., "Reactivity, Mechanism and Structure in Polymer Chemistry," Jenkins, A. D. and Ledwith, A., Eds., Wiley, 1974, Ch. 10.
7. House, H. O., "Modern Synthetic Reactions," Benjamin, 1972, p. 155.
8. Hunter, D. H., "Isotopes in Organic Chemistry," Buncel, E. G. and Lee, C. C., Eds., Elsevier, 1975, Vol. 1, Ch. 4.
9. Schlosser, M., "Structure und Reaktivitat Polarer Organometalle," Springer Verlag, 1973, p. 87, *et. seq.*

10. Szwarc, M., "Ions and Ion Pairs in Organic Reactions," Wiley-Interscience, Vol. 1 and 2.
11. Henderson, J. W., *Chem. Soc. Revs., 2*, 397 (1973).
12. Lambert, J. B., *Topics in Stereochemistry, 6*, 19 (1971).
13. Lehn, J. M., *Fortschr. Chem. Forsch., 15*, 311 (1970).
14. Rauk, A., Allen, L. C., and Mislow, K., *Angew. Chemie, 9*, 400 (1970).
15. Radom, L., in "Methods of Electronic Structure Theory," Schaefer, H. F., ed., Plenum, 1977, Vol. 3, p. 333.
16. Payne, P. W. and Allen, L. C., in "Applications of Electronic Structure Theory," Schaefer, H. F., Ed., Plenum, 1977, Vol. 4, p. 29.
17. (a) $(MeLi)_4$: Weiss, E. and Hencken, G., *J. Organomet. Chem., 21*, 265 (1970).
    (b) MeK: Weiss, E. and Sauermann, G., *Chem. Ber., 103*, 265 (1970).
    (c) MeRb, MeCs: Weiss, E. and Koster, H., *Chem. Ber., 110*, 717 (1977).
18. Stucky, G. D., *Adv. Chem. Ser., 130*, 56 (1974) and references therein.
19. Fraenkel, G., Beckenbaugh, W. E., and Yong, P. P., *J. Am. Chem. Soc., 98*, 6878 (1976).
20. Whitesides, G. M. and Roberts, J. D., *J. Am. Chem. Soc., 87*, 4878 (1965).
21. Applequist, D. E. and Chmurny, G. W., *J. Am. Chem. Soc., 89*, 875 (1967).
22. Pierce, J. B. and Walborsky, H. M., *J. Org. Chem., 33*, 1962 (1968).
23. Glaze, W. H. and Selman, C. M., *J. Org. Chem., 33*, 1987 (1968).
24. Hargreaves, M. K. and Modarai, B., *J. Chem. Soc. C*, 1013 (1971).
25. (a) Abatjoglou, A. G., Eliel, E. L., and Kuyper, L. F., *J. Am. Chem. Soc., 99*, 8262 (1977).
    (b) Kuyper, L. F. and Eliel, E. L., *J. Organomet. Chem., 156*, 245 (1978).
26. Koepple, G. W., Sagatys, D. S., Krishnamurthy, G. S., and Miller, S. I., *J. Am. Chem. Soc., 89*, 3396 (1967).
27. Lehn, J. M. and Wipff, G., *J. Am. Chem. Soc., 98*, 7498 (1976).
28. Csizmadia, I. G., Mangini, A., Schlegel, H. B., Whangbo, M. H., and Wolfe, S., *J. Am. Chem. Soc., 97*, 2209 (1975).
29. Williams, J. E., and Streitwieser, A., *J. Am. Chem. Soc., 97*, 2634 (1975).
30. Radom, L., *Aust. J. Chem., 29*, 1635 (1976).
31. Duke, A. J., *Chem. Phys. Lett., 21*, 275 (1973).
32. Driessler, F., Ahlrich, R., Staemmler, V., and Kutzelnigg, W., *Theor. Chem. Acta, 30*, 315 (1973).
33. Dykstra, C. E., Hereld, M., Lucchese, R. R., Schaefer, H. F., and Meyer, W., *J. Chem. Phys., 67*, 4071 (1977).
34. Marynick, D. S. and Dixon, D. A., *Proc. Natl. Acad. Sci. USA, 74*, 410 (1977).
35. Surratt, G. T. and Goddard, W. A., *Chem. Phys., 23*, 39 (1977).
36. Kollmar, H., *J. Am. Chem. Soc., 100*, 2665 (1978).
37. Streitwieser, A., *J. Organomet. Chem., 156*, 1 (1978).
38. Ellison, G. B., Engelking, P. C., and Lineberger, W. C., *J. Am. Chem. Soc., 100*, 2556 (1978).
39. Dyke, J., Jonathan, N., Lee, E., and Morris, A., *J. Chem. Soc., Farad. Trans. 2, 72*, 1385 (1976).
40. Koenig, T., Balte, T., and Snell, W., *J. Am. Chem. Soc., 97*, 662 (1975).
41. Herzberg, G., *Proc. Roy. Soc. Ser. A, 262*, 291 (1961).
42. Fessenden, R., *J. Phys. Chem., 71*, 74 (1967).
43. Andrews, L. and Pimentel, G. C., *J. Chem. Phys., 47*, 3637 (1967).
44. (a) Zimmerman, A.H., Gigaux, R., and Brauman, J. I., *J. Am Chem. Soc., 100*, 5595 (1978).
    (b) Zimmerman, A. H. and Brauman, J. I., *J. Am. Chem. Soc., 99*, 3565 (1977).
45. Jordan, K. D. and Burrow, P. D., *Accnts. Chem. Res., 11*, 341 (1978).
46. DePuy, C. H., Bierbaum, V. M., King, G. K., and Shapiro, R. H., *J. Am. Chem. Soc., 100*, 2921 (1978).
47. (a) Bowie, J. H., *Chem. Soc. Specialist Period Rpts., Mass. Spectrometry*, Vol. 4, Ch. 11, 1974-6.
    (b) Jennings, K. R., *ibid.*, Ch. 10.
48. Herzberg, G., "Infrared and Raman Spectra," van Nostrand, 1945, p. 222.
49. For a discussion of ammonia and related systems, see Harshberger, W. R., *J. Chem. Phys., 53*, 903 (1970) and *56*, 177 (1972).
50. Note that in principle, one might expect major alteration of the potential energy curve as the change from theoretical energy gap to experimental gap is made. However, the curve shown in Figure 2 correlates best with available experiments and electronic interaction appears to be small.
51. Walsh, A. D., *J. Chem. Soc.*, 2296 (1953).
52. For a review see Gimarc, B. M., *Accnts. Chem. Res., 7*, 384 (1974).
53. Lundgren, J. and Williams, J. M., *J. Chem. Phys., 58*, 788 (1973).

54. Basile, L. J., La Bonville, P., Ferraro, J. R., and Williams, J. M., *J. Chem. Phys., 60*, 1981 (1974).
55. Andrews, L., *J. Chem. Phys., 47*, 4834 (1967).
56. Stille, J. K. and Sannes, K. N., *J. Am. Chem. Soc., 94*, 8489 and 8494 (l972).
57. Babu, T. V. R., Sanders, D. C. and Shechter, H., *J. Am. Chem. Soc., 99*, 6449 (1977).
58. Eigen, M., *Angew. Chemie, 3*, 1 (1964).
59. Jones, J. R., "The Ionization of Carbon Acids," Academic Press, 1973, Ch.8.
60. Ritchie, C. D., "Solute-Solvent Interactions," Dekker, 1969, Ch. 4.
61. Ritchie, C. D., "Physical Organic Chemistry," Dekker, 1975, Ch. 10.
62. Szwarc, M., Streitwieser, A., and Mowery, P. C., "Ions and Ion Pairs in Organic Reactions," Wiley, 1974, Vol. 2, p. 180.
63. Kessler, H., Lecture delivered at the Symposium "Steroeodynamics of Molecular Systems," April 23-24, 1979, State University of New York at Albany.
64. Bridger, R. F. and Russell, G. A., *J. Am. Chem. Soc., 85*, 3754 (1963).
65. New English Bible, Psalms 127, V. 2.
66. Andose, J. D., Lehn, J. M., Mislow, K., and Wagner, J., *J. Am. Chem. Soc., 92*, 4050 (1970).
67. Marchand, A. P. and Allen, R. W., *Tetrahedron Letts.*, 619 (1977) and references therein.
68. Lambert, J. B. and Johnson, D. H., *J. Am. Chem. Soc., 90*, 1349 (1968).
69. Grutzner, J. B. and Winstein, S., *J. Am. Chem. Soc., 99*, 2200 (1972).
70. Moncur, M. V. and Grutzner, J. B., *J. Am. Chem. Soc., 95*, 6449 (1973) and references therein.
71. Kobrich, G., Merkel, D., and Imkampe, K., *Chem. Ber., 106*, 2017 (1973).
72. Knorr, R., and Lattke, E., *Tetrahedron Letts.*, 3969 (1977).
73. Boche, G., Weber, H., Martens, D., and Bieberbach, A., *Chem. Ber. 111*, 2480 (1978).
74. Evans, D. A., Baillargeon, D. J., and Nelson, J. V., *J. Am. Chem. Soc., 100*, 2242 (1978).
75. Fraenkel, G. and Dix, D. T., *J. Am. Chem. Soc., 88*, 979 (1966).
76. Hartmann, J. and Schlosser, M., *Helv. Chim. Acta, 59*, 453 (1976) and references therein.
77. Grovenstein, E., Beres, J. A., Cheng, Y. M., and Pegolotti, J. A., *J. Org. Chem., 37*, 1281 (1972).
78. Jackman, L. M. and Cotton, F. A., "Dynamic NMR," J. Wiley, 1975.
79. Beynon, J. H., Cooks, R. G., Amy, J. W., Baitinger, W. E., and Ridley, T. Y., *Anal. Chem., 45*, 1023A (1973).
80. Cooks, R. G., *Amer. Lab.*, Oct., 1978, p. 111.
81. McClusky, G. A., Kondrat, R. W., and Cooks, R. G., *J. Am. Chem. Soc., 100*, 6045 (1978).
82. Grob, C. A., *Angew. Chemie, 8*, 535 (1969).
83. Booth, H., Bostock, A. H., Franklin, N. C., Griffiths, D. V., and Little, J. H., *J. Chem. Soc. Perkin II*, 899 (1978).
84. Bowman, E. S., Hughes, G. B., and Grutzner, J. B., *J. Am. Chem. Soc., 98*, 8273 (1976).
85. Peoples, P. R., unpublished observations.

Delocalization in Diarylmethyl Anions
Probe of Solution Ion States?

Shelton Bank and John Sturges
Institute of Biomolecular Stereodynamics
Department of Chemistry
State University of New York at Albany
Albany, New York 12222

and

C. H. Bushweller
Department of Chemistry
University of Vermont
Burlington, Vermont 05405

The large difference in acidity between propane ($\sim$60 pKa units) and diphenylmethane (33.4 pKa units) is a direct consequence of the greater stability of the diphenylmethyl anion.[1] While inductive effects no doubt contribute, the major source of stabilization is charge delocalization via resonance into the two aromatic rings. A direct and measurable consequence is that the resultant increase in pi bonding across the benzylic carbon-phenyl bond leads to restricted phenyl rotation.

In this work we discuss first relevant analogies followed by detailed assignments of the high and low temperature ^{1}H and ^{13}C chemical shifts for 4,4'-dideuteriodiphenylmethyl lithium *(1)*, 4,4'-dimethyldiphenylmethyl lithium *(2)* and 4-deuterio, 4'-methyldiphenylmethyl lithium *(3)*. We then consider the results and conclusions of the dynamic nuclear magnetic resonance spectroscopy (DNMR) the symmetrical anions (1 and 2) and conclude with the results and discussion of the DNMR of the unsymmetrical anion 3.

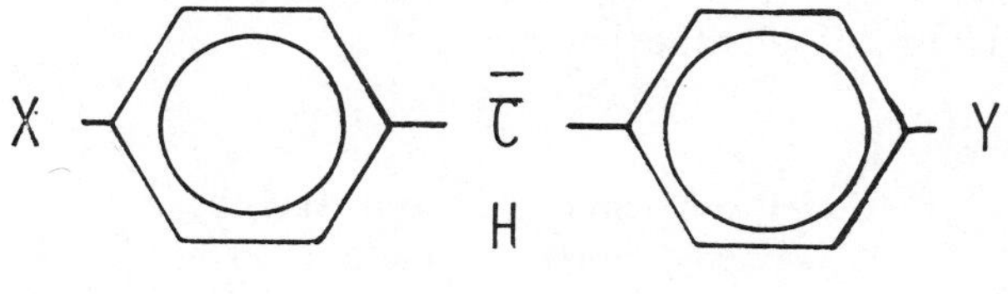

Figure 1. Compounds 1-3.

Static NMR Spectra: Results and Discussion

Nuclear magnetic resonance spectroscopy (NMR) has in recent years been shown to be a powerful tool for investigation of molecular structure and conformation.[2] The advent of improved computer techniques has extended the usefulness of NMR to include the study of kinetic processes in the 4-25 kcal/mol range by DNMR.[3] Appropriately delocalized carbanions display temperature dependent spectra (line shapes) indicative of slowing a unimolecular exchange process at lower temperatures to the point where the exchange rate is very slow. The slow exchange spectra often differ markedly from the rapid exchange spectra. Computer line shape analyses at several intermediate temperatures provide the rates for the exchange process and the activation parameters, ΔG^*, ΔH^*, ΔS^*, are thus obtained.

Several recent examples by Sandel and coworkers[4] for both phenylallyl alkali and naphthylmethyl alkali salts in tetrahydrofuran (THF) and by Fraenkel[5] and Brownstein[6] for benzyl alkali salts in THF reveal conformational barriers in the range of 10-20 kcal/mol. The observed process is restricted rotation about the partially formed double bond as a result of anion delocalization. In this regard, Bushby attempted to observe substituent effects on allyl rotational barriers with groups in the *para* position in 1,3- diphenylpropenyl alkali salts.[7] The barriers however, proved to be too high to be observed.[7] At the same time, diphenylmethyl lithium was observed to have a reasonable barrier in THF in our laboratories.[8] Accordingly, the diphenylmethyl system was chosen as a model and select derivatives were prepared and studied to provide insights into the substituent effects.

The diphenylmethyl lithium salts, 1,2 and 3 (Figure 1) were prepared in THF by reaction of the corresponding hydrocarbons with butyl lithium.[8] For compound 1 X = Y = D, for compound 2 X = Y = methyl and for compound 3 X = D and Y = methyl. The study included the static and dynamic NMR spectra for ^{1}H and ^{13}C. Each salt exhibits temperature dependent line shapes. At room temperature, the spectra are well resolved. However, as the temperature is lowered line shapes broaden and eventually resharpen with separate resonances for each *ortho* and *meta* nuclei, Figure 2. On return to room temperature, the spectra revert to their original appearance. In order to provide a firm basis for the analyses of the DNMR, confident assignments of the static NMR at low and high temperatures are required. The following discussion consider these assignments.

The 270 MHz aromatic ^{1}H low temperature spectra appear as a pair of superimposed AB quartets for each phenyl ring with one quartet upfield of the other (Figure 2B). Consequently, the unsymmetrically substituted anion has four superimposed AB quartets and appears quite complex. The identification of the *ortho* and *meta* resonances are aided by selectively decoupling the *para* proton in diphenylmethyl lithium and comparing the low temperature resonances with those of 1 (Figure 3). The decoupling experiment was performed on a 90 MHz spectrometer and the downfield *ortho* and *meta* resonances appear as a broadened signal at 6.8 ppm. However, the upfield *ortho* and *meta* signals are fairly well resolved and identifica-

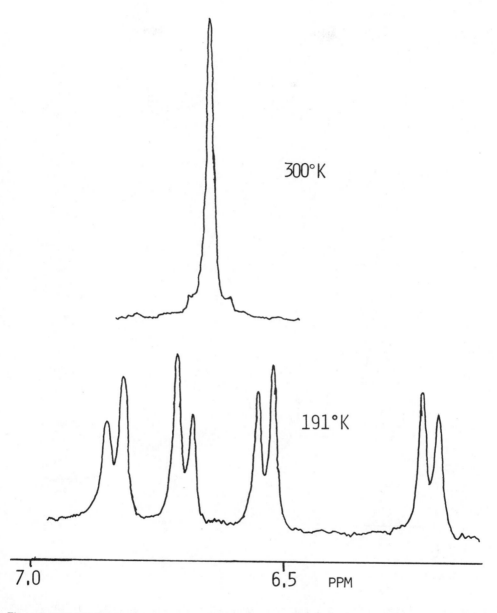

Figure 2. 270 MHz ^{1}H aromatic region of 1 at temperature extremes in THF.

tion of these resonances is straightforward. The *para* proton in diphenylmethyl lithium at 5.7 ppm (Figure 3A) is irradiated by single frequency homo- nuclear decoupling to eliminate coupling to other ring protons (Figure 3B). Accordingly, the *meta* resonance at 6.6 ppm which is a triplet due to nearly equal coupling from the adjacent *ortho* and *para* protons (Figure 3A), becomes a doublet (Figure 3B) upon *para* proton decoupling. Since deuterium coupling is approximately one- seventh the value of a proton, the large *para* coupling (~8.1 Hz) is in effect eliminated by

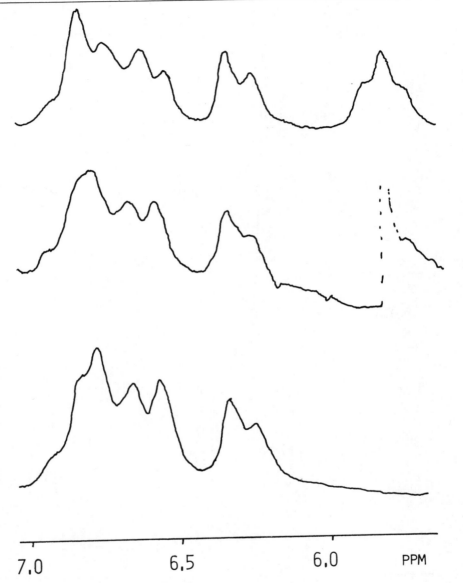

Figure 3. 90 MHz ¹H selective decoupling at 190°K.

deuterium substitution and its spectrum is virtually identical. Importantly, the resonance at 6.2 ppm, an *ortho* resonance, remains a doublet in all instances. The downfield *ortho* and *meta* resonances are then assigned as those which produce the best simulation of experimental line shapes.

Computer simulation of the 270 MHz ¹H DNMR permits only one possible assignment for these resonances; the other assignment gives incorrect computer simulation. Thus the *ortho* protons are assigned the high and low field resonances for each phenyl ring. The other anion spectra are assigned in a similar fashion. Confirmation of the above assignments was obtained from the 270 MHz ¹H low temperature NMR spectrum of diphenylmethyl lithium. *Para - meta* proton coupling clearly identifies the resonances at the *meta* position. Table I reports the 270 MHz ^{1}H anion resonances. We note that some specific assignments remain uncertain due to an incomplete knowledge of factors contributing to chemical shifts.

Table I
270 MHz ^{1}H Anion Resonances[a,b] in THF

Compound	Low Temperature					High Temperature		
	H_o	H_m	H_m	H_o	T°K	H_o	H_m	T°K
1	6.75	6.63	6.49	6.17	195	6.55	6.59	295
2	6.52	6.38	6.24	6.01	190	6.41	6.36	295
3 D-Phenyl	6.56	6.52	6.36	6.03	200	6.46	6.50	300
Methylphenyl	6.63	6.43	6.30	6.08	200	6.50	6.42	300

[a]In ppm downfield from TMS internal standard.
[b]With the following coupling constants determined by fitting $^3J_{HCCH}$ = 8.1 ±0.1 Hz, $^4J_{HCCCH}$ = 1.1 ± 0.1 Hz, $^5J_{HCCCCH}$ = 0.0 ± 0.1 Hz.

Table II
22.64 MHz ^{13}C Resonances[a] of Anions

Compound	Carbon Atom						
6 5 [ring] 4 3 2 1 C—⟨○⟩—C	1	2	3	4	5	6	T°K
1	80.23	146.4	112.63 121.60	127.6 128.8	106.21		223
	76.22	147.13	117.35	127.9	107.53		320
3 (D-Phenyl)	77.91	144.15	111.75 121.18	129.22 128.25	112.54	21.25	195
	72.14	145.83	117.38	128.92	114.70	20.90	205
3 (Methylphenyl)	79.58	146.12	111.77 121.02	126.89 128.54			210
	74.74	147.34	117.04	128.08			320
2	79.58	144.26	112.50 121.84	128.54	113.63	21.14	210
	74.74	147.33	117.92	128.93	115.82	20.9	320

[a]ppm downfield from TMS internal standard.

The anion 22.64 MHz ^{13}C resonances are reported in Table II. The room temperature spectra appear as four well separated resonances for each ring with the central carbon resonance well upfield. Figure 4A is the room temperature ^{13}C spectrum of diphenylmethyllithium in THF. The shift assignments of the symmetrical

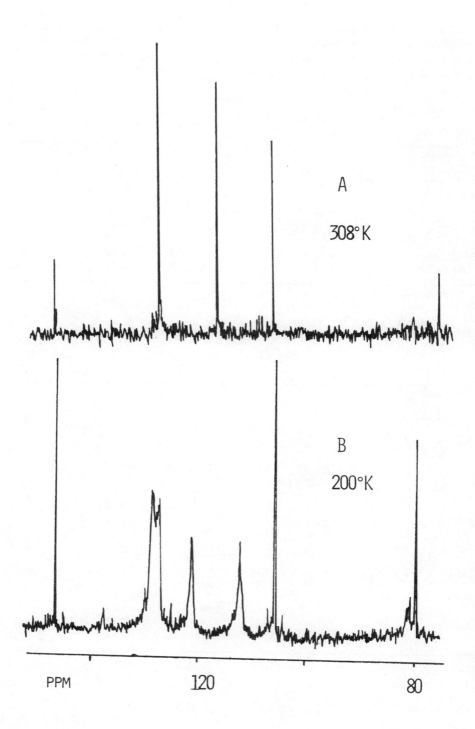

Figure 4. 22.64 MHz ^{13}C spectra of 1 at temperature extremes in THF.

anions are made by comparison with published spectra[9/10] and by comparison with others in the series. However, assignment of the unsymmetrical anion resonances requires the consideration of the low temperature behavior as described below.

At low temperatures individual *ortho* and *meta* [13]C resonances can be discerned. All other resonances are singlets throughout the temperature range. Accordingly, symmetrical anions have six (6) aromatic signals. (Figure 4B is the low temperature spectrum of diphenylmethyl lithium in THF.) The unsymmetrical anion has twelve (12) aromatic signals. As the temperature is raised both *meta* and *ortho* resonances undergo separate coalescence phenomena. The *ortho* carbon resonances show broadening at temperatures some 35° higher than the corresponding *meta* resonances. This is the result of differing separations between the individual *ortho* and *meta* carbon shifts, *ortho* shift separations are ∼9 ppm and the *meta* shift separations are ∼1 ppm.

Since the anion *meta* resonances are quite similar to the respective hydrocarbon resonances, they can be readily assigned. Then, comparison of the relative *ortho* and *meta* coalescence behavior permits assignment of the *ortho* resonances to the correct phenyl ring. The assignment of the *para* resonances are made by comparison with others in the series. Knowledge concerning the coalescence behavior appears to be a straightforward and reliable method for proper assignment. Specific *ortho* and *meta* carbon assignments are not made at this time as in the case of the ^{1}H chemical shifts discussed earlier.

Finally, both the ^{1}H and ^{13}C NMR spectra reveal chemical shift changes of specific nuclei with temperature. These changes are reversible and are related to changes in ion-pairing in the following way. Those systems for which temperature variations are known to bring about changes in ion-pairing[11] also show the greatest changes in chemical shift.

Dynamic NMR Spectra: Results and Discussion

The temperature dependent patterns suggest that certain resonances are undergoing rapid averaging at room temperature but at low temperatures averaging is slow enough to resolve separate ^{13}C and ^{1}H shifts and ^{1}H coupling patterns. A reasonable explanation for this behavior is that a barrier to phenyl ring rotation is present in the diphenylmethyl carbanions. This is supported by the observation that the coalescence phenomenon is restricted to the *ortho* and *meta* nuclei. Nuclei at other positions have single sharp resonances throughout the temperature range studied.

The availability of improved computer programs[12] makes the study of these dynamic processes a straightforward matter. The 270 MHz ^{1}H spectra are particularly suited for computer simulation. The spectra possess sufficient complexity that viscosity related signal broadening can be readily separated from exchange broadening. Chemical shifts and coupling constants are obtained from the slow exchange spectra and an exchange model is assumed. The rate of exchange is varied until the computer generated spectra match the essential aspects of the experimental line shapes.

Examination of the ¹H DNMR spectrum (270 MHz) of the lithium salt of the 4,4'-dideuteriodiphenylmethyl anion (1) (∼0.4M in THF) at 300°K shows a spectrum which can be interpreted as a closely spaced AB or AA′BB′ spectrum (Figure 5).

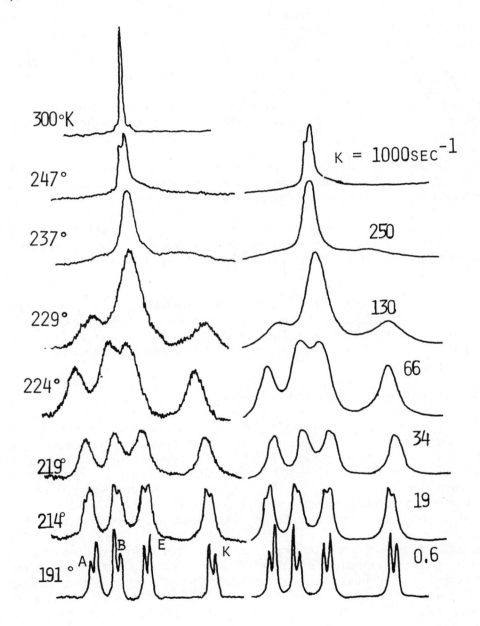

Figure 5. 270 MHz ¹H experimental and simulated line shapes of 1 at selected temperatures.

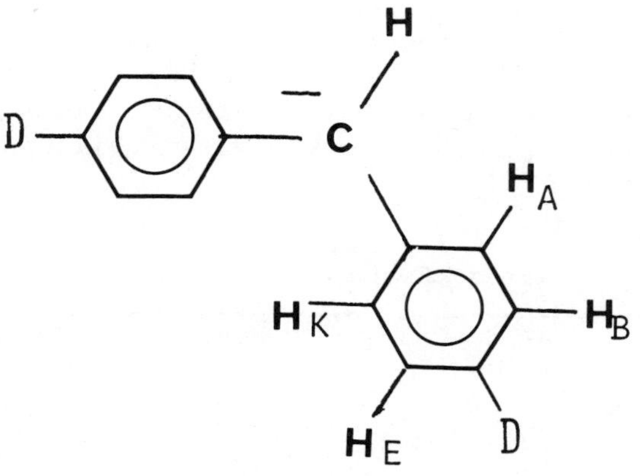

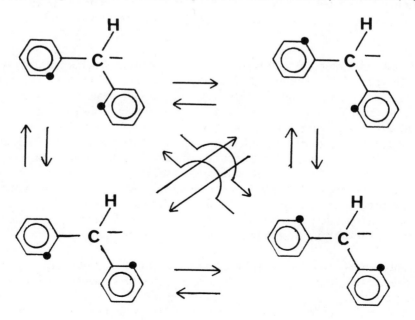

Figure 6. Labeling scheme for 1 phenyl protons.

Below 300°K, the DNMR spectrum of 1 undergoes changes characteristic of slowing a rate process on the DNMR time scale and is sharpened at 191°K into an ABEK spectrum as labeled in Figure 6. The spectrum is simulated theoretically at 191°K (Figure 5) using $\delta_A 6.75$ (TMS reference), $\delta_B 6.63$, $\delta_E 6.49$, $\delta_K 6.17$, $^3J_{AB} = 8.1$ Hz $^5J_{AE} = 0.0$, $^4J_{AK} = 1.1$, $^4J_{BE} = 1.1$, $^5J_{BK} = 0.0$, $^3J_{EK} = 8.1$. The spectrum of 1 at 191°K is consistent with *slow phenyl rotation* on the DNMR time scale and the presence of four different aromatic protons in a static delocalized anion. The assignments of H_A and H_K and H_B and H_E require additional confirmation and such work is in progress. The exchange-broadened DNMR spectra of 1 are accurately simulated (Figure

Figure 7. Itinerary for phenyl rotation.

5) using an ABEK to KEBA exchange model, i.e., H_A and H_K in Figure 6 exchange environments as do H_B and H_E.[4] The itinerary for phenyl rotation in 1 is illustrated in Figure 7. The solid black circles trace the travels of one ortho position on each ring over the rotational itinerary. The activation parameters for phenyl rotation in 1 are compiled in Table III.

Table III
Activation Energies for Phenyl Rotation

Compound	$\Delta G^{\ddagger}$, kcal/mol[a] at 223°K	$\Delta H^{\ddagger}$, kcal/mol[a]	ΔS, eu[b]
1	11.2	11.4	+0.7
2	10.6	11.1	+2.0
3			
4-methylphenyl ring	10.6	11.6	+4.6
4-deuteriophenyl ring	11.4	13.3	+8.4

[a]The error limits for these values are ± 0.2 kcal/mol.   [b]The error limits for these values are ± 3.0 eu.

Similar ^{1}H DNMR behavior (270 MHz) is observed for 2 (∼0.4M in THF). An apparent AB or AA'BB' spectrum observed at 300°K is separated into an ABEK spin system at about 190°K.

The spectrum is simulated using $\delta_A 6.53$ (TMS reference), $\delta_B 6.38$, $\delta_E 6.24$, $\delta_K 6.02$, and $^3J_{AB} = 8.1$ Hz, $^5J_{AE} = 0.0$, $^4J_{AK} = 1.1$, $^4J_{BE} = 1.1$, $^5J_{BK} = 0.0$, $^3J_{EK} = 8.1$. The exchange-broadened spectra for 2 are accurately simulated using an ABEK to KEBA exchange model. The free energy of activation for phenyl rotation is listed in Table III.

The same phenomenon of restricted rate of rotation is demonstrated by the ^{13}C DNMR spectra of 1 (Figure 8). Examination of the ^{13}C DNMR spectrum (22.64 MHz) of 1 (∼0.4M in THF) at 308°K reveals a single peak for both the ortho and meta carbons. The singlet nature suggests that rapid ring rotation brings about rotational averaging for each set that is fast on the DNMR time scale. At lower temperatures, the spectrum undergoes exchange broadening. At 200°K the ortho carbons have sharpened into two distinct singlets and the two meta carbons are easily distinguished. This DNMR change for the ^{13}C spectra of 1 over the same temperature range as the DNMR of the ^{1}H spectra provides additional confirmation for the process assignments and the similarities of the derived kinetic parameters provides independent verification. Accordingly, the experiments provide adequate assurance for the relationship between the data and the interpretations.

The substituent effect on the rotational barrier ($\Delta G^{\ddagger}$) brought about by the change from deuterium to methyl from 11.2 to 10.6 kcal/mol is consistent with the electron-donor properties of methyl. The anticipated decrease in delocalization implies a decrease in the pi bond order. This change of 0.6 kcal/mol in $\Delta G^{\ddagger}$ parallels but is less than the corresponding change of 2.4 kcal/mol in the $\Delta G°$ values derived from the pKa differences for the two hydrocarbons.[13/14] Thus the thermodynamically less

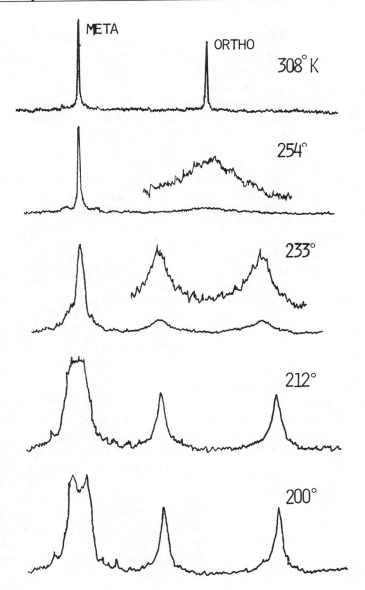

Figure 8. 22.64 MHz ^{13}C *ortho* and *meta* resonances of 1 at selected temperatures.

stable anion has as expected a lower barrier to rotation but the rotational barrier difference is only 25 percent of the anion stability difference. While clearly additional examples are required, a suggested explanation posits a transition state for rotation that involves partial but not complete localization.

While it may be tempting to rationalize these observations solely on the basis of the effect of the electron-donor properties of methyl on the exocyclic π-bond order, the situation with regard to these anions may not be as simple as implied above. In addition to the conformational equilibrations discussed above, very rapid (k $\sim 10^9$ sec^{-1} at 200°K)[15] and temperature-dependent equilibria involving tight and solvent-separated ion pairs exist in solutions of these anions. The state of ion-pairing can vary significantly as a function of anion, cation, and solvent. The origins of the substituent effects on the rate of phenyl rotation discussed above could indeed be found in the electron- donor properties of methyl and/or in changes in the ion-pairing equilibrium. Work on assessing the relative magnitudes of these various effects on anion stereodynamics is in progress.

The anion of 3 shows interesting DNMR behavior. Examination of the ^{1}H DNMR spectrum (270 MHz) of 3 ($\sim$0.4M in THF) at 300°K (Figure 9) shows *two* different overlapping spectra each of which has apparent AB or AA'BB' characteristics ${}^3J_{AB}$ $\sim$ 8 Hz) consistent with the presence of two different rings. The two low-field doublets of each respective phenyl spectrum are *superimposed* while the low-field line of the highest field doublet overlaps with the high field line of the next lower field doublet. At lower temperatures, the spectrum undergoes a series of complex changes and is sharpened at 191°K into overlapping ADFM and BCEN spectra as labeled in Diagram 1. The spectrum is simulated using $\delta_A 6.63$ (TMS reference), $\delta_D 6.43$, $\delta_F 6.30$, $\delta_M 6.08$ ${}^3J_{AD} = 8.1$ Hz, ${}^5J_{AF} = 0.0$, ${}^4J_{AM} = 1.1$, ${}^4J_{DF} = 1.1$, ${}^5J_{DM} = 0.0$, ${}^3J_{FM} = 8.1$ and $\delta_B 6.56$, $\delta_C 6.52$, $\delta_E 6.36$, $\delta_N 6.03$ ${}^3J_{BC} = 8.1$ Hz, ${}^5J_{BE} = 0.0$, ${}^4J_{BN} = 1.1$, ${}^4J_{CE} = 1.1$, ${}^5J_{CN} = 0.0$, ${}^3J_{EN} = 8.1$. The spectrum at 191°K is clearly consistent with rotation of *both* rings being static on the DNMR time scale.

At temperatures above 191°K, complete DNMR line shape analyses reveal a very interesting dynamical situation in 3. Perusal of the DNMR spectra for 3 (Figure 9) at 204°K or 219°K illustrates the situation. It is clear at 204°K or 219°K that the ADFM spectrum is substantially more collapsed than the BCEN spectrum due to the onset of a rate process which is becoming rapid on the DNMR time scale. Indeed, a complete DNMR line shape analysis at 204°K reveals the rate constant for exchange of the ADFM system to an MFDA system is 20 sec^{-1} and that for the BCEN to NECB exchange is 2.5 sec^{-1}, i.e., *one ring is rotating faster than the other*. This observation is very nicely consistent with the *electron-donating* property of methyl leading to reduced π-bonding of the methylated ring to the benzylic carbon of 3 and a reduced barrier to 4-methylphenyl rotation as compared to 4-deuteriophenyl rotation. *Thus, these observations for 3 provide unequivocal evidence for differential rates of rotation in an unsymmetrical 4,4'-disubstituted diphenylmethyl anion.*

The phenomenon of differential rates of rotation is also illustrated very effectively in the ^{13}C[^{1}H] DNMR spectra of 3 (Figure 10). Examination of the ^{13}C[^{1}H] DNMR spectrum (22.64 MHZ) of 3 ($\sim$0.4M in THF) at 313°K reveals two different *ortho* carbon resonances at δ117.9 and 117.0 (TMs reference) consistent with the presence of two different rings. The singlet nature of both signals suggests that the *ortho* carbons are being rotationally averaged at a rate which is fast on the DNMR time scale. At lower temperatures, the spectrum undergoes exchange broadening and the singlet indicated by the asterisk at 313°K is sharpened into two singlets (indicated by asterisks) at 192°K which are separated by 10.25 ppm. The other singlet resonance at 313°K separates into two singlets at 192°K separated by 9.35 ppm. It is obvious from an examination of the ^{13}C[^{1}H] DNMR spectrum of 3 at 218°K that the rate

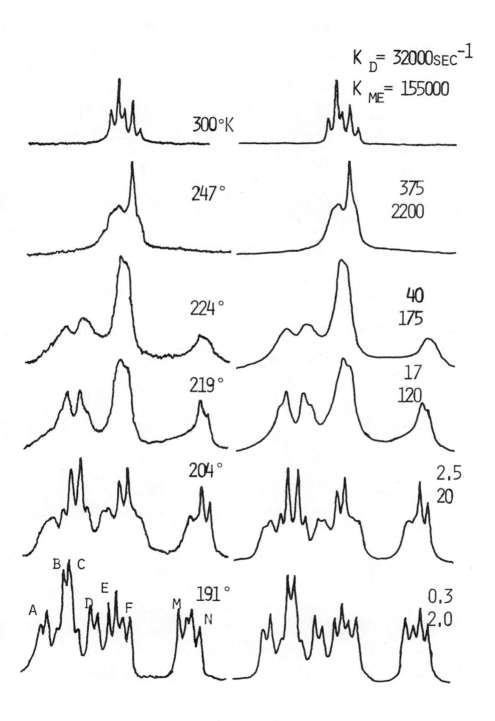

Figure 9. 270 MHz ^{1}H experimental and simulated line shapes of 3 in THF at selected temperatures.

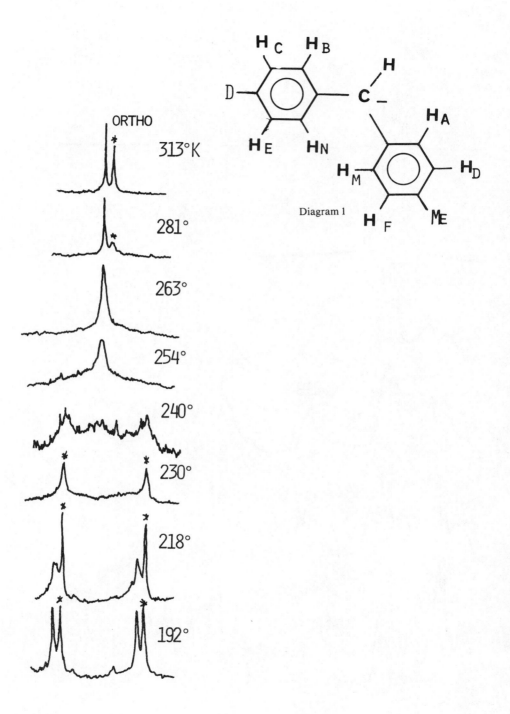

Figure 10. 22.64 MHz ^{13}C *ortho* resonances of 3 in THF at selected temperatures.

process associated with the asterisked peaks is *slow* on the DNMR time scale while the other pair of singlets is subject to exchange broadening, i.e., *one ring is rotating faster than the other.*

Such DNMR spectral behavior is consistent with the asterisked peaks being assigned to the *ortho* carbon of the 4-deuteriophenyl group which is rotating at a *slower* rate than the 4-methylphenyl ring. A complete DNMR line shape analysis at 254°K revealed the rate of rotation of the deuterated ring to be 500 sec^{-1} and that for the methylated ring to be 3000 sec^{-1} in good agreement with the ^{1}H DNMR data (Figure 9). Pertinent energies of activation are compiled in Table III.

The differential barriers reflect the trends observed for the symmetrical compounds. Thus the electron-donor properties of methyl lower the barrier of the 4-methylphenyl ring by some 0.8 kcal/mol relative to the 4-deuteriophenyl ring. This agrees well in both direction and magnitude with the results of the symmetrical compounds.

A final consideration is that the differential rates of rotation provide insight into the mechanism of rotation. For systems such as the diphenylmethyl anion rotation could involve single phenyl group rotation or simultaneous rotation of two phenyl groups. Consideration of Figure 7 reveals several possibilities.

The horizontal and vertical equilibria are visualized as independent one ring flips. The exchange process represented by the diagonals are assumed to occur by correlated rotation or a "gear" mechanism. Hoffman[16] has calculated that correlated rotation clearly offers the least disruption of the overall delocalization and should have the lowest barrier. Further, conrotatory motion is of lower energy than disrotatory motion. The observation of differential rates *demands* that the rotation involve single phenyl group rotation. In view of the similarity of the values for single phenyl group rotation with those observed for the symmetrical compounds it is tempting but by no means certain to conclude that these involve single phenyl rotation as well. In any event, an accessible pathway for single ring rotation has been demonstrated.

Acknowledgements

We are thankful for valuable help from Mr. Terry Aragoni, Mr. John Brennan, Mr. Michael Cipullo, Miss Mona Gabriel, Mr. Martin Goldberg, Mr. Paul Giammatteo, Mr. Dennis Heyer, Dr. Steven Hoogasian, Mr. Richard Marcantonio, Mr. Cyril Migdal, Mr. Mark Mazin and Mr. Mark Mazur. Grateful support is acknowledged by N.S.F., N.I.H., The Research Office of the State University of New York at Albany and the Southern New England High Field NMR Facility.

References and Footnotes

1. Streitwieser, A., Jr., Caldwell, R. A. and Granger, M.R., *J. Am. Chem. Soc., 86*, 3578 (1964); Breslow, R. and Grant, J. L., *J. Am. Chem. Soc., 99*, 7745 (1977).
2. Pasto, D. J. and Johnson, C. R., "Organic Structure Determination," Prentice-Hall, Englewood

Cliffs 1969; Levy, G. C. and Nelson, G. L., "Carbon-13 Nuclear Magnetic Resonance for Organic Chemists," Wiley, New York, 1972.

3. Jackman, L. M. and Cotton, F. A., "Dynamic Nuclear Magnetic Resonance Spectroscopy,"Academic Press, New York, 1975.

4. Sandel, V. R., McKinley, S. V. and Freedman, H. H., *J. Am. Chem. Soc., 90*, 495 (1968); Sandel, V. R. and Kronzer, F. J., *J. Am. Chem. Soc., 94*, 5750 (1972).

5. Fraenkel, G., Russell, J. G., and Chen, Y. H., *J. Am. Chem. Soc., 95*, 3208 (1973).

6. Brownstein, S. and Worsfold, D. J., *Can. J. Chem., 50*, 1246 (1972).

7. Bushby, R. J. and Ferber, G. J., *J. Chem. Soc. Perkins II*, 1688(1976).

8. Bushweller, C. H., Sturges, J. S., Cipullo, M., Hoogasian, S., Gabriel, M. W., and Bank, S., *Tet. Let., 16*, 1359 (1978).

9. Takahashi, K., Kondo, Y., Asami, R., and Inoue, Y., *Org. Mag. Res., 6*, 580 (1974).

10. Von Dongen, C. P. C. M., Von Dijkman, H. W. D. and deBie, M. J. A., *Rec. Trav. Chim., 93*, 29 (1974).

11. O'Brien, D. H., Russell, C. R. and Hart, A. J., *J. Am. Chem. Soc., 101*, 633 (1979).

12. Bushweller, C. H., Bhat, G., Letendre, L. J., Brunelle, J. A., Bilofsky, H. S., Ruben, H., Templeton, D. H., and Zalkin, A., *J. Am. Chem. Soc., 97)*, 65 *(1975).*

13. Streitweiser, A., Jr., Murdoch, J. R., Hafelinger, G. and Chang, C. J., J. Am. Chem. Soc., 95, 4248 (1973).

14. Bank, S. and Sturges, J. S., *J. Organometallic Chem., 156*, 5 (1978).

15. Hirota, N., *J. Am. Chem. Soc., 90*, 3603 (1968).

16. Hoffman, R., Bissell, R. and Farnum, D. G., *J. Phys. Chem., 73*, 1789 (1969).

Steric Effects on Electron-Transfer Reactions of Sulfonamides

W. D. Closson and M. G. Voorhees
Department of Chemistry
State University of New York at Albany
Albany, New York 12222

Introduction

Sulfonamides have long seen great utility in the identification and synthesis of amines and as a protecting group for the amino function in other syntheses. In almost all cases the sulfonyl group is removed from nitrogen, at or near the end of the synthetic sequence, by an electron-transfer reaction in which the S-N bond is reductively cleaved. Typical ways of effecting this have been treatment with alkali metal-liquid ammonia combinations,[1] electrochemical procedures,[2] and treatment with arene anion radicals (such as sodium naphthalene) in tetrahydrofuran (THF) solutions.[3] Studies of the mechanisms of these reductive cleavages have been carried out for both the electrochemical procedure[2-4] and for cleavage with arene anion radicals.[3-4] Some work bearing on the mechanism of the reaction with sodium-liquid ammonia has also been reported.[5]

For the homogeneous reactions involving arene anion radicals in THF, it was found that a typical arenesulfonamide (*1*) reacted rapidly with two moles of the electron donor, producing amide and arenesulfinate ions, and that the rate expression was: rate = k (anion radical)(sulfonamide), (i.e., the initial electron transfer step was rate determining).[3] In an examination of the polar substituent effect on this reaction, using sodium anthracene as the electron donor and a series of sulfonamides, *1* (x = substituent), it was found that for less electronegative substituents (p-dimethylamino through *p*-fluoro), the relative rates of cleavage were correlated moderately well with σ-constants, p = 1.91, (r = 0.987), but that more strongly electron-withdrawing substituents result in rates much slower than expected.[4] This appears to be due to competition of the electronegative substituent with the sulfonyl group for the electron, leading to species such as *2* which neither undergo S-N cleavage nor readily react further with the electron donor. The effect of polar substituents on the amino side of the sulfonamide (e.g., as in *1*, y = substituent) was found to be very small.[6] All of this seems to be best correlated with the mechanism shown in scheme 1, where the slow step is the initial transfer of an electron directly into the sulfonyl group, followed or accompanied by scission of the S-N bond, probably producing arenesulfinate ion and amino radical. Reduction of the amino radical to amide anion by a second molecule of the arene anion radical would be expected to be extremely fast.

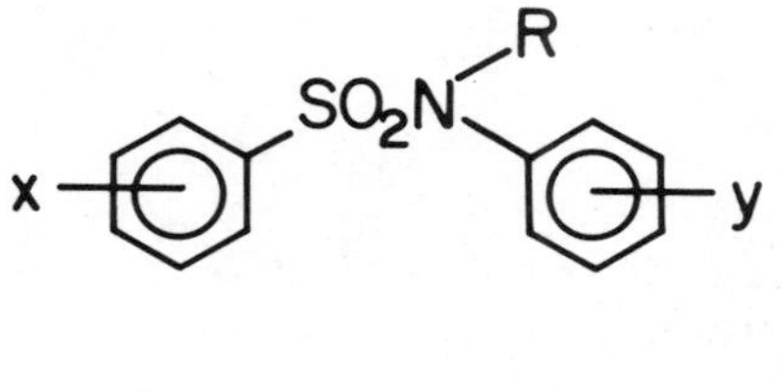

1.

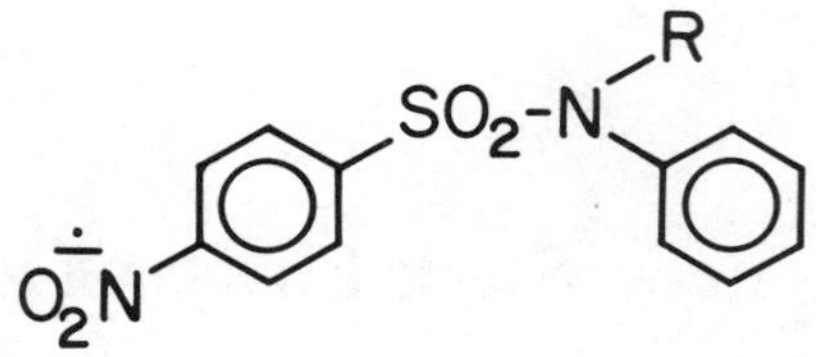

2.

Scheme I

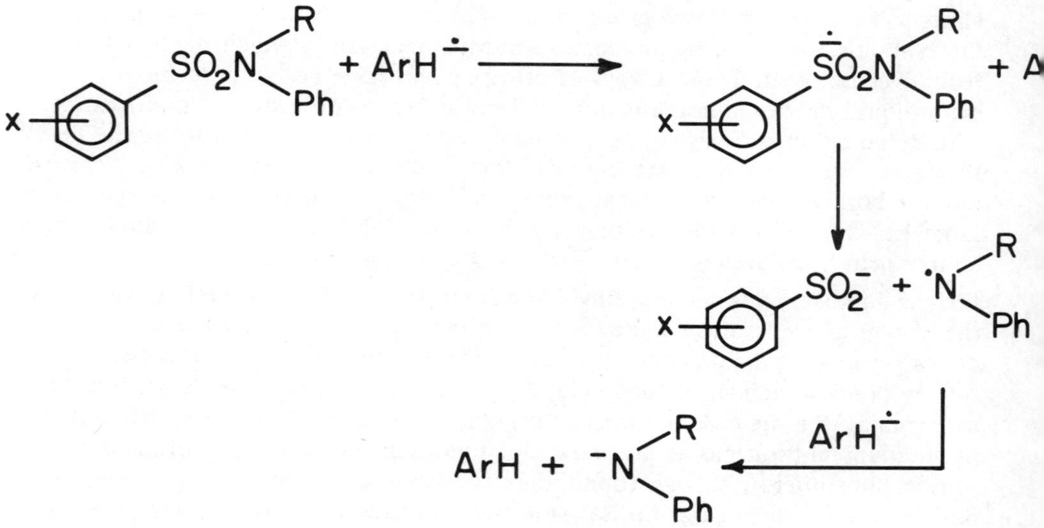

The electrochemistry of sulfonamides has been examined by several groups. Manousek, Exner, and Zuman reported that 4-cyanobenzenesulfonamide undergoes electrochemical cleavage in aqueous solution at the carbon-sulfur bond (eq 1),[4] while Cottrell and Mann observed only S-N cleavage in electrochemical reduction of several arenesulfonamides in acetonitrile.[2] They proposed an irreversible, two-electron reduction followed by rapid cleavage to two anions (eq. 2). Asirvatham and Hawley noted that Cottrell and Mann's results could also be explained by either the *ece* mechanism shown in eq. 3, where the nitrogen or oxygen centered radical would be rapidly reduced at a potential less cathodic than that of the initial reduction, or by a rate determining disproportionation process (eq. 4).[4] Either of these processes would account for the products and *n* values reported.

$$ArSO_2NH_2 + 2e + H^+ \rightarrow ArH + {}^-SO_2NH_2 \rightarrow HSO_3{}^- + NH_3 \qquad (1)$$
$$ArSO_2NR_2 + 2e \rightarrow ArSO_2NR_2{}^{2-} \rightarrow ArSO_2{}^- + R_2N^- \qquad (2)$$
$$ArSO_2NR_2 + e \rightarrow (ArSO_2NR_2)^{\dot{-}} \rightarrow ArSO_2{}^- + R_2N^{\dot{}}, \text{ or } ArSO_2{}^{\dot{}} + R_2N^-,$$
$$\xrightarrow{e} ArSO_2{}^- + R_2N^- \qquad (3)$$
$$2(ArSO_2NR_2)^{\dot{-}} \rightarrow ArSO_2NR_2{}^{2-} + ArSO_2NR_2, \rightarrow ArSO_2{}^- + R_2N^- \qquad (4)$$

We examined the series of sulfonamides (*1*, x = substituent) *via* cyclic voltammetry in acetonitrile solution using a vitreous carbon electrode. Our results agreed well with those of Cottrell and Mann, but we were unable to distinguish between the mechanisms of eqs. 2, 3, and 4.[4] If the electrochemical process is at all similar to homogeneous reduction eq. 3 would be favored, of course. Such similarity is suspect, however. The peak potentials for *all* the sulfonamides (x = *p*-dimethylamino through x = *p*-cyano) correlated very well with σ^n constants. p = 1.07v (r = 0.995). Correlation with normal σ constants, those that gave the best fit with sodium anthracene reduction, was quite poor, r = 0.939. It was felt that this was indicative of important differences in reaction mechanism between the two processes, and it was postulated that while sodium anthracene reacts with a normal, moderately solvated, sulfonamide, where the effect of a para substituent will be the usual mix of resonance and inductive interactions, the electrochemical results may be attributable to heterogeneous surface phenomena at the graphite electrode. If the main site of interaction were the aromatic ring, as might seem reasonable for a vitreous carbon surface, this might disrupt resonance interactions between the substituent and sulfonyl group, while still allowing modest inductive effects.[4] Recently, the possibility that the large difference in substituent effects might be due to the differences in solvent and counter ion (THF and sodium *vs.* acetonitrile and tetrabutylammonium) was tested. Using the method of Nicholson and Shain[7] the rates of reaction of a series of sulfonamides with electrochemically generated anthracene anion radical in acetonitrile (tetrabutylammonium perchlorate as supporting electrolyte) were determined. The rates so obtained correlated excellently with σ, p = 1.95 (r = 0.982), clearly ruling out solvent and counter ion as the cause.[8]

From the above discussion, it is clear that two key questions about the mechanism remain. Is there an intermediate, and does S-N cleavage occur at a one-electron or two-electron stage of reduction? The existence of metastable anion radical intermediates during reductive cleavage of several similar types of compounds, e.g., alkyl[9] and aryl methanesulfonates,[10] trifluoromethanesulfonamides of aryl amines,[8] and aryl phosphate esters,[11] has been proved, and cleavage at both one- and two-electron stages of reduction demonted for some of them. It was felt that a study of sterically hindered sulfonamides might prove helpful in answering these questions. In addition, information useful in improvement of the reaction for synthetic purposes might result.

Results and Discussion

Initially, we felt that preparation of several series of sterically hindered sulfonamides should present little problem considering the ready availability of a wide variety of highly substituted amines and sulfonyl halides, even though few highly hindered sulfonamides had been previously reported in the literature. It was quickly determined, however, that there was good reason for the scarcity of references. The reaction of amines and sulfonyl chlorides is very sensitive to steric hindrance, particularly in the vicinity of the amino nitrogen. The reaction is also complicated by the fact that the desired products are clearly less thermally stable than are normal sulfonamides. The result was that we often had to be content with meager yields of compounds obtained after long reaction times at moderate temperatures.

A short series of N-alkyl-N-phenyl tosylamides (3) was obtained, as well as a larger series of N-alkyl-N-benzyl derivatives (4). In addition, the highly hindered compounds 5, 6, and 7 were obtained without too much difficulty.

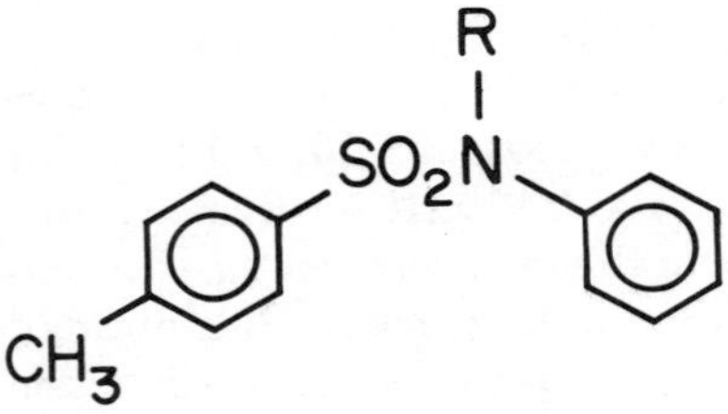

3. a, R = CH$_3$
    b, R = CH$_3$CH$_2$
    c, R = (CH$_3$)$_2$ CH

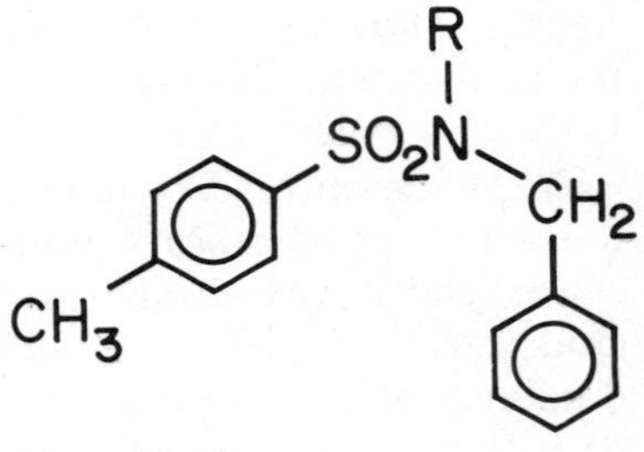

4. a, R = CH$_3$
    b, R = CH$_3$CH$_2$
    c, R = (CH$_3$)$_2$CH
    d, R = (CH$_3$)$_3$C
    e, R = (CH$_3$)$_2$ CH CH$_2$
    f, R = (CH$_3$)$_3$C CH$_2$

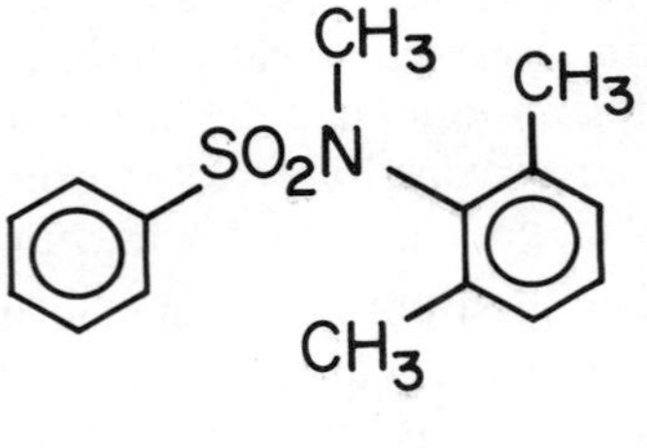

5.

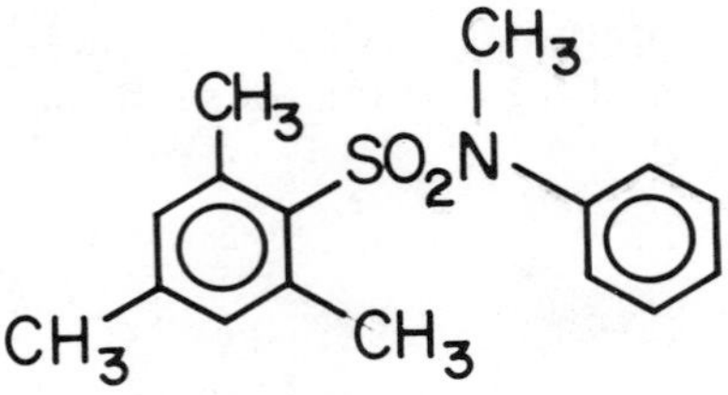

6.

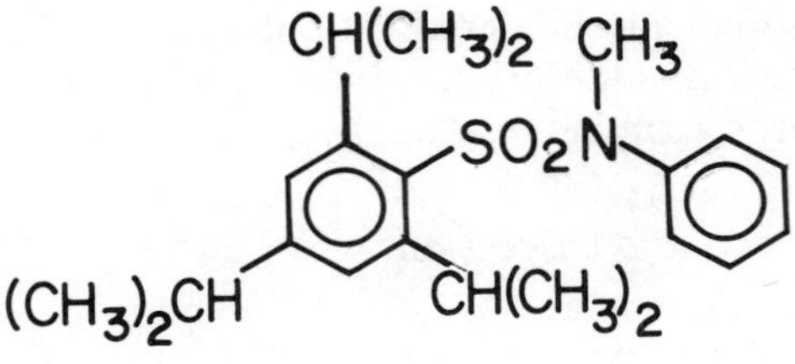

7.

The compounds were first examined by cyclic voltammetry, using a vitreous carbon electrode in acetonitrile, with $0.2M$ tetrabutylammonium perchlorate as supporting electrolyte. The technique was essentially the same as described previously.[4] Most showed an irreversible reduction wave in the vicinity of -2.5 v *vs.* SCE, with an *n* value of two. On reversal of the scan oxidation peaks typical of arenesulfinate ion could usually be observed. There are interesting exceptions. This data is presented in Table I. In addition, the rates of reaction of most of the sulfonamides with elec-

Table I

Cyclic Voltammetric Data for Sulfonamides

Sulfonamide	m.p.[1]	Reduction Peak[2]	Oxidation Peaks[2]
3a	65°	-2.56	+0.90[3], +0.46[4]
b	86°	-2.52	+0.80[3], +0.46[4]
c	96°	-2.52	+0.43[4]
4a	94°	-2.52	+0.45[4]
b	50°	-2.52	+0.45[4]
c	48-49°	-2.51[5]	+0.66[6]
d	100-102°	-2.46[5]	+0.70[7]
e	66-67°	-2.56	+0.30[8]
f	74-76°	-2.51	+0.48[4]
5	112-113°	-2.35	+0.82[3], +0.49[4]
6	94-96°	-2.46	+0.87[3], +0.38[4]
7	71-72°	-2.58	+0.36[4]

[1] Compounds either had m.p.s. corresponding to lit. values, or had spectroscopic properties (ir, nmr) in accord with proposed structure.
[2] Volts *vs.* SCE.
[3] Aniline anion oxidation.
[4] Arenesulfinate anion oxidation.
[5] Value of *n* is *ca.* one.
[6] N-isopropyl tosylamide oxidation.
[7] Probably N-t-butyl tosylamide anion oxidation.
[8] Probably arenesulfinate ion oxidaton.

trochemically generated anthracene anion radical were determined in the same solvent-electrolyte system, using the method of Nicholson and Shain.[7] The relative rates of reaction are presented in Table II.

The cyclic voltametric data in Table I show that reduction occurs in all cases fairly close to where one would expect, considering only polar effects in the arenesulfonyl portion of the molecule. Possible exceptions are N-t-butyl-N-benzyl tosylamide (*4d*) and the hindered compounds *6* and *7*, where reduction appears to occur at a somewhat lower potential than one might expect. The differences are not very large, however. More striking is the fact that *4d* and N-isopropyl-N-benzyl tosylamide (*4c*) undergo 1-electron reductions rather than the usual 2-electron process. Also for

Table II

Relative Rates of Reaction of Sulfonamides with
Electrochemically Generated Anthracene Anion Radical

Sulfonamide	Relative Rate
3a	13.2
3b	4.0
4a	4.1
4c	1.6
4d	1.4
4f	3.6
5	10.3
6	5.4
7	1.4

these compounds, reversal of the scan shortly after passage through the reduction peak (at *ca.* -2.65v) shows neither reversibility of this reduction process nor the presence of any other oxidizable species out to *ca.* +1.0v. If the scan is continued to about -2.9v and then reversed, an oxidation peak is observed near +0.7v. This is far too positive for toluenesulfinate ion (which normally undergoes oxidation at *ca.* +0.45v under these conditions), and corresponds best with oxidation of the corresponding debenzylated N-alkyl tosylamide anions. (It should be noted that N-benzyl tosylamide anion is oxidized at +0.71v under these conditions, so that the identity of the anion in the case of 4d is in some doubt). Apparently, an *eec* process, shown in eq. 5, is being followed by these hindered, benzyl substituted, sulfonamides. The benzyl anion produced in the second step would probably react rapidly with either the acenonitrile or tetraalkylammonium cation, producing toluene which would not be detectable electrochemically.

$$ArSO_2NRCH_2Ph \xrightarrow{e} (ArSO_2NRCH_2Ph)^{-} \xrightarrow{e} ArSO_2N\bar{R} + PhCH_2^{-} \qquad (5)$$

Why formation of the intermediate sulfonamide anion radical is not reversible is puzzling, as is the fact that it gives rise to no easily oxidizable products. Apparently, it must react fairly rapidly with solvent or electrolyte to produce non-electrochemically active species. The answer to this must await preparation of larger amounts of material and controlled potential electrolysis experiments.

The homogeneous rates of reaction of sulfonimides with anthracene anion, listed in Table II, show only moderate sensitivity to steric hindrance. All rates measured were within a factor of ten. Even so, there is a clear pattern of increased steric hindrance near the sulfonyl group resulting in a slower rate. Yajima, *et al.*, previously noted that mesitylenesulfonamides, similar to compound *6*, were rather less reactive toward sodium-liquid ammonia than less hindred arenesulfonamides,[11] and *6* is indeed slower than 3a. Unfortunately, since the design of these kinetic experiments does not lend itself to identification of products, little more can be said about the homogeneous reduction.

In conclusion, this preliminary examination of simple steric effects on electron transfer reactions of sulfonamides has failed to provide much insight into the basic questions remaining. It has shown, however, that for certain structures steric effects may divert normal electrochemical reduction to an entirely different path. This finding will certainly be explored further.

References and Footnotes

1. duVigneaud, V. and Behrens, O. K., *J. Biol. Chem.*, *117*, 27 (1937); Birch, A. J. and Smith, H., *Quart. Rev.*, (London), *12*, 17 (1958).
2. Cottrell, P. T. and Mann, C. K., *J. Am. Chem. Soc.*, *93*, 3579 (1971).
3. (a) Ji, S., Gortler, L. B., Waring, W., Battisti, A., Bank, S., Closson, W. D., and Wriede, P., *ibid.*, *89*, 5311 (1967).
(b) Closson, W. D., Ji, S., and Schulenberg, S., *ibid.*, *92*, 650 (1970).
4. (a) Manousek, O., Exner, O., and Zuman, P., *Collect Czech. Chem. Commun.*, *33*, 4000 (1968).
(b) Asirvatham, M. R., and Hawley, M. D., *J. Electroanal. Chem. 53*, 293 (1974).
(c) Quaal, K. S., Ji, S., Kim, Y. M., Closson, W. D., and Zubieta, J. A., *J. Org. Chem.*, *43*, 1311 (1978).
5. Kovacs, J. and Ghatak, U. R., *ibid.*, *31*, 119 (1966).
6. Ji, S., Ph. D. Thesis, State University of New York, Albay, 1970.
7. Nicholson, R. S., and Shain, I., *Anal. Chem.*, *36*, 706 (1964).
8. Saboda Quaal, K., Ph. D. Thesis, State University of New York, Albany, 1978.
9. Ganson, J. R., Schulenberg, S., and Closson, W.D., *Tetrahedron Lett.*, 4397 (1970).
10. Carnahan, J. C., Jr., Closson, W. D., Ganson, J. R., Juckett, D. A., and Saboda Quaal, K., *J. Amer. Chem. Soc.*, *98*, 2526 (1976).
11. Shafer, S. J., Closson, W. D., van Dijk, J. M. F., Piepers, O., and Buck, H. M., *ibid.*, *99*, 5118 (1977).
12. Yajima, H., Takeyama, M., Kanaki, J., and Mitani, K., *J. Chem. Soc. Chem. Commun.*, 482 (1978).

The Effect of Isotopic Perturbation
on NMR Spectra
A New Method for Distinguishing Rapidly Equilibrating
Molecules from Symmetric Molecules

Martin Saunders
Sterling Chemistry Laboratory
Yale University
New Haven, Connecticut 06520

Introduction

Molecules translate, rotate and vibrate but in addition, many undergo motions over internal barriers. Knowledge of these barriers and rates of such motions is essential to the understanding of the properties, interactions and reactions of molecules. We obtain such knowledge from experimental observations and from theories of bonding .[1]

NMR spectroscopy has aided in this task through its ability to investigate rapid torsion, inversion and other molecular rearrangement processes through observation of changes in lineshape.[2] Measurable uncertainty broadening occurs when the rates are greater than a few times a second.

However, when a process which interchanges the NMR shifts becomes faster than a limit which is roughly equal numerically to the square of the frequency separation in Hz, only sharp peaks are seen at positions which are the weighted averages of the separate NMR frequencies. Depending on these frequencies and on the temperature range accessible for observation, barriers to internal motion as high as 20 kcal/mole can be measured. In a few cases where exceptionally large NMR shifts are present, barriers as low as 3 kcal/mole have been measured.[3]

If the barrier to rearrangement goes to zero, we no longer have a set of equilibrating structures but a single *intermediate* structure. In this way the dynamics of molecular rearrangement can be connected to questions of molecular structure. The fundamental question is whether a rearrangement is occurring over a barrier on a double or multiple minimum surface or whether there is a single energy minimum with a hybrid structure. This talk will describe a new method of employing NMR spec-

$$A \underset{}{\overset{\text{fast}}{\rightleftharpoons}} B \qquad K_{eq} = [A]/[B]$$

$$f_{AVR} = \frac{f_A[A] + f_B[B]}{[A] + [B]} = \frac{f_A K + f_B}{1 + K}$$

Figure 1. Weighted average of NMR frequencies.

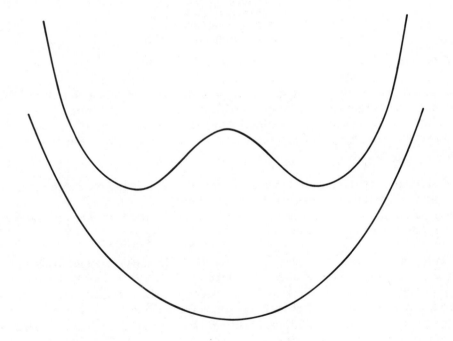

Figure 2. Single and double energy minimum surfaces.

troscopy to help decide this kind of question.

Carbonium Ions and their Rearrangement

While carbonium ions were first studied as *intermediates*[4] in reactions, in the past fifteen years it has become possible to prepare stable solutions of many simple acyclic, monocyclic and polycyclic tertiary cations. Many ions with stabilizing groups (substituted allyl, benzyl, dienyl, anti-7-norbornenyl, 2-norbornyl, 2-bicyclo[2.1.1]hexyl, etc.) and a few simple secondary cations (isopropyl, sec-butyl and cyclopentyl) have also been made.[5] The most widely used reaction for the preparation of these stable solutions is the following:

$$RCl + SbF_5 \rightarrow R^+ + SbF_5Cl^-$$

We have carried out this reaction to form all of the ions described here using a new preparative technique which we have called the "molecular beam" method.[6] Carbonium ions[7/8] are electron deficient species since trivalent C^+ has only six bonding electrons. Rearrangement processes are very common and ready in these ions, since when groups migrate toward the C^+, they are forming bonds to the empty orbital on carbon.[9] The most common and facile carbonium ion rearrangement is the 1-2 shift of hydride or methide. A number of stable carbonium ions rearrange by

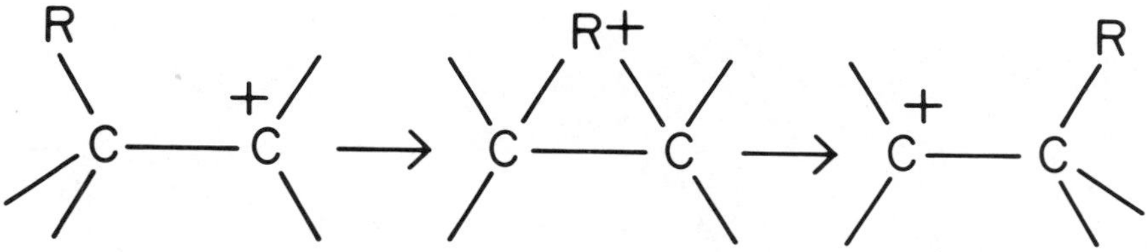

Figure 3. Carbonium ion 1,2 shift rearrangement.

first undergoing 1,2- shifts uphill in energy (i.e., tertiary to secondary or secondary to primary rearrangements). Such transformations are reversible and return to the starting ion occurs. However, if the complete cycle of steps leads to carbon or hydrogen scrambling, the reactions can be observed. Isotopic tracers can be used to follow these processes, but it is more convenient to study them in stable ion solutions at temperatures where the reactions are fast enough to produce nmr lineshape changes.[5d/10/11] From lineshape analysis, the rates of the overall reactions can be found and detailed mechanistic information can often be obtained.[12]

Rapid, Degenerate 1,2-Hydride and Methide Shifts in Carbonium Ions

If the product of a rearrangement step is chemically indistinguishable from the starting material, we call the process a degenerate rearrangement. In the simplest case, the activation energy may just be that required to form a symmetrically bridged transition state. Many simple acyclic and monocyclic carbonium ions, capable of undergoing degenerate 1,2-shifts of hydride and methide, give sharp, averaged proton and carbon peaks in their nmr spectra with the usual methods of measurement. Using CMR spectroscopy on a high field instrument and going down to -140°C, it has been possible to measure appreciable line-broadening and obtain rates of and barriers toward rearrangement in the first three of the following cases:[3] Failure to observe line broadening at low temperature (as in the fourth case) for ions capable

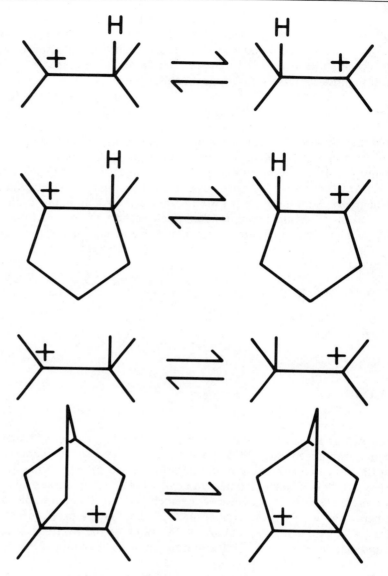

Figure 4. Ions undergoing fast, degenerate, 1,2 shifts.

of undergoing degenerate rearrangements may mean that we have a double minimum energy surface with a barrier in the range *below* 3 or 4 kcal/mole. However, the possibility remains that the barrier is zero and that instead of a double-minimum energy surface we have a single energy minimum where the stable structure is now the *bridged* ion. Such bridged carbonium ions have commonly been termed non-classical ions.[13] The nmr *lineshape* cannot, in general, distinguish between these two situations.

If the NMR peaks observed are far away from the positions predicted by averaging ones chosen from suitable model systems, often the conclusion is drawn that the system is not the equilibrating set of ions and is probably the bridged ion. However, the factors which influence carbon shifts, in particular, are not well understood. Therefore, a small discrepancy between predicted and observed chemical shifts is *not* convincing evidence for a non-classical structure. In cases where very unusual peak positions are found, valid structural conclusions can be obtained this way. In cases where the bridged ion is *very much* more stable than the unbridged ion, the rates of solvolysis reactions leading to these ions often provide compelling evidence. Reaction rate increases [14] of up to 10^{11} compared with the rates in model substances have been seen and in such cases are convincing. However, when such rate increases are small, it is possible to dispute endlessly the suitability and use of the models.

The Effect of Deuterium Perturbation on Rapidly Equilibrating Systems

The introduction of deuterium into ions undergoing rapid degenerate rearrangemens breaks the symmetry. Equilibrium constants between the structures which are rapidly interchanging are no longer unity, since the structures and hence the energies are

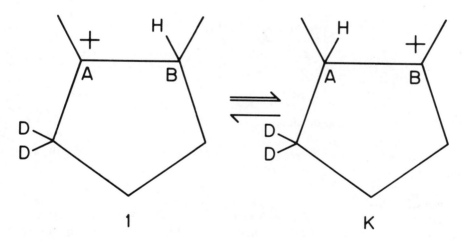

Figure 5. Deuterated 1,2-dimethylcyclopentyl cation. K and 1 refer to concentration.

made slightly different by the isotope. As a consequence of this, pairs of nuclei which gave averaged, single peaks in the non-isotopic substance now are found to give separate lines.[15-18] The observed splitting between these lines δ is related to the chemical shift difference which would be seen if there were *no* exchange Δ, and the isotopic equilibrium constant K. The equation for this splitting if just two nuclei are interchanged by the reaction may be derived as follows: In the case shown in Figure 5, the observed cmr splitting between carbon atoms 1 and 2 (δ) was found to be 81.8 ppm at -142°C as a result of an equilibrium deuterium isotope effect,[18] K = 1.91. Since the positively charged and neutral carbons in carbonium ions are separated by

$$\text{Let } f_{CH} = 0 \, , \quad f_{C^+} = \Delta$$

$$\text{averaged frequencies:} \quad f_A = \frac{\Delta}{K+1} \, , \quad f_B = \frac{\Delta K}{K+1}$$

$$f_B - f_A = \delta = \frac{\Delta(K-1)}{K+1}$$

$$\frac{\delta}{\Delta} = \frac{K-1}{K+1}$$

$$K = \frac{\Delta + \delta}{\Delta - \delta}$$

Figure 6. Derivation of splitting equation.

particularly large frequency differences, their ^{13}C spectra show exceptionally large splittings.

In the cases described above, deuterium isotope effects are due to hyperconjugation which changes the force constants for stretching and bending CH bonds adjacent to a carbonium center and are a function of the dihedral angle.[18/19] Splitting in simple monocyclic systems is typically around 25 ppm per deuterium on methyl and 50 ppm per deuterium on methylene. As already mentioned, the rearrangement barriers in several of these cases have been measured and been found[3] to be 3-4 kcal/mole. However, large splitting has *also* been observed in several cases where it has not been possible to measure the rate using the nmr lineshape method since the barriers appear to be lower than 3 kcal/mole.[20f]

Deuterium Perturbation of the NMR Spectra of Symmetric (Single Energy Minimum) Cases

When deuterium is used to break the symmetry of symmetrical substances, nuclei which are equivalent in the non-isotopic case are also split in frequency. Cyclohexenyl and cyclopentenyl cations, with *no* rapid rearrangement processes occurring

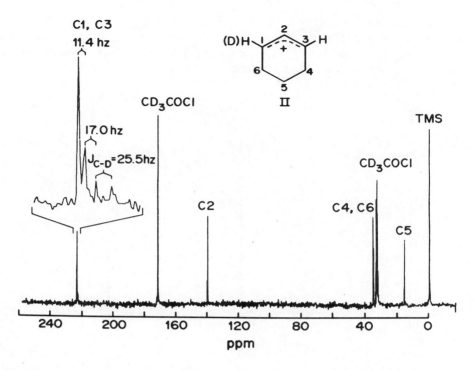

Figure 7. Deuterated cyclohexenyl cation.

(single energy minimum), show splittings in the ^{13}C spectrum upon the introduction of deuterium ($\delta = 0.5$ ppm).[21] The splitting in these and similar substances is smaller than in the rapidly equilibrating cases by about two orders of magnitude.

NMR frequencies are always averages over the zero point motion. The small splittings observed in these cases are believed to be due to such motions in the isotopically labeled compound averaging the NMR peaks over portions of the single-minimum energy surface different from those in the unlabeled material. In the tricyclononyl

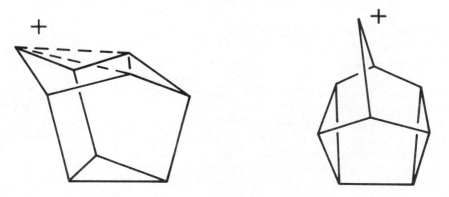

Figure 8. Non-classical and classical structures for Coates' tricyclononyl cation.

cation reported by Coates[22] the proton and carbon spectra establish that the ion *must* have a nonclassical structure (rapidly equilibrating C_s or C_{2v} alternatives are ruled out since they must lead to more extensive degeneracy than is observed). The introduction of deuterium results in a spectrum where the splitting is only 0.1 ppm or less.[20f]

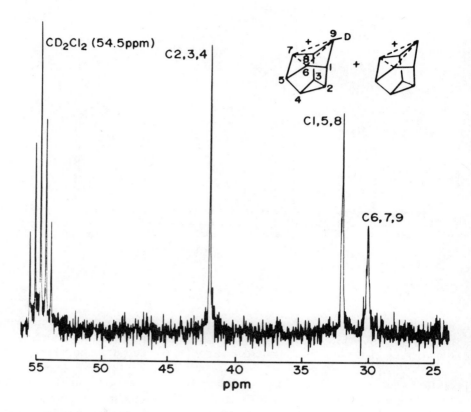

Figure 9. Spectrum of deuterated Coates' cation.

The Deuterium Perturbation Method for Distinguishing Rapidly Equilibrating from Symmetric Cases

To summarize, in species where rapid, degenerate equilibria are known to occur, deuterium substitution results in characteristically large CMR splittings between nuclei which are identical in the unlabeled molecules. On the other hand, when a system possesses *true* symmetry rather than time-averaged molecular symmetry, the corresponding splittings are found to be very much smaller as might be expected since there is no equilibrium to upset. The observation of these model cases suggests that the magnitude of perturbation of the carbon spectrum on introduction of deuterium might therefore distinguish rapidly equilibrating systems from symmetrical ones.

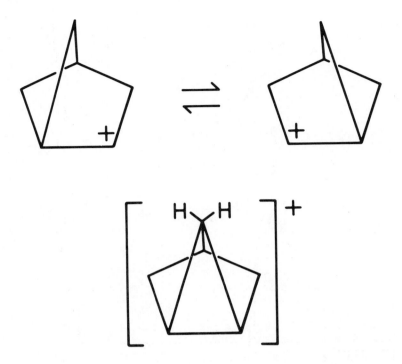

Figure 10. Classical and nonclassical structures for bicyclo [2.1.1] hexyl cation.

This new method has been applied, so far, to three cases where the answer had not been previously known. In the bicyclo[2.1.1]hexyl cation the inroduction of deuterium as shown resulted in a splitting in the ^{13}C spectrum of 1.2 ppm, a value far smaller than that to be expected for a rapidly equilibrating system.[23] The conclusion is that the energy surface is single-minimum and that the ion is bridged (nonclassical).

The problem of deciding whether the norbornyl cation is a rapidly equilibrating or a non-classical ion is the oldest and most intensively discussed case of this kind.[13/20] There is a complication in applying this new method.[20] Besides the Wagner-Meerwein rearrangement, a further rapid process, the 6,2-hydride shift, has a barrier[20d] of only 5.9 kcal/mole and results in a certain amount of unavoidable line broadening. Even in the ion with *no* deuterium, the downfield carbon line (due to the former 2 and 6 carbons) is found to be $\sim$ 2 ppm wide.[20f]

Nevertheless, no *additional* isotopic splitting or broadening can be discerned in either the 2-monodeutero or the 3,3-dideutero ions and therefore the isotopic splitting can be no more than 2 ppm. This result can be compared with those for methylene deuterated dimethylcyclopentyl and dimethylnorbornyl cations, models for rapidly equilibrating classical ions.

The < 2 ppm splitting found for the norbornyl cation is far smaller than the values

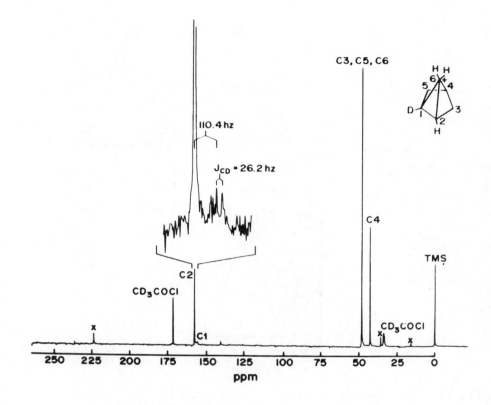

Figure 11. Spectrum of deuterated bicyclo [2.1.1]hexyl cations.

measured in the cases of equilibrating, classical ions and strongly suggests a symmetric, nonclassical ion.

The 1,2-dimethylnorbornyl cation is an interesting case. The cation deuterated on one of the methylene groups yields an intermediate value for the splitting ($\delta = 24$ ppm) reduced by about a factor of 4 from the splitting observed in the analogous

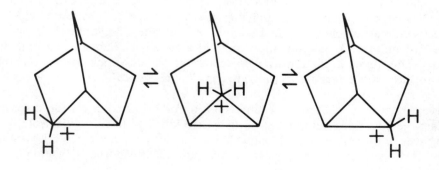

Figure 13. 6,2-Hydride shift in norbornyl cation.

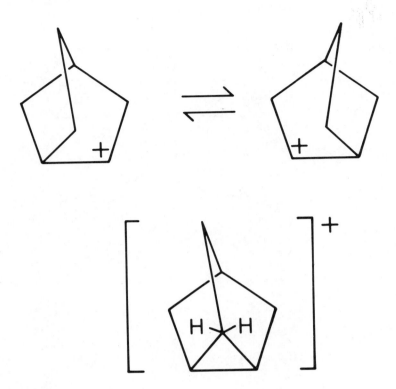

Figure 12. Classical and nonclassical structures for norbornyl cation.

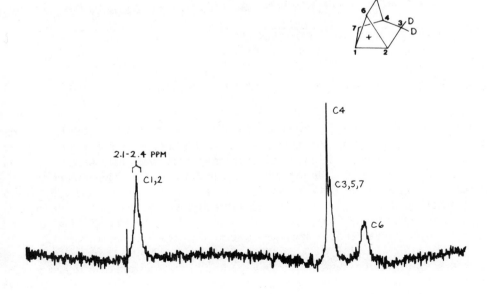

Figure 14. Spectrum of deuterated norbornyl cation.

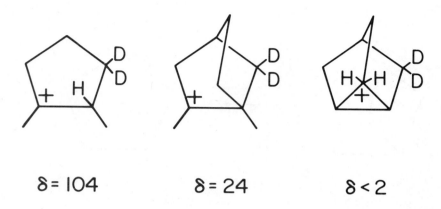

$$\delta = 104 \qquad\qquad \delta = 24 \qquad\qquad \delta < 2$$

Figure 15. Comparison of δ for norbornyl and model cations.

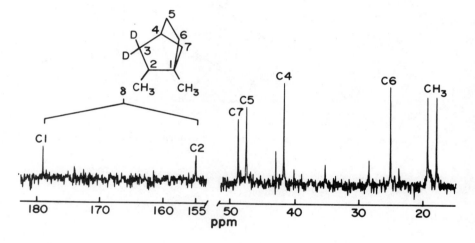

Figure 16. Spectrum of deuterated 1,2- dimethylnorbornyl cation.

open system, 1,2-dimethylcyclopentyl cation (δ = 104 ppm). This case may represent the *intermediate* situation of a double minimum surface with a very low barrier between the two structures. These structures may be *partially* delocalized as the ion approaches the bridged structure. The smaller the barrier is, the smaller the difference will be between these structures. The equilibrium isotope effect K would become closer to unity and the chemical shift difference between interchanging carbons Δ would decrease, both changes reducing δ. Thus, the nmr splitting could be continuously reduced as a result of a continuous decrease in the barrier.

Recently, Sorenson has applied this new method to the question of whether the 1,6-dimethylcyclodecyl cation is classical or bridged. The ^{13}C spectrum showed a *single* peak for the 1 and 6 carbons resulting either from a very rapid 1,6- hydride shift or from the true symmetry of the bridged, non-classical ion. Sorenson found that the

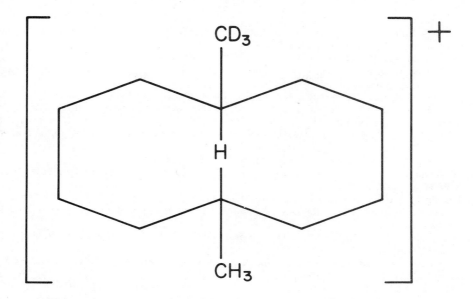

Figure 17. Deuterated 1,6-dimethylcyclodecyl cation.

ion prepared with one deutromethyl group showed a splitting of only 0.5 ppm, clearly supporting the bridged structure.[24] An additional, important piece of evidence is that a peak of area one was found in the proton spectrum 2 ppm *above* TMS.

Since in simple 1,2 shifts of hydride and methide, barriers of 3-4 kcal/mole are found it should be no surprise that in more complex carbonium ions that analogous barriers are reduced through factors which can be generally described as strain. When such a barrier to a rearrangement process becomes zero, we have a single minimum energy surface and the ion can be appropriately described as a non-classical ion. Thus, the existence of non-classical ions, first proposed as a result of studies of the kinetics of solvolysis reactions has now been confirmed through the observation of the NMR spectra of stable solutions of the ions themselves. The new method described here for attacking such problems, may well be useful in many other, non-carbonium ion, systems.

Acknowledgements

I should like to gratefully acknowledge the collaboration of Mark Jaffe, Linda Telkowski, Dr. Pierre Vogel and especially Mandes Kates who did the research leading to the method described here. I should also like to acknowledge the support of this work by the National Science Foundation.

References and Notes

1. a) Radom, L., Poppinger, D. and Haddon, R.C., *in* "Carbonium Ions,"Vol. V, (Olah, G.A. and Schleyer, P.v.R., ed.), Chapt. 38, p. 2303, Wiley- Interscience, New York, 1976; b) Hehre, W. J. in "Applications of Electronic Structure Theory," Schaefer, H. F., III, ed.), Vol. 4 of "Modern Theoretical Chemistry," Plenum Press, New York, 1977.

2. Saunders, M. in "Magnetic Resonance in Biological Systems" (Ehrenberg, A., Malstrom, B. G., and Vanngard, T., eds.) Pergamon, Oxford, 1967, p. 85.

3. Saunders, M. and Kates, M. R., *J. Am. Chem. Soc. 100*, 7082 (1978).

4. Nenitzeschu, C. D., *in* "Carbonium Ions," Volume 1 (Olah, G. A. and Schleyer, P. v. R., ed.), Chapt. 1, p. 1, Wiley- Interscience, New York, 1968.

5. a) See Olah, G. A., "Carbocation and Electrophilic Reactions," J. Wiley and Sons, 1974, for an historical account of the development of this area b) Meerwein, H., Bodenbenner, K., Borner, P., Kunert, F., and Wunderlich, H., *Ann.632*, 38 (1960); c) Brouwer, D. M., McLean, S., and Mackor, E. L., *Disc. Farad. Soc. 39*, 121 (1965); d) Saunders, M. and Hagen, E. L., *J. Am. Chem. Soc. 90*, 2436 (1968).

6. Saunders, M., and Lloyd, J. R., submitted for publication, *J. Am. Chem. Soc.*.

7. "Carbonium Ions" (Olah, G. A. and Schleyer, P. v. R., ed.), Vols. I-V, Wiley Interscience New York, 1968-1976.

8. Bethell, D., and Gold, V., "Carbonium Ions, an Introduction," Academic Press, New York, 1967.

9. "Molecular Rearrangements" (de Mayo, P., ed.), Parts I and II, Interscience Publishers, New York, 1963 and 1964.

10. a) Saunders, M., and Hagen, E. L., *J. Am. Chem. Soc. 90*, 6881 (1968); b) Saunders, M., Hagen, E. L. and Rosenfeld, J., *J. Am. Chem. Soc. 90*, 6882 (1968).

11. a) Saunders, M., and Rosenfeld, J., *J. Am. Chem. Soc. 91*, 7756 (1969); b) Saunders, M., and Budiansky, S., *Tetr., 34*, 0000 (1979).

12. Saunders, M., Vogel, P., Hagen, E. L., and Rosenfeld, J., *Acc. Chem. Res. 6*, 53 (1973).

13. Brown, H. C. (with comments by Schleyer, P. v. R.), "The Nonclassical Ion Problem," Plenum Press, New York, 1977).

14. Winstein, S., Shatavsky, M., Norton, C. and Woodward, R. B., *J. Am. Chem. Soc. 77*, 4183 (1955).

15. Saunders, M., Jaffe, M. H., and Vogel, P., *J. Am. Chem. Soc. 93*, 2558 (1971).

16. Saunders, M. and Vogel, P., *J. Am. Chem. Soc. 93*, 2559 (1971).

17. Saunders, M. and Vogel, P., *J. Am. Chem. Soc. 93*, 2561 (1971).

18. Saunders, M., Telkowski, L. and Kates, M. R., *J. Am. Chem. Soc. 99*, 8070 (1977).

19. Sunko, D. E., Szele, I. and Hehre, W. J., *J. Am. Chem. Soc. 99*, 5000 (1977).

20. a) Schleyer, P. v. R., Watts, W. E., Fort, Jr., R. C., Comisarow, M. B., and Olah, G. A., *J. Am. Chem. Soc. 86*, 5679 (1964) b) Saunders, M., Schleyer, P. v. R., and Olah, G. A., *ibid., 86*, 5680 (1964) c) Jensen, F. R. and Beck, B. H., *Tetrahedron Lett.*, 4287 (1966); d) Olah, G. A., White, A. M., De Member, J. R., Commeyras, A. and Lui, C. Y., *J. Am. Chem. Soc. 92*, 4627 (1970); e) Olah, G. A., Liang, G., Mateescu, G. D. and Riemenschneider, J. L., *ibid. 95*, 8698 (1973); f) Kates, M. R., Ph.D. Thesis, Yale University, 1978.

21. Saunders, M. and Kates, M. R., *J. Am. Chem. Soc. 99*, 8071 (1977).

22. Coates, R. M. and Fretz, E. R. *J. Am. Chem. Soc. 97*, 2538 (1975).

23. Saunders, M., Kates, M. R., Wiberg, K. B. and Pratt, W., *J. Am. Chem. Soc. 99*, 8072 (1977).

24. Sorensen, T. S., private communication.

Part III
Medium Size Molecular Systems

Conformational Mobility of the Backbone of Cyclic Tripeptides[1]

Horst Kessler
Institut fur Organische Chemie
der Universitat Frankfurt a. M.,
Niederurseler Hang, D-6000
Frankfurt a M. 50, Germany

Introduction

Small cyclic peptides are of special interest for conformational investigations. Restrictions of their molecular mobility lead to a relative small number of structures—a necessary precondition for drawing definite conclusions about their conformation in solution. This report deals with cyclic tripeptides, which have not yet been found in nature, but nevertheless are important due to the above mentioned reasons.

Cyclic tripeptides form nine-membered rings which contain three cis-peptide bonds. Due to steric reasons a trans-peptide bond is energetically disfavored. Formation of a cyclus requires cis-peptide bonds in the linear precursors, too. These are only populated in N-alkyl amino acids (Pro[2-6], Hyp[3-5], Sar[7/8], N-Benzyl-Gly[9/10], o-Nitrobenzyl-Gly[11]). Therefore so far no cyclic tripeptide with a "free" NH-group in an amide bond is known.

Double bonds and amide bonds have similar structures. Thus the cis,cis, cis-nonatriene-(1,4,7) can be regarded as an excellent model for cyclic tripeptides. The most stable conformation of the hydrocarbon is the "crown" with C_{3v} symmetry.[12] It can be derived from the cyclohexane chair conformation by substitution of the three CH_2-groups in 1,3,5-position with three cis double bonds. The "crown" was previously observed for cyclo[L-Pro$_3$][3-6], cyclo[Sar$_3$][7/8], and cyclo[Bzl.Gly$_3$][13].

We have found that in all cyclic tripeptides containing at least one achiral amino acid (Sar, N-Bzl.Gly, 2-Nitrobenzyl-Gly) a conformational equilibrium exists between "crown" and "boat"[9/10/13]. This equilibrium is analogous to the chair-boat interconversion of cyclohexane (Figure 1).

If a cyclic tripeptide contains only chiral amino acids no boat-crown equilibrium is possible but only two alternative structures depending on the relative chirality of the amino acids. In case of the same chirality only the C_3-symmetric "crown" is allowed (example: cyclo[L-Pro$_3$][3-6]). In case of different chirality the "boat" is the only possible conformation (example: cyclo[L-Pro-L-Pro-D-Pro][1]).

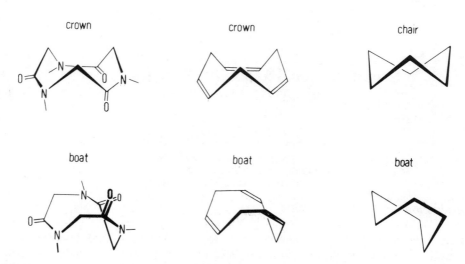

Figure 1. Crown and boat conformation of cyclic tripeptides and comparison with the conformational analogues cyclononatriene- (1,4,7) and cyclohexane.

The "Crown"

The characteristic feature of the rigid "crown" conformation is the inward orientation of three α-protons, which is comparable to the axial orientation of the 1,3,5-hydrogens in cyclohexane. The Dreiding model for cyclic tripeptides shows an even stronger steric interaction of these positions. Hence, substitution of one "inward" hydrogen by an alkyl grup destabilizes the "crown" dramatically. For that reason, a complete ring inversion of the "crown" is only possible in the case of cyclic tripeptides containing only achiral amino acids (cyclo[Sar$_3$][12/14], cyclo[N-Bzl.Gly]$_3$[9/13]).

In homogeneous cyclic tripeptides (cyclo[Pro$_3$], cyclo[Sar$_3$] and cyclo[N-Bzl.Glyc$_3$]) assignment of the crown is directly evident from the equivalence of the three amino acids in their ^{1}H and ^{13}C NMR spectra. In other cases the analogy of the ^{13}C NMR chemical shifts to those above mentioned "crowns" determines assignment.[1] The x-ray analysis of cyclo[Pro-N-Bzl.Gly$_2$] crystallized from polar solvents yields the "crown."[15] A solution of these crystals at low temperatures exhibits only NMR signals of the "crown." At higher temperatures the equilibration to a mixture of both conformations is observed.[16]

The orientation of all three amide groups to one side of the ring leads to higher dipole moments for the "crown" than for the "boat." Therefore the crown-boat-equilibrium is shifted towards the "crown" with increasing solvent polarity. This is demonstrated for cyclo[Pro$_2$Bzl.Gly] in Table I.

The characteristic downfield shift of all inward orientated proton signals of the "crown" during titration of a CDCl$_3$ solution with DMSO-d$_6$ is helpful for assignment, too.[9/16]

Table I

Crown/Boat Equilibrium in cyclo[Pro_2-Bzl.Gly]

Solvent	Polarity	parameter	% boat
	E_T[a]	DK	
$CDCl_3$	39.1	4.7	90
CD_2Cl_2	41.1	8.9	79
CD_3CN	46.0	37.5	67
acetone-d_6	42.2	20.7	59
DMSO-d_6	45.0	48.9	58
DMF-d_7	43.8	36.7	42

[a] C. Reichard, Angew. Chem. *91*, 119 (1979).

The "Boat"

The boat conformation differs from the crown by inverting one α-carbon to the opposite side of the ring plane. The flexibility of the boat corresponds to that of the cyclohexane boat. Overall there are six different boat conformations which are degenerated in case of cyclo[Sar_3] and cyclo[Bzl.Gly_3] (Figure 2).

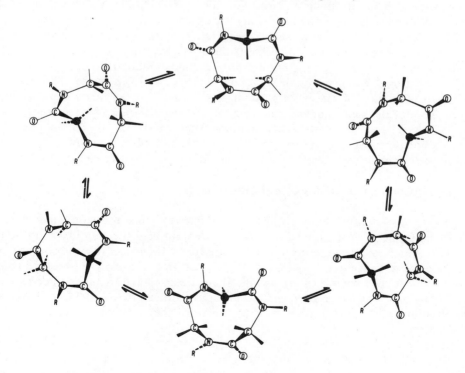

Figure 2. Pseudorotation of the six boat conformations in cyclic tripeptides.

The interconversion between two boats occurs via a twist boat by a pseudorotational pathway. This process is fast on the NMR time scale[17] and leads to a singlet for all α-protons in cyclo[Sar$_3$][10] and cyclo[Bzl.Gly$_3$].[9] Insertion of a chiral amino acid destroys this degeneracy. This will be discussed later. The "ideal" boat of the model cyclononatriene-(1,4,7), has a symmetry plane. Cyclic tripeptides certainly do not have such a symmetry but still the angles have nearly the same absolute value (pseudosymmetry with $\phi_1 \sim -\psi_2$; $\psi_1 \sim -\phi_2$; $\phi_3 \sim -\psi_3$, Figure 3). Twisting reduces the interaction between the inward orientated α-protons at carbon 1 and 2 (see Figure 3) and between the α-proton at carbon 3 with the opposite amide bond. On the other hand, the Dreiding model shows some increasing ring strain during this process. It is difficult to determine the amount of twisting by NMR spectroscopy. X-ray analysis of the only known boat conformation so far (cyclo[L-Pro-L-Pro-D-Pro], Figure 4) demonstrates an intermediate conformation between the ideal boat and the twist-boat (Figure 3).

The Dynamics of the Backbone

Cyclic tripeptides with at least one achiral amino acid show an equilibrium of both backbone conformations. Interconversion between the crown and the boat conformation is slow on NMR time scale at room temperature. Coalescence occurs at temperatures higher than 100°C. Interconversion of both conformations has also been demonstrated by saturation transfer experiments[16] at somewhat lower temperatures. Thus meaning a barrier of about 90 kJ/mol. A similar barrier is observed for the complete ring inversion of the crown conformation of cyclo[Sar$_3$][14] and cyclo[Bzl.Gly$_3$].[9/13] Theoretically[12] one could consider an analogy to the ring inversion of cyclohexane which proceeds via the boat conformation as indicated in the isomerization graph in Figure 5. For example, a small amount of "boat" form in the CDCl$_3$ solution spectra of cyclo[Sar$_3$] and cyclo[Bzl.Gly$_3$] was found. The free enthalpy of activation of the process "boat" → "crown" (determined from the line broadening of the signals of the boat singlet) plus the difference in free energy between the boat and the crown conformation (determined from the population difference boat/crown) is equal to the barrier of the complete ring inversion of the crown (determined from the line shape of the "crown"- signals) (see Figure 5).

Cyclic Tripeptides Containing Chiral Amino Acids

The rigid crown conformation is chiral (C$_3$-symmetry). Ring inversion interconverts both enantiomers. When introducing a chiral amino acid the ring inversion is no longer degenerated, because diastereomers are formed. Obviously, a substitution of the three inward orientated hydrogens of a N-substituted cyclotriglycine by bigger groups leads to considerable steric strain. There are no such difficulties when replacing the outer C$_\alpha$-hydrogens by one, two or three other groups. If more than one chiral amino acid is introduced into the "crown" they necessarily must have the same chirality.

The boat conformation also allows the introduction of chiral amino acids, although the two inward orientated C$_\alpha$-hydrogens as is the position opposite to the amide bond is substantially hindered. Thus the outside positions are the best places for substituents to be put in. The cyclic structure of a proline residue restricts the allow-

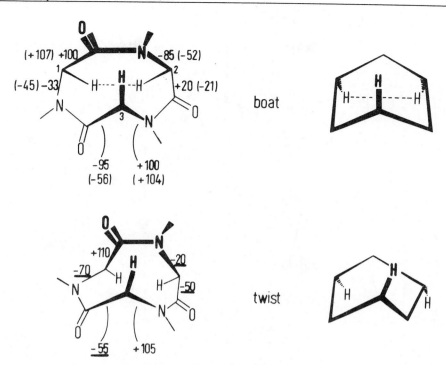

Figure 3. The boat conformation of cyclic tripeptides and cyclohexane. The "ideal boat" (above) can be twisted (below) to reduce steric interactions. Conformational angles are given. The values in parentheses are the experimentally observed ones for cyclo[L-Pro-L-Pro-D-Pro].[1/15]

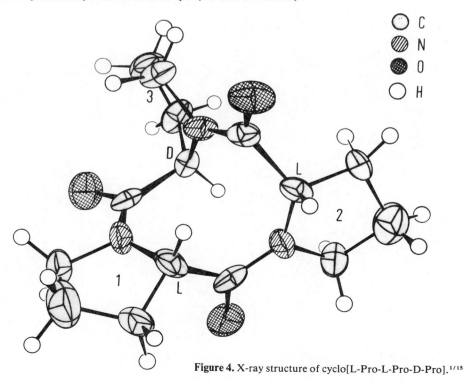

Figure 4. X-ray structure of cyclo[L-Pro-L-Pro-D-Pro].[1/15]

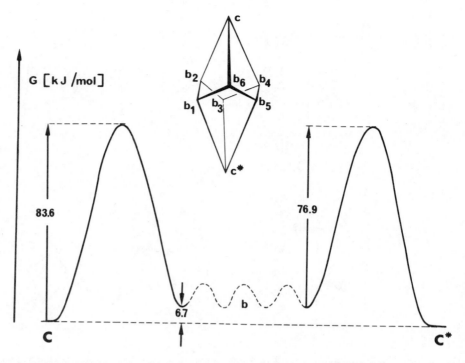

Figure 5. Isomerizational graph and experimentally determined energy profile in cyclo[Sar₃] (see text).

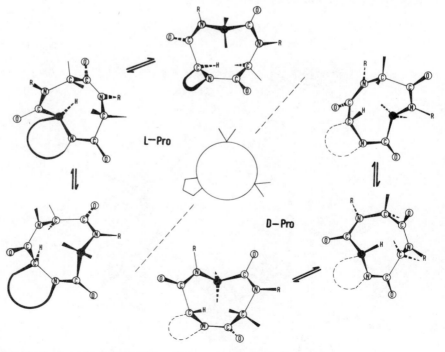

Figure 6. The boat conformation of cyclo[L,D-Pro-Xxx-Xxx], Xxx = Sar, N-Bzl.Gly. D-Pro is indicated by dashed lines.

ed positions even more because of the outward orientation of the substituents at nitrogen. This is shown in Figure 6. Altogether three boat conformations are in a rapid equilibrium for peptides of the type cyclo[Pro-Xxx₂][15] (Xxx = Sar, N-Bzl.Gly).

Further reduction of the number of possible boat conformations occurs if a second prolyl residue is introduced (Figure 7). If both prolyl residues have the same chirality (example: cyclo[L-Pro-L-Pro-Bzl.Gly][9]), there is only one boat conformation favored by steric effects. In case of having different chiralities, there should be two boat conformations in fast equilibrium. So far no example of this type exists.

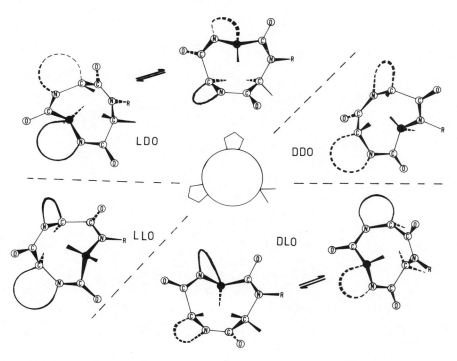

Figure 7. The boat conformations of cyclo[D,L-Pro-D,L-Pro-Xxx], Xxx = Sar, N-Bzl.Gly. D-Pro is indicated by dashed lines.

Three chiral amino acids which differ in chirality only form a distinct boat conformation. This was proven for cyclo[L-Pro-L-Pro-D-Pro] by NMR spectroscopy and x-ray analysis.[1]

The conformational possibilities of all cyclic tripeptides known till now are summarized in Table II.

The remaining question which of the rapidly equilibrating boat conformations is the thermodynamically most stable one was solved for cyclo[L-Pro-Bzl.Gly₂] by C-13 NMR spectroscopy.[16] Comparison with the completely assigned signals of

Table II

Possible Conformations in Cyclic Tripeptides
Depending on the Chirality of the Amino Acid Residues
(O means an achiral amino acid; the tripeptides so far synthesized are given).

SSS/RRR	SSO/RRO	SOO/ROO	OOO
crown	crown + boat	crown + 3 boats	2 crowns + 6 boats
c[Pro$_3$][2-6]	c[Pro$_2$Bzl.Gly][9/10/13]	c[ProBzl.Gly$_2$][9/10/13]	c[Bzl.Gly$_3$][9/10/13]
c[Pro$_2$Hyp][3-5]	c[Pro$_2$Sar][22]	c[ProSar$_2$][22]	c[Sar$_3$][7/8]

SSR/RRS	SRO
boat	2 boats
c[Pro$_2$D-Pro][1/21]	unknown

Table III

Population of the Crown Conformation
in Cyclic Tripeptides in Percent (in CDCl$_3$)

c[L-Pro$_3$]	c[Pro$_2$Bzl.Gly]	c[ProBzl.Gly$_2$]	c[Bzl.Gly$_3$]
100	10	64	90

c[L-Pro$_2$D-Pro]	c[Pro$_2$Sar]	c[ProSar$_2$]	c[Sar$_3$]
0	15	?	94

cyclo[L-Pro-L-Pro-D-Pro] and cyclo[L-Pro-L-Pro-Bzl.Gly] leads immediately to the solution of the problem. Starting with the ''boat'' of cyclo[L-Pro[1]-L-Pro[2]-D-Pro[3]] and replacing the D-proline by N-benzyl-glycine, the carbon-13 signals of the ring 3 diminish (Figure 8). This result corresponds with the expectation, because the Bzl.Gly residue can only occupy the position of the D-proline in the boat conformation. Further replacement of one L-proline leaves only the signals of proline-1.

Conclusions

Several times we have pointed out the conformational analogy between cyclic tripeptides and cyclohexane, following Dunitz' and Waser's concept of the analogy between cis double bonds and methylene groups.[18] The conformations of cyclic tripeptides are much easier to study than those of cyclohexane, because—the barrier of the crown-boat interconversion in tripeptides is about twice that of the chair-boat process in cyclohexane—the thermodynamic stability of the flexible boat conformation in cyclic tripeptides is similar to the crown. On the other hand, the stability of the cyclohexane chair exceeds that of the boat by about 23 kJ/mol.[19]

One fact which destabilizes the crown is the stronger steric interaction of the three inward orientated hydrogens which come in closer contact than the three axial hydrogens of cyclohexane.

There is another trend in the equilibrium between crown and boats in peptides of the general structure cyclo[Pro$_m$X$_n$] *(m + n = 3)*. With decreasing m the crown conformation becomes more and more populated (Table III). The small amount of less

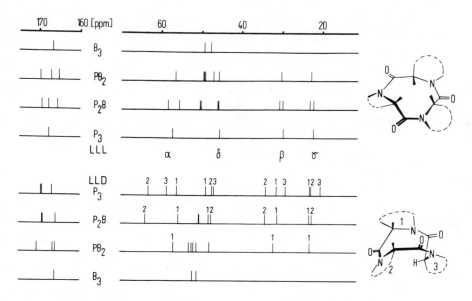

Figure 8. Comparison of the C-13 spectra of proline containing cyclic tripeptides for conformational assignments in cyclo[L-Pro-Bzl.Gly-Bzl.Gly].

than 10 percent boat conformation in cyclo[Sar₃] and cyclo [Bzl.Gly₃] has been overlooked in the past. So far we do not see an easy way to explain these trends.

We have focused our interest in this paper only to the backbone conformation of cyclic tripeptides. There are interesting features concerning the mobility of the proline rings, too. These processes are too fast to observe distinct species in NMR spectra. Carbon-13 relaxation measurements[5/20] and x-ray analysis[1] show a higher mobility of the γ-carbon at the proline residues of cyclo[L-Pro₃] and cyclo[L-Pro-L-Pro-D-Pro].

Acknowledgements

I am very much indebted to my coworkers Dr. P. Kramer, A. Friedrich, P. Kondor, and G. Krack, who were involved in these studies. I am also very grateful to Professor Dr. J. Dale and Dr. K. Titlestad, who provided us with a sample of cyclo[Sar₃] and to Professor Dr. M. Rothe and Dr. W. Mastle for the sample of cyclo[L-Pro-L-Pro-D-Pro]. This work was supported by the Fonds der Chemischen Industrie and the Deutsche Forschungsgemeinschaft.

References and Notes

1. Peptide conformations, VIII, Part VII; Bats, J.W., Friedrich, A., Fuess, H., Kessler, H., Mastle, W., and Rothe, W., *Angew, Chem.,* in press.
2. Rothe, M., Steffen, K. D., Rothe, I., *Angew. Chem. 77,* 347 (1965)   *Angew. Chem. Intern. Ed., 4,* 356 (1965).
3. Deber, C. M., Torchia, D. A., Blout, E. R., *J. Am. Chem. Soc., 93,* 4893 (1971).
4. Kartha, G., Ambady, G., Shankar, P. V., *Nature, 247,* 204 (1974); Kartha, G., Ambady, G., *Acta Cryst. B 31,* 2035 (1975).
5. Deslauriers, R., Smith, I. C. P., Rothe, M., Peptides, Eds. R. Walter, J. Meienhofer, *Ann Arbor Science Publ.,* 1975, p. 91.
6. Druyan, M. E., Coulter, C. L., Walter, R., Kartha, C., Ambady, G. K., *J. Am. Chem. Soc. 98,* 5496 (1976).
7. Dale, J., Titlestad, K., *Chem. Comm. 1969,* 656.
8. Dale, J., Titlestad, K., *Acta Chem. Scand. B29,* 153 (1975).
9. Kessler, H., Kondor, P., Krack, G., and Kramer, P., *J. Am. Chem. Soc. 100,* 2548 (1978).
10. Kessler, H., Krack, G., and Kramer, P., "Peptides," 1978, Eds. J. Z. Siemion and G. Kupryszewski, Wrozlaw.
11. Kessler, H. and Siegmeier, R., unpublished results.
12. Dale, J., "Topics Sterochem." (Eds. E. L. Eliel and N. C. Allinger) *9,* 199 (1976) and references cited therein.
13. Kramer, P., Doctoral Thesis, Frankfurt, 1976.
14. Schaug, J., *Acta Chem. Scand. 25,* 2771 (1971).
15. Bats, J. W., Fuess, H., *J. Am. Chem. Soc.,* in preparation.
16. Kessler, H. and Krack, G., unpublished results.
17. The low temperature measurements of cyclo[Sar₃] in CD_2Cl_2 down to -100°C did not show any broadening of the singlet of the boat form.
18. Dunitz, J. D. and Waser, J., *J. Am. Chem. Soc. 94,* 5645 (1972).
19. Squillacote, M., Sheridan, R. S., Chapman, O. L., and Anet, F. A. L., *J. Am. Chem. Soc. 97,* 3244 (1975).
20. Deslauriers, R. and Smith, I. C. P., "Topics in Carbon-13 NMR Spectroscopy," Ed. G. Levy, *2,* 1 (1976), Wiley- Interscience, New York, 1976.
21. Rothe, M. and Mastle, W., "Peptides 1978," Eds. J. Z. Siemion and G. Kupryszewski, Wrozlaw.
22. Rothe, M.. Theysohn, R., Muhlhausen, D., Eisenbeib, F., and Schindler, W., "Chemistry and Biology of Peptides," Ed. J. Meienhofer, Science Publ., Ann Arbor, 1972, p. 51.

Multinuclear NMR Studies of Crown and Cryptand Complexes

Alexander I. Popov

Department of Chemistry
Michigan State University
East Lansing, Michigan 48824

Introduction

The popularity of macrocyclic complexes of alkali elements has grown rapidly since the synthesis of crown ethers by Pedersen[1] in 1967 and of cryptands two years later by Lehn and co-workers.[2] During the last decade numerous macrocyclic ligands have been synthesized and their interactions with the alkali cations have been studied by a large variety of physicochemical techniques such as potentiometry, electrical conductance measurements, polarography, cyclic voltametry, liquid-liquid extraction, electronic spectroscopy, vibrational spectroscopy, electron spin resonance and last, but certainly not least, nuclear magnetic resonance,[3] which is the subject of this presentation.

Nearly all naturally occurring elements have at least one isotope with a non-zero nuclear spin and thus, in principle, nearly every element can be studied by NMR. However, an NMR-active isotope may exist only in low natural abundance (for example, ^{13}C, ^{17}O, and ^{43}Ca). In addition, many nuclei have very low sensitivity compared to that of the familiar proton.

Recent advent of superconducting solenoids and of Fourier transform spectroscopy has increased the sensitivity of the NMR measurements to the point where the resonance of many NMR active nucleus can be observed at reasonable concentrations. Thus, it seems, that the era of multinuclear NMR is only beginning.

It is natural that nuclear magnetic resonance has been used extensively in the studies of macrocyclic complexes. It is impossible to cover all such studies in a limited time and only some representative examples will be given here. Some macrocyclic ligands discussed below are shown in Figure 1.

Proton Magnetic Resonance

Many of the most commonly used macrocyclic polyethers and cryptands have very symmetrical structures and, therefore, give simple ^1H and ^{13}C NMR spectra. The ultimate case is that of unsubstituted crown ethers, *i.e.* 18C6 (Figure 1), where all

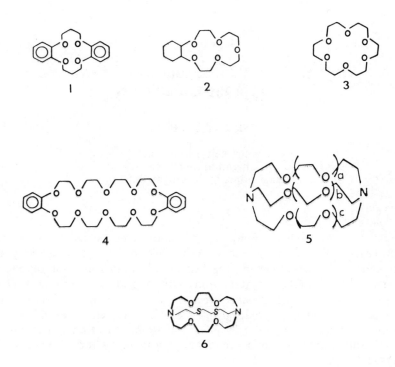

Figure 1. Crown ether and cryptands. 1. Dibenzo-12-Crown-4, DB12C4; 2. Benzo-15-Crown-5, B15C5; 3. 18-Crown-6, 18C6; 4. Dibenzo-30-Crown-10, DB30C10; 5. [2]-cryptands, a=b=c=1—C222; a=b=1, c=0—C221; a=1, b=c=0—C211; 6. C22S₂.

protons and the carbons of the free ligand are equivalent, and only one resonance is observed for each nucleus.

The resonance of the polyether chain hydrogens is moderately sensitive to the complexation reaction. For example, Wong, et al.[4] found that the addition of sodium fluorenyl to dimethyldibenzo-18C6 in tetrahydrofuran solution resulted in an upfield shift of the ring protons of about 0.75 to 1.0 ppm while the protons of the aromatic rings and of the methyl groups were affected to a much smaller extent. Lockhart, et al.[5] reports proton magnetic resonance measurements on B15C5 and B21C7 in the free state and on their alkali complexes. They noted that in cases where both 1:1 and 2:1 ligand to metal ion complexes were formed, the pmr moves initially downfield but after the formation of the first complex the direction of the chemical shift is reversed. Similar results were observed by Live and Chan in their studies of DB18C6 complex with the cesium ion in acetone solutions.[6] Figure 2 shows the variations in the chemical shift of the ether proton during the formation of the 1:1 and of the 2:1 "sandwich" complex.

Live and Chan observed that, in general, the pattern of the ¹H chemical shifts upon complexation is similar for the potassium and cesium complexes of DB18C6 and DB30C10. However, the addition of sodium ion to DB30C10 results in a quite dif-

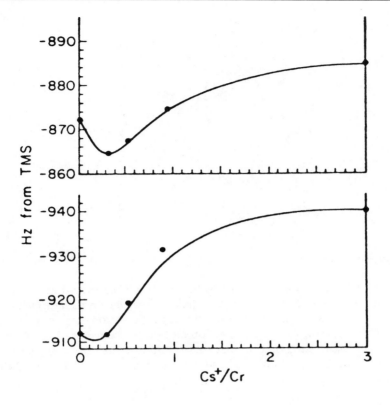

Figure 2. Chemical shift behavior of the ether protons (H₁ lower, H₂ upper graph) on complexation at DB18C6 with Cs⁺ ion. Reference 6. Reproduced with the permission of the Journal of the American Chemical Society.

ferent spectrum from that of the larger alkali cations indicating that the structure of Na⁺•DB30C10 complex is different from those of the K⁺•DB30C10 and Cs⁺•DB30C10 complexes. It seems, therefore, that a conformational change of the DB30C10 occurs upon formation of the sodium complex.

Carbon-13 NMR

Numerous examples of the use of ^{13}C NMR measurements for the study of macrocyclic ligands and their complexes can be found in the literature. A typical example is the protonation of a spherical [3]-cryptand (Figure 3), and the formation of an anionic complex.[7] The free ligand has two carbons in different environments and ^{13}C resonances are observed at 70.42 and 56.97 ppm respectively (with respect to TMS). When one equivalent of hydrochloric acid is added to a methanol solution of the ligand, half of the ligand present is diprotonated while the other half remains as the free base (Figure 4). Two equivalents of the acid convert all of the ligand to the diprotonated form. Further addition of one equivalent of HCl converts *half* of the diprotonated ligand to a new species resonating at 69.87 and 54.00 ppm. Finally, when four equivalents of HCl are present all of the ligand is converted into this new species which is the anionic complex $LH_4 \cdot Cl^{3+}$.

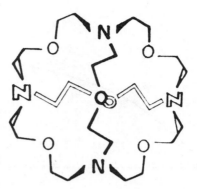

Figure 3. Spherical [3]-cryptand.

Alkali Metal NMR

Studies of the resonance of alkali nuclei by us and others[8] showed that the magnetic resonances of these nuclei are very sensitive indicators of the immediate chemical environment of the alkali ions in solution. In fact, the resonance frequencies are dependent of the solvent, on the counter ion and on the concentration of the solution. It is interesting to note that in the case of ^{23}Na nucleus the chemical shifts are proportional to the donor abilities of the solvents[9] as expressed by the Gutmann donicity scale.[10]

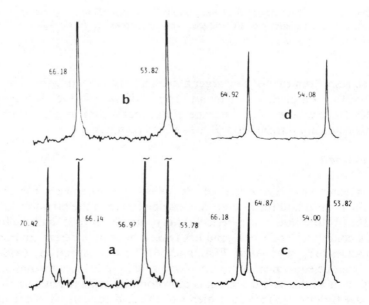

Figure 4. Carbon-13 spectra of the spherical [3]-cryptand in methanol solution upon addition of HCl. (a) One equivalent of HCl; (b) two equivalents; (c) three equivalents, (d) four equivalents. Reference 7. Reproduced with the permission of the Journal of the American Chemical Society.

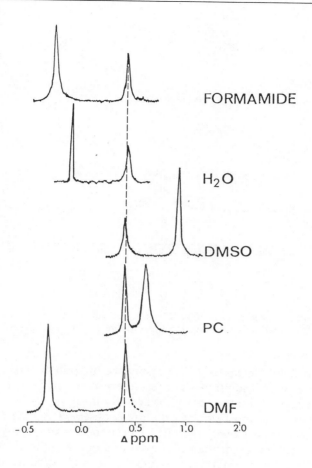

Figure 5. Lithium-7 NMR spectra of lithium—C211 cryptates in various solvents. [C211] = 0.25 M, [Li$^+$] = [0.50 M]. Reference 12. Reproduced with the permission of the Journal of Physical Chemistry.

It is well known that the stability of cryptates (as well as of the crown complexes) depends on the consonance between the sizes of the macrocyclic activities and the ionic radii. Cryptand C211 is tailor-made for the lithium ion[11] and the resulting complexes are quite stable. In the presence of an excess of the metal ion the exchange between the two lithium sites is slow by the NMR time scale and two resonance signals are observed[12] (Figure 5). It is evident that while the resonance frequency of the free cation depends strongly on the solvent, the encapsulated cation is completely insulated from the solvent.

Dye and co-workers studied Na$^+$•C222 complex in ethylenediamine solution[13] and found that two ^{23}Na resonances are observed when the solution contains an excess of the sodium ion. The addition of the cryptand C222 to a solution of metallic sodium in ethylammine solutions results in the formation of the sodium cryptate salt of the sodium anion [C222•Na$^+$] Na$^-$ (Figure 6) and the sodium-23 resonance of the sodium anion is at 62.8 ppm upfield from the sodium signal of a saturated aqueous sodium chloride solution used as the reference.[14]

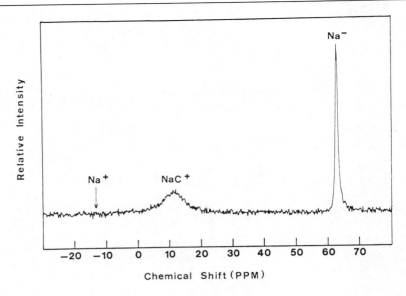

Figure 6. Sodium-23 NMR spectra of [C222•Na⁺] Na⁻ salt in ethylenediamine solution at 1.4°C. Reference 14. Reproduced with the permission of the Journal of Chemical Physics.

Kintziger and Lehn[15] measured ^{23}Na resonance of sodium cryptates in 95 percent MeOH - 5 percent D_2O solutions. The results are shown in Table I. It is seen that the resonance frequency and the linewidths are very much dependent on the size of the cryptand cavity. The most downfield shift is observed for the Na⁺C211 complex where one would expect a very tight fit between the ion and the ligand resulting in a large overlap between the orbitals of the donor atoms and the other p-orbitals of the cation, but where the stability of the complex is small as compared to the sodium complexes of C221 and of C222. Thus the magnitude of the paramagnetic shift on complexation is not an indication of the stability of the alkali complex.

In the case of the cesium-133 resonance the natural linewidth is narrow (< 1 Hz), the range of the chemical shifts is quite large and, therefore, the sensitivity of the measurement is relatively high.

Table I
Sodium-23 Chemical Shifts of Some Sodium Cryptates
in 95 Percent MeOH - 5 Percent D_2O

Cryptate	δ(ppm)[a]	$\Delta\gamma$ (Hz)
C211•Na⁺	+11.15	132
C221•Na⁺	-4.25	46
C222•Na⁺	-11.40	29
C22S₂•Na⁺[b]	-6.20	49

[a]With respect to 0.25 M aqueous NaCl solution.
[b]See Figure 1.

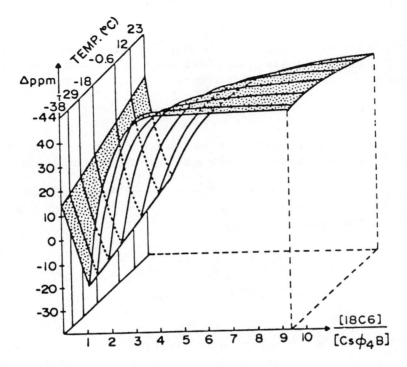

Figure 7. A three-dimensional plot of the ^{133}Cs chemical shift *vs.* mole ratio and temperature for solutions of CsBPh$_4$ and 18C6 in pyridine. Reproduced with the permission of the Journal of the American Chemical Society.

Crown complexes of Cs$^+$ ion in a variety of solvents have been studied by Mei *et al.*[16/17] A plot of the ^{133}Cs resonance as a function of the 18C6:Cs$^+$ mole ratio at different temperatures is shown in Figure 7. It is evident that the behavior is very similar to that observed by Live and Chan[6] for the DB18C6•Cs$^+$ system by proton NMR in acetone. The change in the direction of the chemical shift indicates stepwise formations of the 18C6•Cs$^+$ and of (18C6)$_2$Cs$^+$ complexes. The analysis of the data gave the complexation constants at these temperatures as well as the enthalpy and the entropy of complexation (Table II).

Table II
Limiting Chemical Shift of the Cs$^+$•2(18C6) Complex
and the Thermodynamic Parameters
for the Reaction Cs$^+$•18C6 + 18C6 ⇌ Cs$^+$(18C6)$_2$ in Pyridine[a]

Temp. K	$K_{1(min)}$, M^{-1}	K_2, M^{-1}	δ_2, ppm
297	10^5	79 ± 2	47.8
285	10^6	121 ± 5	49.4
272	10^6	218 ± 14	49.9
255	10^6	432 ± 58	51.4
244	10^6	623 ± 35	51.9
235	5×10^6	1173 ± 160	51.2

[a]$(\Delta G°_2)_{298}$ = -2.58 ± 0.02 kcal mol^{-1}; $\Delta H°_2$ = -5.8 ± 0.2 kcal mol^{-1}; $\Delta S°_2$ = -10.7 ± 0.6 cal mol^{-1}deg^{-1}.

It has been observed that in the case of crown complexes the cation exchange between the two sites usually is fast by the NMR time scale and for a 1:1 complex only
one resonance signal is observed whose chemical shift is given by the expression:

$$\delta_{obs} = X_f \delta_f + X_c \delta_c \tag{1}$$

where X_f and X_c are the respective mole fraction of the cation in the free and in the
complexed state while δ_f and δ_c are the cation chemical shifts in the two states.

Using the above equation and the equilibrium constant expression one can derive[18]
the following equation which relates the observed chemical shift to the total concentration of the metal ion and of the ligand,

$$\delta_{obs} = [(K_f C_t^M - K_f C_t^L - 1) \pm (K_f^2 C_t^{L^2} + K_f^2 C_t^{M^2} - 2K_f^2 C_t^L C_t^M + \tag{2}$$
$$2K_f C_t^L + 2K_f C_t^M + 1)^{\frac{1}{2}}] [(\delta_f - \delta_c)/(2K_f C_t^M)] + \delta_c$$

where C_t^M and C_t^L are the total concentrations of the metal ion and of the ligand
while K is the formation constant. The usual procedure is to measure δ_{obs} at a number
of different (L)/(M) mole ratios, then substitute the experimental parameters δ_{obs}, δ_f,
C_t^M and C_t^L and determine K_f and δ_c by an iterative procedure. This method fails
when the formation constants are $> 10^5$. Very similar treatment of data was used by
Live and Chan.[6] In the Cs$^+$•18C6 case described above, a curve-fitting method was
used to *estimate* the magnitude of K_1.

Large cryptands have somewhat flexible skeletons, for example, has been shown by
Weiss, *et al.* by X-ray studies[19] that C222 can accommodate cesium ion inside the
cavity although the latter, in the free state, appears to be somewhat smaller than the
size of the Cs$^+$ ion. Both potentiometric and ^{133}Cs NMR measurements[20/21] have
shown that in nonaqueous solvents cesium forms quite stable complexes with C222.

A cesium-133 NMR study of the Cs$^+$-C222 system (containing excess ligand) in three
different solvents showed that at room temperature the resonance of the complexed
Cs$^+$ ion is solvent-dependent.[22] Thus the cation cannot be completely enclosed inside
the ligand cavity. When the temperature is lowered, however, the resonance frequencies approach each other (Figure 8) and, at about -100°C they coalesce. It seemed to us that this behavior is indicative of an equilibrium between two forms of the
C222•Cs$^+$ complex, an *exclusive* complex (Figure 9) in which partially solvated cation only partially fills the cavity of the ligand, and the inclusive complex where the
cation is inside the cavity and is effectively insulated from the solvent. The
equilibrium between the two forms

$$Cs^+•C222_{exclusive} \rightleftharpoons Cs^+•C222_{inclusive} \tag{3}$$

is temperature dependent and shifts to the right at lower temperatures.

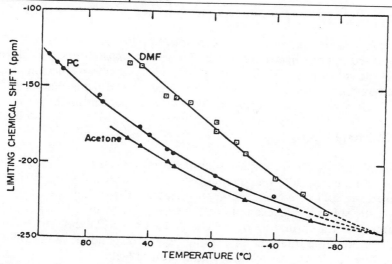

Figure 8. Limiting [133]Cs chemical shift at high mole ratio [C221][Cs+] *vs.* temperature in acetone, propylene carbonate and dimethylformamide solution. Reference 22. Reproduced with the permission of the Journal of the American Chemical Society.

It is interesting to note that in 1978 Weiss *et al.*[23] reported crystal structures of sodium and a potassium complexes with cryptand C221. The results clearly indicate the formation of an *inclusive* complex in the first case and of an *exclusive* one in the second. Thus in the crystalline state the existence of an exclusive cryptate has been clearly demonstrated.

Recent studies of the Cs+•C222 in tetrahydrofuran solutions containing small amounts of water[24] at ~-100°C show slow exchange between the two forms of the cesium cryptate. The [133]Cs resonance of the exclusive complex is at -70 ppm while that of the inclusive complex is at -240 ppm. These results confirm the existence of the two forms of the complex in solutions.

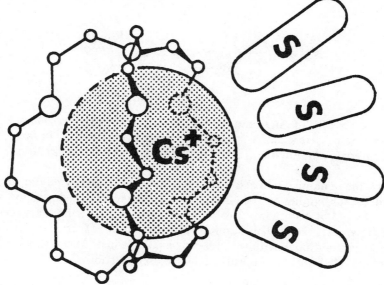

Figure 9. Schematic illustrations of an "exclusive" C222•Cs+ complex. Reference 21.

It is interesting to note that the formation of macrocyclic complexes is very often enthalpy stabilized but entropy destabilized. Thus Izatt, *et al.* found by calorimetric studies that the entropy of complexation of alkali ions by crown ethers in methanol-water mixtures becomes more negative as the size of the macrocyclic ring increased.[25] Thus $\Delta S°$ of complexation for Na+ in 20 percent methanol are -3.7, -3.8,-18.9, and -32.5 cal.deg^{-1}mole^{-1} with benzo-15C5, 18C6, dibenzo-24C8 and dibenzo-27C9 respectively.

A decrease in entropy upon complexation has also been observed by us for cesium complexes with dibenzo-30C10 and dibenzo-24C8 in various solvents.[26] While in all cases studied the $\Delta S°$ of complexation was negative the magnitude of the change varied with the solvent. For example, $\Delta S°$ values of -5.9 and -1.4 cal.deg^{-1}mole^{-1} were obtained for the formation of the DB30C10•Cs+ and DB24C8•Cs+ complexes respectively in pyridine solution. In acetone, the corresponding values for the two complexes were -26.2 and -20.1 cal.deg^{-1}mole^{-1} respectively.

While there are several factors which affect the entropy change of a complexation reaction, it seems that the overall negative $\Delta S°$ is due to a decrease in the conformational entropy of the ligand upon the formation of a metal complex. The ligands should be rather flexible in the free state and the flexibility should increase with the size of the macrocyclic ring. The formation of a complex with a metal ion probably locks the ligand into a more or less rigid configuration with a resulting decrease of the conformational entropy.

It is interesting to note that a recent study[27] of the sodium complex of a linear macrocycle shown below

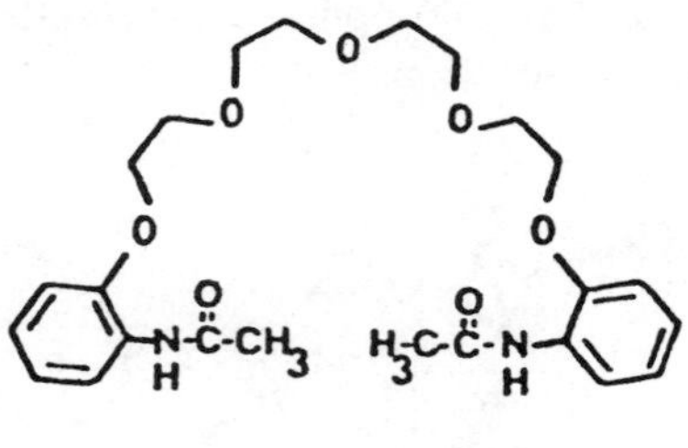

by ^{23}Na NMR in pyridine solutions showed enthalpy change to be -48 cal.deg^{-1}mole^{-1}. Here again, it was assumed that the decrease in the entropy results from a transformation of a flexible structure into a rigid conformation.

Finally, the remaining two alkali elements, potassium and rubidium, should be mentioned. So far, studies on [39]K NMR have been rather sparse due to low sensitivity of the nucleus (5×10^{-4} that of proton at constant field). Preliminary measurements[28] have shown that the nucleus is suitable for the studies of the macrocyclic complexes of potassium although with the presently available instrumentation the sensitivity of the results is below that of the alkali nuclei discussed above.

Rubidium seems to be the black sheep of the alkali family, from the point of view of NMR, that is. both ^{85}Rb and ^{87}Rb nuclei have broad lines. In nonaqueous solvents the linewidths at half-height are 600-800 Hz.[29] It seems that at this time studies of macrocyclic rubidium complexes by rubidium nucleus NMR do not look too promising.

References and Footnotes

1. C. J. Pederson, *J. Am. Chem. Soc. 89*, 2495, 7017 (1967).
2. B. Dietrich, J.-M. Lehn and J.-P. Sauvage, *Tetrahedron Letters, 1969*, 2885, 2889.
3. A. I. Popov and J.-M. Lehn, "Physicochemical Studies of Crown and Cryptate Complexes," in "Chemistry of Macrocyclic Complexes," G. A. Melson, ed., Plenum Publishing Co., New York, 1979.
4. K. H. Wong, G. Konizer, and J. Smid, *J. Am. Chem. Soc., 92*, 666 (1970).
5. J. C. Lockhart, A. C. Robson, M. E. Thompson, P. D. Tyson and I. H. M. Wallace, *J. C. S. Dalton, 1978*, 611.
6. D. Live and S. I. Chan, *J. Am. Chem. Soc., 98*, 3769 (1976).
7. E. Graf, and J.-M. Lehn, *J. Am. chem. Soc., 98*, 6903 (1976).
8. A. I. Popov, *Pure Appl. Chem., 51*, 101 (1979).
9. R. H. Erlich and A. I. Popov, *J. Am. Chem. Soc., 93*, 5620 (1971).
10. V. Gutmann and E. Wychera, *Inorg. Nucl. Chem. Lett., 2*, 257 (1966).
11. J.-M. Lehn, *Structure and Bonding, 16*, 1 (1973).
12. Y. M. Cahen, J. L. Dye and A. I. Popov, *J. Phys. Chem., 79*, 1289 (1975).
13. J. M. Ceraso and J. L. Dye, *J. Am. Chem. Soc., 95*, 9932 (1973).
14. J. M. Ceraso and J. L. Dye, *J. Chem. Phys., 61*, 1585 (1974).
15. J. P. Kintzinger and J.-M. Lehn, *J. Am. Chem. Soc., 96*, 3312 (1974).
16. E. Mei, J. L. Dye and A. I. Popov, *J. Am. Chem. Soc., 98*, 1619 (1976); *99*, 5308 (1977).
17. E. Mei, A. I. Popov, and J. L. Dye, *J. Phys. Chem., 81*, 1677 (1977).
18. Y. M. Cahen, R. F. Beisel, and A. I. Popov, *J. Inorg. Nucl. Chem.*, Supplement, 1976, p. 209.
19. D. Moras, B. Metz, and R. Weiss, *Acta. Cryst., B29*, 388 (1973).
20. J.-M. Lehn and J.-M. Sauvage, *J. Am. Chem. Soc., 97*, 6700 (1975).
21. E. Mei, L. Liu, J. L. Dye and A. I. Popov, *J. Solution Chem., 6*, 771 (1977).
22. E. Mei, A. I. Popov and J. L. Dye, *J. Am. Chem. Soc., 99*, 6532 (1977).
23. F. Mathieu, B. Metz, D. Moras, and R. Weiss, *J. Am. Chem. Soc., 100*, 4912 (1978).
24. E. Kauffmann, J.-M. Lehn, J. L. Dye and A. I. Popov, in press.
25. R. M. Izatt, R. E. Terry, D. P. Nelson, Y. Chan, D. J. Eatough, J. S. Bradshaw, L. D. Hansen, and J. J. Christensen, *J. Am. Chem. Soc., 98*, 7626 (1976).
26. M. Shamsipur and A. I. Popov, *J. Am. Chem. Soc.*, in press.
27. J. Grandjean, P. Laszlo, F. Vogtle and H. Sieger, *Angew. Chem., Int. Ed. 17*, 856 (1978).
28. J. S. Shih and A. I. Popov, to be published.
29. E. T. Roach and A. I. Popov, unpublished results.

Frontier Orbital Models for and Rearrangements of Organotransition Metal Compounds

C. Peter Lillya
Department of Chemistry
University of Massachusetts
Amherst, Massachusetts 01003

Introduction

The Dewar-Chatt-Duncanson (D-C-D)[1,2] model has provided the basis for almost all qualitative thinking about bonding in transition metal pi-complexes.[3] It ascribes bonding to charge transfer interactions between occupied ligand orbitals and unoccupied metal orbitals (donation) *and* between occupied metal orbitals and empty ligand orbitals (back donation). Difficulties arose when the model was applied, owing to the large number of potential charge transfer interactions between ligand and metal valence shell orbitals permitted by the low symmetry of most complexes. The unclear picture of bonding which resulted was pointed out by Green, for example.[3]

The frontier orbital concept[4] offers a solution to the above problem by providing an orbital energy criterion for identification of the most important charge transfer interactions. Examples of this approach have appeared in literature[5,6] the most extensive being a series of papers in which Hoffmann and his coworkers have developed a treatment of interactions between the frontier orbitals of submolecular fragments, e.g., CpM or M(CO)$_3$, and additional ligands.[7,8] Our interest in the frontier orbital approach arose because we felt the need for a model which was more illuminating than the D-C-D model alone, yet stopped short of the complexity of molecular orbital calculations.[9,10] Use of the frontier orbital approach within the D-C-D framework provides an incisive picture of metal carbon bonding which is conceptually simple enough for routine use by synthetic chemists. We present below examples of the frontier orbital method for analysis of bonding in organotransition metal pi complexes.

Structures of Organotransition Metal Pi Complexes

In the interest of simplicity we treat bonding in terms of one or, at the most, two donor-acceptor interactions. This approach involves some serious approximations. We use an approach based upon perturbation theory to treat strong bonding interactions, and we run the risk of neglecting important interactions when we fix our attention upon one or two. Nevertheless, as the discussion below will show, this approach successfully accounts for the structures of a wide variety of metal pi-complexes. Optimum cases for this simple approach will be those in which the donor and acceptor roles of ligand and metal fragment are clear and in which the metal fragment possesses relatively few frontier orbitals. Thus, in proceeding from metal fragments which lack only an electron or two of the 18-electron inert gas configuration to metal fragments with higher "electron demand," we will tend to move from the simpler to the more complex cases. We will focus our attention on prediction of structure for complexes of ligands which offer several non-equivalent coordination sites.

Anthracenetetrakis (silver perchlorate) (1)[11] and the mono silver perchlorate complexes of indene[11] and acenaphthene[25] crystallize in structures consistent with donation from the arene HOMO to an unoccupied Ag(I) orbital. Complexation occurs where the HOMO coefficients are largest, avoiding nodes.

Properties of cyclopentadienyl(olefin)dicarbonyliron cations are completely in accord with symmetric pi-bonding between the two sp^2 olefin carbons and iron (2)[12]. The $Cp(CO)_2Fe^+$ fragment, which can be constructed from a $CpFe^+$ fragment,[22] has a single low-lying unoccupied orbital, as expected for a 16 electron fragment. Bonding in an olefin pi-complex such as 2 is the result of donation from the olefin pi-MO to this LUMO. Kerber and Entholt have reported the complex with heptafulvene (3),[13] a ligand which offers 3 non-equivalent coordination sites. Donation from the heptafulvene HOMO (4)[14] should give a $1,8-\eta^2$ structure with a much stronger interaction with C_8 than with C_1. An X-ray diffraction study revealed an ion with strong $Fe-C_8$ bonding ($r_{Fe-C_8} = 2.16$Å) and little if any $Fe-C_1$ bonding ($r_{Fe-C_1} = 3.00$Å).[15] The comparatively high Huckel energy (0.216β) for the HOMO suggests that heptafulvene should be a more effective donor than simple olefins, a conclusion reached by Kerber and Entholt by interpretation of nmr and infrared data.[13]

The $Cr(CO)_5$ and $Fe(CO)_4$ fragments, both of which fall short of an 18 electron (valence shell) configuration by 2 electrons, possess an acceptor orbital of suitable symmetry for interaction with a σ-donor and a lower lying donor orbital (a degenerate pair for the Cr octahedral fragment) with pi-symmetry.[7] These are shown right for the $Fe(CO)_4$ case, 5 and 6. Thus, when these fragments interact with an olefin ligand, both donation, from the olefin π (to 5) and backdonation (from 6) to the olefin π^* orbital, may be important. The $Fe(CO)_4$ fragment affords the better case for backdonation because the energies of the donor and acceptor frontier orbitals are closer. Rossi and Hoffmann[16] have pointed out that preference of olefin ligands for the $eq_\perp$ conformation in olefin $Fe(CO)_4$ and related ML_4 (olefin) complexes finds its origin in donation from 6 to the olefin π^* orbital.

Acrylonitrile (AN) offers two non-equivalent sites for pi-coordination, the $C=C$ and $C\equiv N$ bonds. Examination of the CNDO/2 coefficients for the acrylonitrile HOMO and LUMO in Table 1 reveals that both donation and backdonation should occur preferentially at the carbon-carbon double bond. The complexes $ANFe(CO)_4$,[17a] $FNCr(CO)_5$,[17b] $(AN)_3M(CO)_3$ where $M = Mo, W$,[17c] and $(TCNE)M(CO)_5$ where $M = Cr, Mo, W$[17d] all exhibit $C=C$ coordination, as do the closely related complexes $CpMn(CO)_2AN$,[17e] $CpCo(PPh_3)FN$,[17f] cobaloxime AN^-,[17g] $(\eta$-arene$)Cr(CO)_2AN$,[17h] $(PPh_3)_2IrX(CO)(ol)$ where $X = Cl, Br$ and $ol = AN$,[17i] FN,[17i] $TCNE$[17j] and $[P(O$-o-$tol)_3]_2NiAN$[17k] ($FN = fumaronitrile,

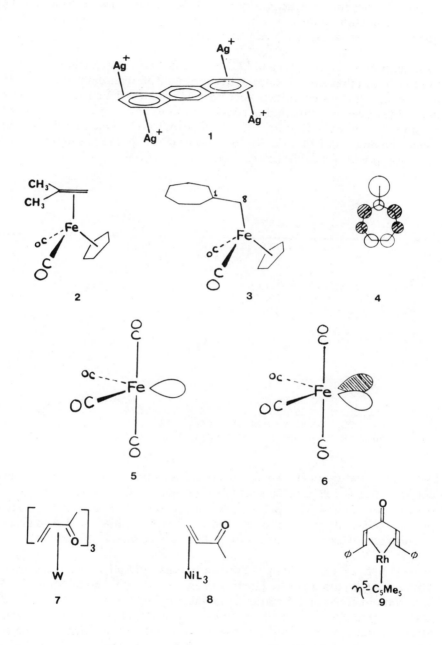

TCNE = tetracyanoethylene, Cp = η^5-cyclopentadienyl). No substantiated example of η^2 coordination to the $C \equiv N$ bond of acrylonitrile exists though similar coordination to saturated nitriles has been demonstrated.[18]

Examination of the HOMO and LUMO coefficients for acrolein in Table I leads to the conclusion that enones too should coordinate preferentially at their $C = C$ bonds in η^2 complexes. η^4- complexes such as cinnamaldehydeFe(CO)$_3$[19] and (3-buten-2-one)$_3$W (7)[20] demonstrate that pi complexation to both the $C = C$ an $C = O$ bonds is possible. Yet only $C = C$ complexation is known for η^2-enone complexes.[18] Examples include complexes of Mn, Fe, Rh, Ni, and Pt (Cf. 8 and 9)[21] as well as complexes of α,β-unsaturated carboxylic acids, esters, and anhydrides.

Table I

CNDO/2 Coefficients for the HOMO and LUMO of Acrylonitrile and Acrolein	
.66 -.54 -.30 .43 C === C --- C ≡≡≡ N LUMO	.59 -.39 -.48 .51 C === C --- C === O LUMO
.60 .49 -.35 -.54 C === C --- C ≡≡≡ N HOMO	.58 .48 -.30 -.58 C === C --- C === O HOMO
Acrylonitrile[a]	Acrolein[b]

a. K. N. Houk and L. L. Munchausen, *J. Amer. Chem. Soc.*, **98**, 937 (1976).
b. K. N. Houk and R. W. Strozier, *ibid.*, **95**, 4096 (1973).

A typical metal fragment which is 4 electrons short of the 18 electron, closed shell, configuration is Fe(CO)$_3$. This octahedral fragment (C_{3v} symmetry) possesses 3 frontier orbitals: a low-lying e pair of e symmetry, one a donor and one an acceptor 10, plus a higher lying a orbital of a symmetry.[7,22] Thus, it is a 4-electron acceptor but will also tend to donate electrons to the ligand from one of its e orbitals.

η^3-Allyl ligands are formally 4-electron donors, and allyl complexes are common. They have been discussed by Hoffmann and Hofmann[23] and by Mingos.[10] The most important interaction is donation from the non-bonding allyl orbital (HOMO of the allyl anion) to the unoccupied e orbital of the metal, 11. η^3-Bonding to ligands which offer alternative sites should occur where the HOMO coefficients are largest for the type of overlap depicted in 11. For example, η^3-benzyl complexes always exhibit 1,2,7-η^3 coordination, e.g., 12[24] as predicted on the basis of HOMO coefficients for benzyl anion, 13[25] and compound 14 possesses the predicted structure.[26] Owing to the fluxional character of these compounds, there is no question that these are the most stable structures.

Of special interest are the fluorenyl complexes, 15 and 16. Stucky, *et al.*[6] have

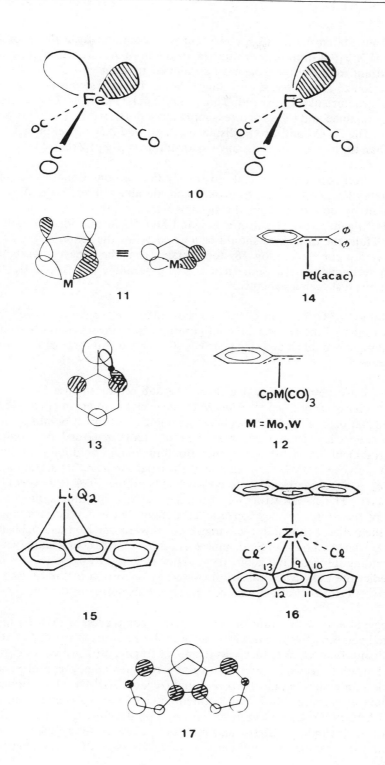

10

11 ≡

14

Pd(acac)

13

CpM(CO)₃

M = Mo, W

12

Li Q₂

15

16

17

pointed out that structure 15 is consistent with donation from the fluorenyl anion HOMO (17)[25] to a lithium p orbital. In 16 the normal tetrahedral bond angles are distorted and one fluorenyl ligand has slipped to what is described as 9,10,13-η^3 coordination to relieve steric crowding.[27] Were this true η^3 coordination, we would expect a structure like 15 instead. Thus, significant Zr-C$_{11}$ and Zr-C$_{12}$ interaction is probably retained and the structure is more aptly described as a strongly distorted η^5 complex. The Zr-C$_{11}$ and Zr-C$_{12}$ distances, 2.801 and 2.807Å are, in fact, only 0.30Å longer than the average Zr-C distance in unstrained $(\eta^5$-Cp)$_2$ZrCl$_2$.[28]

The question of direct metal to exocyclic carbon bonding in cyclobuta-dienylmethylFe(CO)$_3$cations has evoked considerable interest.[29] Bonding should be dominated by donation from the occupied iron e orbital to the LUMO of the C$_4$H$_3$CH$_2^+$ ligand (18). The iron e′ - ligand HOMO interaction (19) will also contribute. Thus, the iron atom should be located below the center of the ring to maximize overlap for interaction 18, too distant from the exocyclic carbon to bond through secondary interactions such as 19. Crystalline [C$_4$H$_3$Fe(CO)$_3$]$_2^+$CPh BF$_4^-$ (20) exhibits just such a structure.[30]

The common η^5-CpFe$^+$ and Cr(CO)$_3$ fragments[22] are typical of those which fall 6 electrons short of the 18 electron configuration. Both possess a low lying e set plus a higher lying a orbital as described for Fe(CO)$_3$ above. All three of these orbitals are unoccupied.

The ferrocenylmethyl cation has been the subject of intense interest and considerable controversy since 1959. As with the cyclobutadienylmethylFe(CO)$_3$ cation the principal point in dispute has been the question of direct bonding between iron and the exocyclic carbon.[31] The parent cation can be regarded as a complex of the CpFe$^+$ fragment and a fulvene ligand. Bonding should be dominated by donation from the fulvene HOMO (π^3)[32] to one of the e pair on iron (21) accompanied by the lesser ($\pi_2 \rightarrow$ e) interaction (22). (see Figure 1) No direct Fe-C (exocyclic) interaction is predicted. The crystal structure of ion 23[33] reveals iron atoms directly below the centers of their respective cyclopentadienyl rings. The exocyclic carbon is displaced out of the planes of the adjacent rings 0.54Å toward one iron and 0.34Å away from the other. The resulting C-Fe distances are 2.69 an 2.87Å, indicating very weak, if any, bonding. Since ligand geometry is assumed as a starting point, this model cannot predict the rich structural detail caused by distortion of the fulvene unit from planarity.[34] Yet it goes clearly and accurately to the essence of the bonding question.

The unsymmetrical structure of the dinuclear complexes 25 (M = Fe or Ru)[35] are predicted correctly by consideration of both donation from the ligand HOMO and to the ligand LUMO. We start with the metal fragment (CO)$_2$Fe^+_a - Fe^-_b(CO)$_3$. Iron a, which needs 6 electrons, interacts to give a cyclopentadienyl complex intermediate 24. This interaction should involve mainly ligand HOMO which possesses finite amplitude at carbons 1 - 4 only.[32] Interaction of the remaining iron(b) with the ligand LUMO (26) should clearly give the 6,7,8-η structure (25) and not the 6,7,13-η alternative. A similar analysis accounts for preference by a dinuclear anthracene iron complex for the symmetrical structure 27[36] over the unsymmetrical 5,6,7-η alternative.

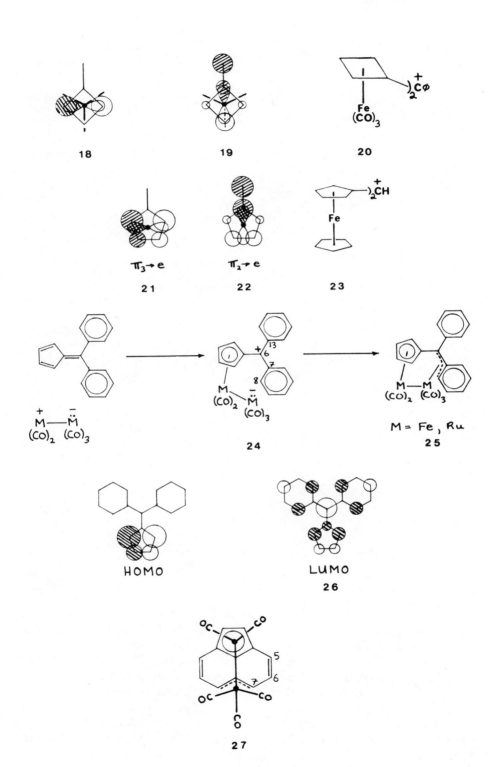

18

19

20

$\pi_3 \rightarrow e$

21

$\pi_2 \rightarrow e$

22

23

24

M = Fe , Ru

25

HOMO

LUMO

26

27

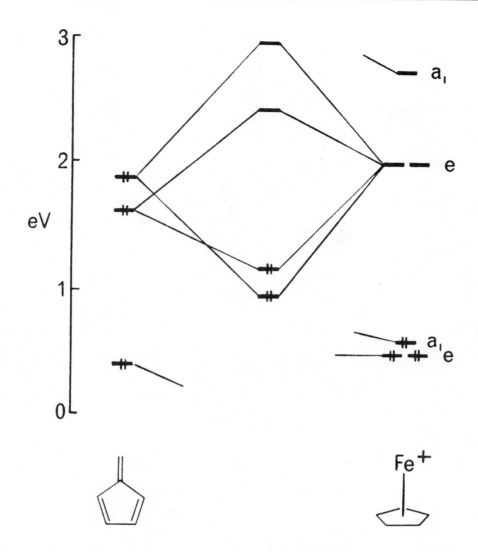

Figure 1. Frontier orbital interactions in ferrocenylmethyl cation.

Molecular Rearrangements
DienylFe(CO)₃ Cations

We have studied a series of these cations (e.g., 28) generated from dienolFe(CO)₃ precursors by treatment with FSO₃H in SO₂ using ¹³C nmr spectroscopy.[37] The ¹³C chemical shifts show that carbons 1,3, and 5 are significantly more shielded than carbons 2 and 4. Simple semiempirical all valence electron MO calculations using the INDO[38] and extended Huckel[37/39] methods predict that these carbons (1,3 and 5) are electron-rich relative to carbons 2 and 4 and suggest strongly that this is the cause of the shielding. Frontier orbital analysis suggests that the strongest bonding interaction should involve donation from an e orbital of the Fe(CO)₃ fragment to the dienyl cation LUMO (29).[39/40] Strong donation to this LUMO could lead to just such an alternating charge distribution.

At low temperatures the cations exhibit non-equivalent low field signals for terminal carbonyl cations. Cations which lack a plane of symmetry, e.g., 28b, generally exhibit 3 signals, one in the 206-208 ppm range and a closely spaced pair in the 197-199 ppm region. The carbonyl region of the ^{13}C spectra for 28a and b recorded over a broad temperature range are shown in figures 2 and 3, respectively. Computer simulation of these spectra (see figures) has led to the conclusion that the site exchange is either an unrestricted pairwise exchange process or a simultaneous exchange equivalent to that produced by $Fe(CO)_3$ rotation. The pairwise process has been ruled out at least in the case of cation 31. Owing to the intrinsic asymmetry of the ion, the two carbonyls can never become equivalent as the result of $Fe(CO)_2PPh_3$ rotation; but pairwise exchange would interconvert them. No broadening of the carbonyl ^{13}C signals for 31 is apparent up to 0° to +10°, above which the sample decomposes.

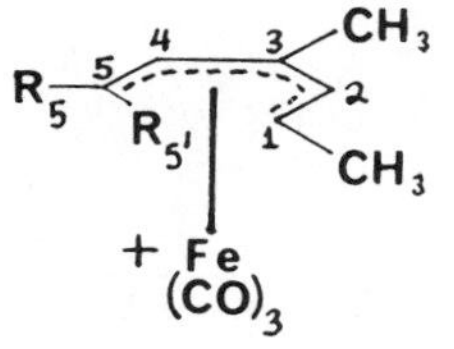

28

a $R_5 = CH_3$, $R_5' = H$

b $R_5 = H$, $R_5' = CH_3$

29

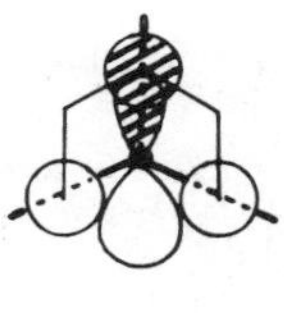

30

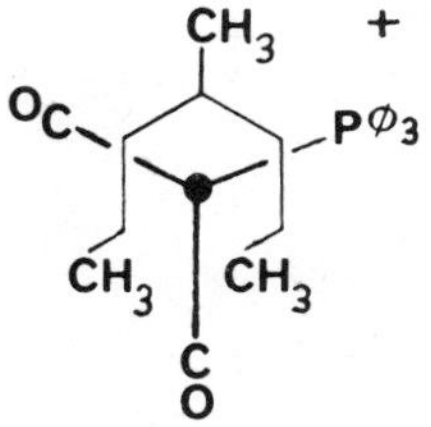

31

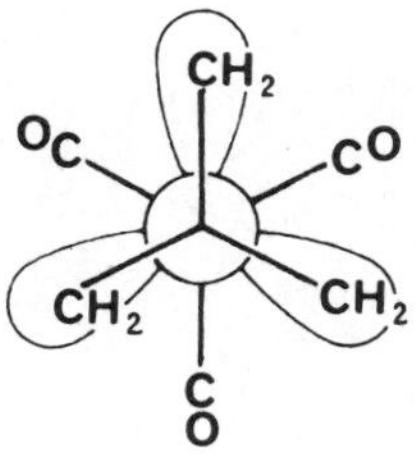

32

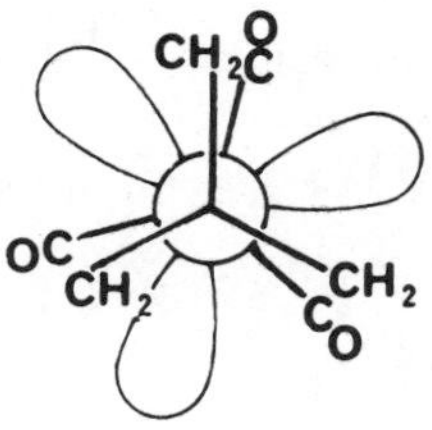

33

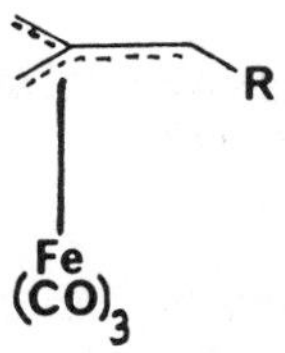

34

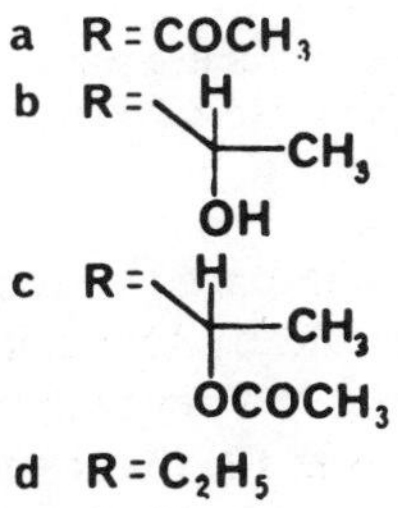

a R = COCH₃

b R =

c R =

d R = C_2H_5

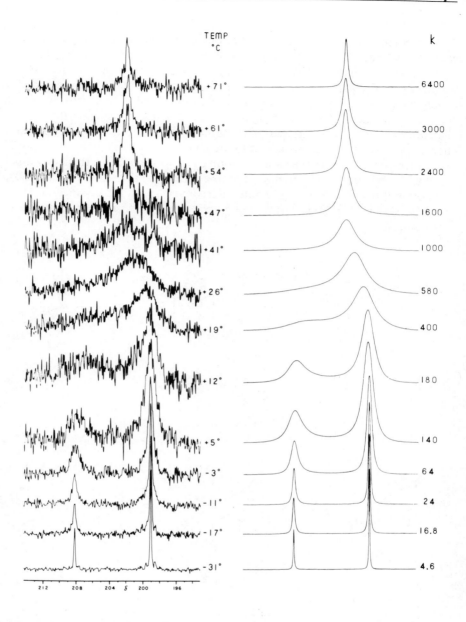

Figure 2. Observed (left) and calculated (right) line shapes for the carbonyl ^{13}C nmr signals of cation 28a in flurosulfonic acid. (From reference 37b, reproduced with permission.)

Barriers to the exchange based on computer lineshape fitting of the experimental spectra are given in Table II. Albright, Hofmann, and Hoffmann have investigated the source of the barrier.[40] They conclude that owing to polarization of the e orbitals of the $Fe(CO)_3$ group (see 10) that overlap for the HOMO-LUMO interaction is better in conformatin 29 than in 30.[41] Their calculated barrier to $Fe(CO)_3$ rotation is in good agreement with our experimental values (Table II).

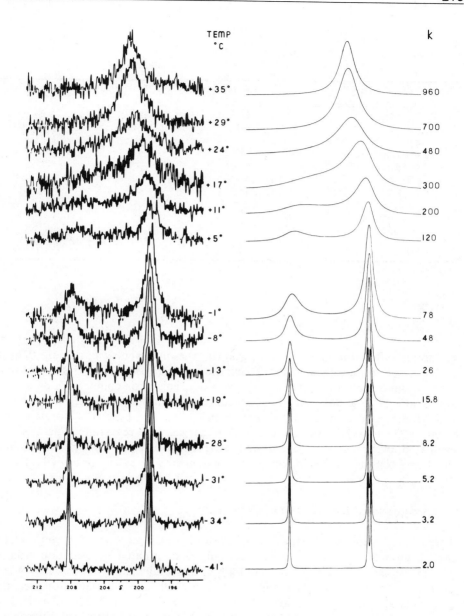

Figure 3. Observed (left) and calculated (right) line shapes for the carbonyl ^{13}C nmr signals of cation 28b in fluorosulfonic acid. (From reference 37b, reproduced with permission.)

TrimethylenemethaneFe(CO)₃

The barriers to $Fe(CO)_3$ rotation in trimethylenemethaneFe(CO)₃ compounds should be significantly larger than those for the dienylFe(CO)₃ cations. Elian and Hoffmann have pointed out that the frontier orbitals of an octahedral $M(CO)_3$ fragment can be viewed as 3 equivalent hybrids which project in the direction of the missing ligands.[7] Interaction of such an $Fe(CO)_3$ unit with a trimethylene methane ligand leads to the conclusion that bonding should be strong in the staggered con-

Table II
Activation Parameters for Carbonyl Site Exchange

Compound	$\Delta G^{\ddagger}$(kcal/mole)	$\Delta H^{\ddagger}_{277.8}$(kcal/mole)	$\Delta S^{\ddagger}_{277.8}$(e.u.)
28a	13.6 ± 0.1^{a}	11.1 ± 0.2	-9.0 ± 0.9
28b	13.6 ± 0.1^{a}	11.4 ± 0.1	-7.7 ± 0.6
Parent dienylFe(CO)$_3$ cation		10.9^{b}	
34a	18 ± 1^{c}		
	18 ± 1^{d}		
34b	17 ± 2^{e}		
Trimethylene-methaneFe(CO)$_3$		20.8^{b}	

a) at 277.8°K
b) Calculated, Ref. 40
c) 332°K
d) 344°K
e) 330°K

formation where the metal hybrids point at the methylene groups (and their p orbitals) (32) but much weaker in the eclipsed conformation (33).[42] Using ^{13}C nmr we have found 3 separate signals in the carbonyl region for compounds 34 a-d.[43] Examination of 34a and 34b over a temperature range revealed successive peak broadening, coalescence, and development of a sharp singlet as the temperature range +27° to +88° was traversed. All charges were reversible. The $\Delta G^{\ddagger}$ values estimated by the coalescence method (Table II) are significantly higher than those of the dienyl complexes and in reasonable agreement with calculations of Hoffman, *et al.*.[40]

Cross-conjugated DienylFe(CO)$_3$ Cations[5/44]

These cation were conceived as a severe test of the simplest, one-interaction frontier orbital model.[5/44] If metal-carbon bonding is dominated by donation from an e HOMO on iron to the non-bonding LUMO of the dienyl cation (35), maximization of overlap for this interaction will lead to the η^3-allyl structure 36. However, the second orbital of the e pair on iron (the LUMO of Fe(CO)$_3$), fails to find a dienyl partner with which it can interact effectively in this geometry. The consequence is that the allyl structure has a 16 electron, coordinatively unsaturated, iron atom. Alternatively, two coordinatively saturated structures, 37 and 38, can be envisaged for which η^4-diene[45] and η^4-trimethylenemethane[46] complexes offer structural precedent. Thus, existence of structure 36 would be impressive evidence for domination of bonding by a single frontier orbital interaction. The characteristics which should distinguish 36 from 37 and 38 are coordinative unsaturation and a low barrier to rotation about the C$_2$-C$_3$ bond.

Evidence for C$_2$-C$_3$ rotation was first provided by isolation of the rearranged methyl ether 40 by low-temperature methanolysis of trimethyl cations.[5] However, no

35

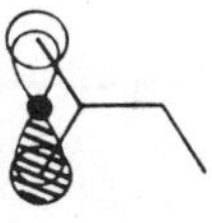

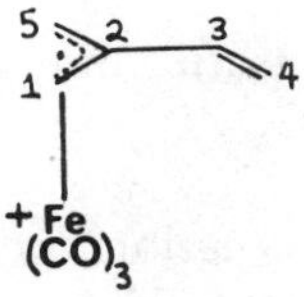

36

$+Fe(CO)_3$

37

$Fe(CO)_3$

38

$Fe(CO)_3$

39 $\xrightarrow[\text{or}]{\text{FSO}_3\text{H}/\text{SO}_2\ -65°}$ Cations $\xrightarrow{\text{CH}_3\text{OH}\ -78°}$ **40**

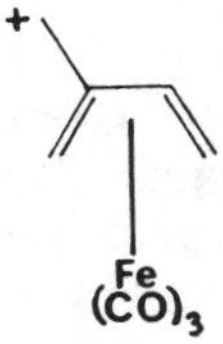

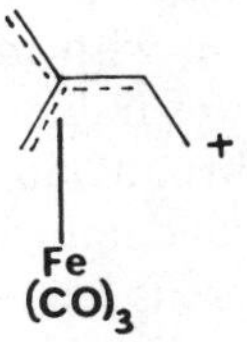

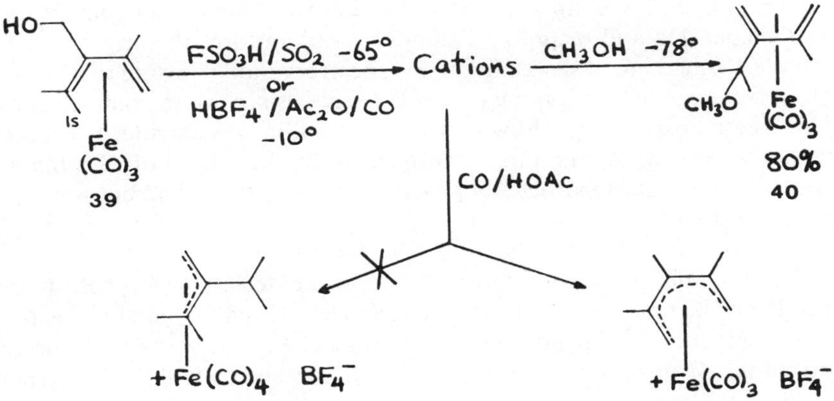

$HBF_4 / Ac_2O/CO$
$-10°$

CH_3O $Fe(CO)_3$

80%

CO/HOAc

$+ Fe(CO)_4\ BF_4^-$

$+ Fe(CO)_3\ BF_4^-$

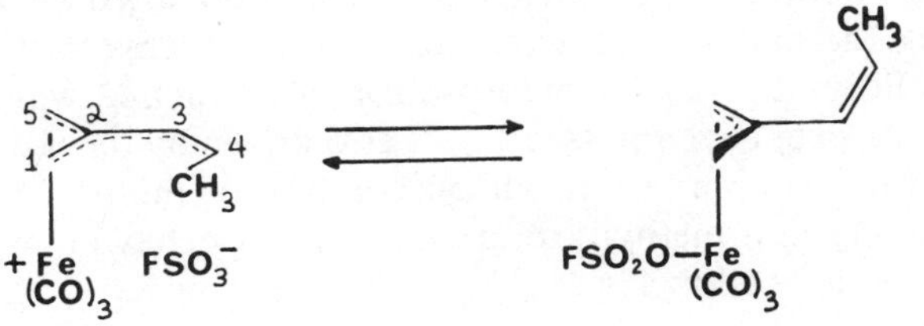

41 $+Fe(CO)_3$ FSO_3^-

42 $FSO_2O-Fe(CO)_3$ CH_3

43 $+Fe(CO)_3$

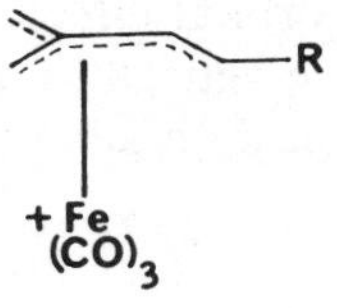

a $R = CH_3$

b $R = H$

evidence for coordinative unsaturation of the intermediate cations could be obtained by generation under 1 atm CO.[47]

To avoid irreversible loss of cation via rearrangements at higher temperatures, we have prepared cations which lack the 1s-methyl (see 39). Cations 41 and 43 have been thoroughly characterized by quenching and by variable temperature ^1H and ^{13}C nmr.[44] They exhibit no tendency to coordinate with added CO. The anti methyl cation, 41, exists in equilibrium with the fluorosulfonate adduct 42. ^1H and ^{13}C nmr spectra of 42 are temperature independent over the range 0° to -75°C. Cation 41 however exhibits selective broadening of the signals for C_1, C_5 and their attached hydrogens. For example, broadening of the ^1H signals becoms apparent at -48°. At -27° all coupling fine structure is lost while at -2° the 4 signals have almost vanished into the baseline. Estimation of ΔG^+ from lineshapes[48] at -31°, -17°, and -2° all gave values in the range 12.8 ±0.4 kcal/mole. Further warming causes conversion of 41 to its syn isomer 43a and recooling from -2 to -60° restores the original spectrum. Comparable changes are observed in the ^{13}C spectrum though owing to a larger (18.9 ppm) slow exchange separation the exchanging signals are not so completely averaged at -2°. These observations show that C_2-C_3 rotation is occurring. It is important to note that when a 41/42 mixture is warmed to -2° the signals of 42 remain sharp and distinct. Thus, rotation does *not* proceed via formation of 42 but is an intrinsic property of cation 41.

In contrast to 41, the syn-methyl cation 43a forms no detectable fluorosulfonate adduct. Its ^1H an ^{13}C nmr spectra show non-equivalent C_1 and C_5 methylene units and exhibit no temperature dependence in the range -10 to -70°C. The occurrence of C_2-C_3 rotation, albeit slow on the nmr time scale, has been established by generation of 43a-5-d_1, but clearly the barrier to rotation is substantially higher than that in 41. Rapid C_2-C_3 rotation in anti methyl cation 41 is probably driven by relief of steric strain involving the anti methyl group as the cis-propenyl unit rotates out of the $C_1C_2C_5$ plane. Formation of the fluorosulfonate adduct 42 which allows a similar rotation of the cis-propenyl unit (see 42), is also driven by relief of steric strain. This interpretation finds support from our observation of the parent cation 43b which lacks the offending anti-methyl group and exhibits behavior which is identical to that of the syn-methyl cation 43a.[44]

In the absence of the anti-methyl steric effect it is clear that cross-conjugated dienyl cations do not behave as coordinatively unsaturated species and possess barriers to C_2-C_3 rotation substantially larger than 13 kcal/mole. Thus, the η^3-allyl structure predicted on the basis of one dominant frontier orbital interaction is not the structure of lowest energy. The η^3- allyl structure is energetically accessible however, for it must be just such a structure which is the transition state (or intermediate) for C_2-C_3 rotation.

Summary

These results serve to define some limits for application of a simple frontier orbital model for metal carbon bonding. As a number of the examples discussed earlier illustrate, bonding can be approximated as one or two frontier orbital charge transfer interactions with success when the geometrical consequences of this analysis do not involve sacrifice of other important interactions. But in a case like tricarbonyl (cross-conjugated dienyl)iron cations, in which maximization of overlap for the most important frontier orbital interaction leads to sacrifice of other frontier orbital interactions and to sacrifice of a filled metal valence shell, the simple model fails. Like any simplified approximate theory, this one must be applied with due regard to the approximations being made. With this caveat the frontier orbital model provides a valid and fruitful way for the practicing organometallic chemist to form a first picture of bonding and structure in a pi-complex.

Acknowledgements

My coworkers have made vital contributions to the work described here and have been a constant source of stimulation. The most important have been Dr. Benedict R. Bonazza, Prof. Paul A. Dobosh, Mr. Douglas G. Gresham, Dr. David J. Kowalski, Dr. Elaine Magyar, and Dr. Gary Scholes. Prof. C. Hackett Bushweller provided his modified version of the Binsch DNMR-3 nmr line shape program. Financial support from the donors of the Petroleum Research Fund, administered by the American Chemical Society, the National Science Foundation and the University of Massachusetts, Amherst are gratefully acknowledged.

References and Footnotes

1. M. J. S. Dewar, *Bull Soc. Chim. Fr.*, C71 (1951).
2. J. Chatt and L. A. Duncanson, *J. Chem. Soc.*, 2939 (1953).
3. Its exclusive use for presentation of bonding in general monographs on organotransition metal chemitry serves as a measure of its acceptance. *Cf.* (a) M. L. H. Green, "The Transition Elements Vol. 2 of Organometallic Compounds," Methuen, London, 1968 (b) R. B. King, "Transition Metal Organometallic Chemistry, An Introduction,"Academic Press, New York, N.Y., 1969 (c) M. Tstusui, M. N. Levy, A. Nakamura, M. Ichikawa, and K. Mori, "Introduction to Metal π-Complex Chemistry," Plenum Press, New York, N.Y., 1970 (d) P. L. Pauson, "Organometallic Chemistry,"St. Martin's Press, New York, N.Y., 1967 (e) E. O. Fischer and H. Werner, "Metal π-Complexes,"Vol. 1, Elsevier, Amsterdam, 1966 (f) R. F. Heck, "Organotransition Metal Chemistry, A Mechanistic Approach," Academic Press, New York, N.Y., 1974.
4. *Cf.* K. Fukiu in "Molecular Orbitals in Chemistry, Physics, and Biology," P.-O. Lowdin and B. Pullman, ed., Academic Press, New York, N.Y., 1964, p. 513 and R. Hoffmann, *Accts. of Chem. Res.*, *4*, 1 (1971)
5. B. R. Bonazza and C. P. Lillya, *J. Amer. Chem. Soc.*, *96*, 2298 (1974).
6. Explicit use of orbital energy criteria within a Dewar-Chatt-Duncanson framework is exemplified by (a) E. O. Greaves, C. J. L. Lock, and P. M. Maitlis, *Can. J. Chem.*, *46*, 3879 (1968), (b) P. M. Maitlis, "The Organic Chemistry of Palladium," Vol. 1, Academic Press, New York, N.Y., 1971, (c) J. J. Brooks, W. Rhine, and G. D. Stucky, *J. Amer. Chem. Soc.*, *94*, 7339 (1972), and (d) W. E. Rhine and G. D. Stucky, *ibid.*, *97*, 737 (1975).
7. M. Elian and R. Hoffman, *Inorg. Chem.*, *14*, 1058 (1975) and related (mostly subsequent) papers.
8. Construction of extensive MO energy level schemes for metal pi-complexes has always involved consideration of the relative energies of interacting orbitals, *Cf* F. A. Cotton, "Chemical Applications of Group Theory," 2nd ed., Wiley, New York, New York, 1963.
9. Mingos has stated this need clearly, D. M. P. Mingos in *Advances in Organometal. Chem.*, *15*, 1(1977).
10. For a non-perturbation theory approach to this problem, see Mingos' topological Huckel model, D. M. P. Mingos, *J. Chem. Soc., Dalton*, 20, 26, and 31 (1977).
11. (a) E. A. H. Griffith and E. L. Amma, *ibid.*, *96*, 5407 (1974). (b) P. F. Rodesiler, E. A. Hall Griffith and E. A. Amma, *Ibid.*, *94*, 761 (1972). (c) P. F. Rodesiler and E. A. Amma, *Inorg. Chem.*, *11*, 388

(1972). describe acenaphthalene-AG⁺ complex which does not have the structure predicted for the mono-coordinated acenaphthalene but exhibits a structure in which each acenaphthalene has 2 coordinated Ag⁺ ions.

12. A. Cutler, D. Entholt, P. Lennon, K. Nicholas, D. F. Marten, M. Madhavarao, S. Raghu, A. Rosan, and M. Rosenblum, *J. Amer. Chem. Soc.* **97**, 3149 (1975).

13. D. J. Entholt, G. F. Emerson, and R. C. Kerber, *ibid.*, *91*, 7547 (1969) and R. C. Kerber and D. J. Entholt, *ibid.*, *95*, 2927 (1973).

14. M. J. S. Dewar and N. Trinajstic, *Coll. Czech. Chem. Comm.*, *35*, 3484 (1970).

15. M. R. Churchill and J. P. Fennessey, *Chem. Comm.*, 1056 (1970).

16. A. R. Rossi and R. Hoffmann, *Inorg. Chem*, *14*, 365 (1975).

17. (a) A. R. Luxmoore and M. R. Truter, *Acta. Cryst.* *15*, 1117 (1962).

(b) S. N. Avakyan and R. A. Karapetyan, *Izv. Akad. Nauk. S.S.R. Khim. Nauki*, *18*, 15 (1965).

(c) D. P. Tate, J. M. Augl, and A. Buss, *Inorg. Chem.*, *2*, 427 (1963) and B. L. Ross, J. G. Grasselli, W. M. Ritchey, and H. D. Kaesz, *ibid.*, *2*, 1023 (1963).

(d) M. Heberhold, *Angew. Chem. Int. Ed. Eng.*, *7*, 305 (1968).

(e) M. L. Ziegler and R. K. Sheline, *Inorg. Chem.*, *4*, 1230 (1965).

(f) H. Yamazaki and N. Hagihara, *J. Organometal. Chem.*, *7*, P22 (1967) and *ibid.*, *21*, 431 (1970).

(g) G. N. Schrauzer, J. H. Weber, and T. J. Beckham, *J. Amer. Chem. Soc.*, *92*, 7078 (1970).

(h) J.-F. Guttenberger, W. Strohmeier, *Chem. Ber.*, *100*, 2807 (1967).

(i) J. A. McGinnety and J. A. Ibers, *Chem. Comm.*, 235 (1968).

(j) W. H. Baddley, *J. Amer. Chem. Soc.*, *90*, 3705 (1968).

(k) K. Sumitani and K. Tamao, *J. Organometal. Chem.*, *50*, 311 (1973).

18. *Cf.* (a) S. Zecchini, G. Zotti, and G. Pilloni, *Inorg. Chim. Acta*, *33*, L117 (1979).

(b) M. Heberhold, "Metal π-Complexes," Vol. II, part 1, Elsevier, Amsterdam, 1972.

19. K. Stark, J. E. Lancaster, H. D. Murdoch, and E. Weiss, *Z. Naturforsch.*, *196*, 284 (1964).

20. R. B. King and A. Fronzaglia, *Chem. Comm.*, 274 (1966) and *Inorg. Chem.*, *5*, 1837 (1966).

21. Mn - C. R. Jablonski and M. Heberhold, *Chem. Ber.*, *102*, 767, 778 (1969). Fe - E. Weiss, K. Stark, J. E. Lancaster, and H. D. Murdoch, *Helv. Chim. Acta*, *46*, 288 (1963) and K. Stark, J. E. Lancaster, H. D. Murdoch, and E. Weiss, *Z. Naturforsch.*, *B*, *19*, 284 (1964). A. N. Nesmeyanov, K. Ahmed, L. V. Rubin, M. I. Rubinskaya, and Yu. A. Ustynyuk, *Dokl. Akad. Nauk. S.S.S.R.*, *175*, 1070 (1967) and *J. Organometal. Chem.*, *10*, 121 (1967). V. G. Andrianov and Yu. T. Struchkov, *Chem. Comm.*, 1590 (1968). Ni - C. A. Tolman, *J. Amer. Chem. Soc.*, *96*, 2780 (1974). Pt - S. Cenini, R. Ugo, F. Bonati, and G. LaMonica, *Inorg. and Nucl. Chem. Lett.*, *3*, 191 (1967). Rh - J. A. Ibers, *J. Organometal. Chem.*, *73*, 389 (1974).

22. M. M. L. Chen, D. M. P. Mingos, M. Elian, and R. Hoffmann, *Inorg. Chem.*, *15*, 1148 (1976).

23. R. Hoffmann and Peter Hofmann, *J. Amer. Chem. Soc.*, *98*, 598 (1976).

24. R. B. King and A. Fronzaglia, *ibid.*, *88*, 709 (1966); M. D. La Prade and F. A. Cotton, *ibid.*, *90*, 5418 (1968); and F. A. Cotton and J. J. Marks, *ibid.*, *91*, 1339 (1969).

25. C. A. Coulson and A. Streitwieser, Jr. with help from M. A. Poole and J. I. Brauman, "Dictionary of π-Electron Calculations," Pergamon Press, London, 1965.

26. P. M. Maitlis, A. Sonoda, and B. E. Mann, *Chem. Comm.*, 108 (1975).

27. C. Kowala, P. C. Wailes, H. Weingold, and J. A. Wunderlich, *Chem. Commun.*, 993 (1974) and C. Kowala and J. A. Wunderlich, *Acta Cryst. B*, *32*, 820 (1976).

28. K. Prout, T. S. Cameron, R. A. Forder, S. R. Critchley, B. Denton, and G. V. Rees, *Acta Cryst. B30*, 2290 (1974).

29. J. D. Fitzpatrick, L. Watts, and R. Pettit, *Tetrahedron Lett.*, 1299 (1966).

30. R. E. Davis, H. D. Simpson, N. Grice, and R. Pettit, *J. Amer. Chem. Soc.*, *93*, 6688 (1971).

31. In support of direct Fe-C bonding - E. A. Hill and J. H. Richards, *J. Amer. Chem. Soc.*, *81*, 3483 (1959) *83*, 3840, 4216 (1961) and M. Cais, *Organometal. Chem. Rev.*, *1*, 435 (1966). Opposed - J. C. Ware and T. G. Traylor, *Tetrahedron Lett.*, *18*, 1295 (1963) and T. G. Traylor and J. C. Ware, *J. Amer. Chem. Soc.*, *89*, 2304 (1967).

32. A. Streitwieser, Jr. and J. I. Brauman, "Supplemental Tables of Molecular Orbital Calculations," Vol. 1, Pergamon Press, London, 1965.

33. S. Lupan, M. Kapon, M. Cais, and F. H. Herbstein, *Angew. Chem. Int. Ed. Eng.*, *11*, 1025 (1972) and M. Cais, S. Dani, F. H. Herbstein and M. Kapon, *J. Am.Chem. Soc.*, *100*, 5554 (1978). See also R. L. Sime and R. J. Sime, *ibid.*, *96*, 892 (1974).

34. See the extended Huckel treatment of this cation by R. Gleiter and R. Seeger, *Helv. Chim. Acta*, *54*, 1217 (1971).

35. U. Behrens and E. Weiss, *J. Organometal. Chem.*, *73*, C67 (1974); *ibid.*, *96*, 399, 435 (1975).
36. M. R. Churchill and J. Wormald, *Chem. Comm.*, 1597 (1968); *Inorg. Chem.*, *9*, 3239 (1970).
37. (a) P. A. Dobosh, D. G. Gresham, C. P. Lillya, and E. S. Magyar, *Inorg. Chem.*, *14*, 2311 (1976).
(b) P. A. Dobosh, D. G. Gresham, D. J. Kowalski, C. P. Lillya, and E. S. Magyar, *ibid.*, *17*, 1775 (1978).
38. D. W. Clark, M. Monshi, and L. A. P. Kane Maguire, *J. Organometal Chem.*, 107, C40 (1976).
39. R. Hoffmann and P. Hofmann, *J. Am. Chem. Soc.*, *98*, 598 (1976).
40. T. A. Albright, R. Hoffmann and P. Hofmann, *ibid.*, *99*, 7546 (1977).
41. They have also identified an added repulsive interaction which destabilizes 30.
42. See also referenc 40.
43. E. S. Magyar and C. P. Lillya, *J. Organometal. Chem.*, *116*, 99 (1976).
44. (a) B. R. Bonazza, C. P. Lillya, E. S. Magyar, and G. Scholes, Abstracts of Papers, 172nd ACS National Meeting, San Francisco, Aug. 29 - Sept. 3, 1976, INOR. 109
(b) B. R. Bonazza, C. P. Lillya, E. S. Magyar, and G. Scholes, *J. Am. Chem. Soc.*, *101*, 0000 (1979).
(c) P. A. Dobosh, C. P. Lillya, E. S. Magyar, and G. Scholes, submitted for publication.
45. B. F. Hallam and P. L. Pauson, *J. Chem. Soc.*, 642, 646 (1958) and O. S. Mills and G. Robinson *Acta Cryst.*, *16*, 758 (1963).
46. G. F. Emerson, K. Ehrlich, W. P. Giering, and P. C. Lauterbur, *J. Amer. Chem. Soc.*, *88*, 3172 (1966), and K. Ehrlich and G. F. Emerson, *ibid.*, *94*, 2464 (1972).
47. G. Scholes, unpublished work.
48. *Cf.* F. A. Bovey, "Nuclear Magnetic Resonance Spectroscopy," Academic Press, New York, New York, 1969, pp. 183-188 and E. D. Becker, "High Resolution Nuclear Magnetic Resonance," Academic Press, New York, New York, pp. 214-219.

Spectroscopic and X-ray Crystallographic Structural Studies of Mo(IV), Mo(V), and Mo(VI) Complexes with Linear Thiolate Ligands

J. Hyde, L. Magin, P. Vella and J. Zubieta
Institute of Biomolecular Stereodynamics
Department of Chemistry
Center for Biological Macromolecules
State University of New York at Albany
Albany, New York 12222

The coordination chemistry of molybdenum in its higher oxidation states (IV to VI) has received a great deal of attention, particularly because of the possible relationship of these species to the redox-active molybdoenzymes.[1,2] In particular, the iron-molybdenum protein of nitrogenase from *Azotobacter vinelandii*,[3] *Clostridium pasteurianum*,[4] and *Klebsiella pneumoniae*,[5] and xanthine oxidase[6] all contain two molybdenum atoms per mole of enzyme. The proposal has been made that the two molybdenum atoms are at the active site and function in a concerted fashion during catalysis.[7] Although recent EXAFS studies[8] on lyophylized samples of nitrogenase provide the first direct evidence as to the nature of the molybdenum-binding site and suggest the absence of the nearly ubiquitous oxo-coordination, the extensive model studies of Schrauzer[9] clearly implicate oxo-coordination in synthetic catalytic systems. Although the nature of the detailed geometry at molybdenum remains problematical, cysteine-sulfur coordination is a common theme of both EXAFS and model studies. In the case of xanthine oxidase, oxo-coordination has been established,[8] and a Mo(V)-cysteinato-complex has been proposed for the metal site.

The importance of molybdenum in these redox-active proteins and the speculation that dimeric species involving sulfur coordination may be relevant to the active site geometry[1,7] have prompted a number of studies on the chemical[8,10,11] electrochemical[12] and structural aspects[13-34] of dimeric and monomeric molybdenum complexes with sulfur-containing ligands.

The structural chemistry of molybdenum in the higher oxidation states is dominated by oxo- or sulfido-coordination,[2,35] with fairly regular coordination polyhedra. A common feature of these complexes is the presence of 1,1 dithioacid ligands, providing an extensively delocalized four-membered chelate ring on coordination to the metal. It is clear that this type of sulfur coordination is quite distinct from that provided by either cysteinato- or methionato-sulfur donors available for coordination in enzyme systems.

227

X = CNR$_2$, OR, SR, PR$_2$, C=CR$_2$.

1,1 dithioacid-metal chelate

As part of our general studies of the chemistry of higher oxidation state complexes of molybdenum with sulfur-donor ligands, we have sought to extend the investigations to the interactions of molybdenum with facultative and fully-saturated thiolate ligands. We initiated our studies with the preparation and structural characterization of complexes containing potentially bidentate, tridentate, and quadridentate thiolate ligands.

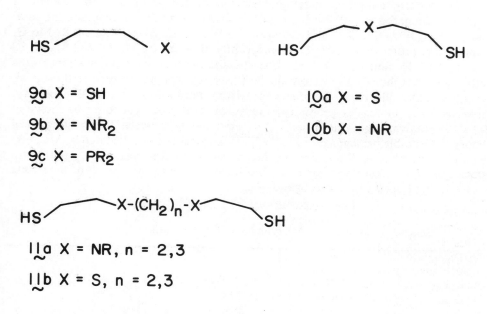

$\underset{\sim}{9}a$ X = SH

$\underset{\sim}{9}b$ X = NR$_2$

$\underset{\sim}{9}c$ X = PR$_2$

$\underset{\sim}{10}a$ X = S

$\underset{\sim}{10}b$ X = NR

$\underset{\sim}{11}a$ X = NR, n = 2,3

$\underset{\sim}{11}b$ X = S, n = 2,3

Complexes with Bidentate Thiolate Ligands

Reaction of a number of starting materials, $[MoO_4]^{2-}$, $[MoS_4]^{2-}$, and $[MoO_2(acac)_2]$, with ethanedithiol, 9a, yielded products of dubious analytic purity. The major products were disulfide and molybdenum oxides from the redox reaction:

$$Mo(VI) + HSC_2H_4SH \rightarrow Mo(V) + \tfrac{1}{2} HSC_2H_4SSC_2H_4SH$$

With $[MoCl_5]$ as precursor a 1 percent yield of the complex $[Mo_2S_4(SC_2H_4S)_2]^{2-}$ was obtained. This species had been previously shown to possess a structure of the type 7a.[30]

On the other hand, 1,2 aminoethanethiol, 9b, reacts readily with Mo(VI) starting materials to give the complex $[MoO_2(SC_2H_4NH_2)_2]$, which displays infrared bands, 899 and 875 cm^{-1}, consistent with a *cis*-dioxo structure 3. Reaction of this species with PPh$_3$ yields a purple complex characterized as $[Mo_2O_3(SC_2H_4NH_2)_4]$ of common structural type 4. The complexes display no unusual reactivity patterns or electrochemical behavior.

The ligand $HSC_2H_4PPh_2$, 9c, was chosen in order to effect the stabilization of the Mo(IV) oxidation state, a recurrent theme in model studies and generally a useful intermediate in further reactions with small donor molecules.[9/37/38] Reaction of two equivalents of 9c with $[MoO_2(acac)_2]$ gave a bright green complex

[MoO(SC$_2$H$_4$PPh$_2$)$_2$], (I), a relatively rare example of a stable, monomeric Mo(IV) complex which additionally enjoys formal coordinative unsaturation.

The crystal structure of (I), Figure 1, (Table I) consists of monomeric molecules [MoO(SC$_2$H$_4$PPh$_2$)$_2$] of geometry intermediate between trigonal bipyramidal and tetragonal pyramidal, exhibiting approximate C$_{2v}$ local symmetry.[56] The distortions from regular trigonal bipyramidal geometry are evident in the P1-Mo-P2 angle of 154.7(1)° compared to 180° for the regular polyhedron and in the angles in the basal plane, 122.3(3)°, 117.3(3)°, and 120.4(3)° for S1-Mo-S2, S1-Mo-O, and S2-Mo,O, respectively, rather than 120°. The significantly different P1-Mo-P2 and S1-Mo-S2 angles (154.7(1)° and 122.3(1)°, respectively) and the concomitant deviations of the sulfur and phosphorus atoms from the best least squares planes through these positions, ±0.30Å, illustrate the distortion from tetragonal pyramidal geometry. Analysis of the molecular geometry in terms of ideal polytopal forms as shown in Table II confirms that the structure is intermediate between the limiting ideal shapes.[39] The table presents not only the ideal dihedral angles calculated by Mueterrties and Guggenberger,[39] but also the interplanar angles for idealized C$_{2v}$ polyhedra which take into account the bond length distortion inherent as a result of the strong π-substituent effects of the oxo-ligand.

Table I
Selected Bond Lengths and Angles for MoO[SC$_2$H$_4$P(C$_6$H$_5$)$_2$]$_2$

a) Bond Lengths

Mo-P1	2.481(4)	P2-C15	1.85(1)
Mo-P2	2.486(3)	P2-C21	1.88(1)
Mo-S1	2.372(4)	P2-C27	184(1)
Mo-S2	2.348(4)	S2-C28	187(1)
Mo-O	1.733(9)	C-C av, ring I	1.43(2)
P1-C1	1.85(2)	C-C av, ring II	1.42(3)
P1-C7	1.83(1)	C-C av, ring III	1.43(2)
P1-C13	1.91(2)	C-C av, ring IV	1.42(2)
S1-C14	1.91(2)	C13-C14	1.54(2)

b) Bond Angles

P1-Mo-P2	154.7(1)	C15-P2-Mo	121.9(4)
S1-Mo-S2	122.3(1)	C21-P2-Mo	114.8(5)
O-Mo-P1	101.6(3)	C27-P2-Mo	105.3(4)
O-Mo-P2	103.6(3)	C14-S1-Mo	110.6(5)
O.Mo:S1	117.3(3)	C28-S2-Mo	109.1(5)
O-Mo-S2	120.4(3)	P1-C13-C14	103(1)
P1-Mo-S1	81.9(1)	C13-C14-S1	109(1)
P1-Mo-S2	87.4(1)	P2-C27-C28	106(1)
P2-Mo-S1	84.6(1)	C27-C28-S2	110(1)
P2-Mo-S2	81.9(1)	C-C-C, ring I av	120(2)
C1-P1-Mo	123.5(5)	C-C-C, ring II av	120(2)
C7-P1-Mo	112.1(5)	C-C-C, ring III av	120(2)
C13-P1-Mo	107.7(5)	C-C-C, ring IV av	120(1)

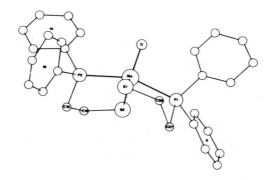

Figure 1. Perspective view of the structure of [MoO(SC$_2$H$_4$PPh$_2$)$_2$], showing the atom labelling scheme.

The distorted geometry found for MoO[SC$_2$H$_4$P(C$_6$H$_5$)$_2$]$_2$ appears to be atypical for five-coordinate molybdenum complexes. The monomeric Mo(IV) species [MoO(S$_2$CN-(C$_3$H$_7$)$_2$)$_2$][13] and [MoO(h^2-S$_2$CSC$_3$H$_7$)(h^3-S$_2$CSC$_3$H$_7$)][37] exhibit local C$_{4v}$ symmetry, i.e., tetragonal pyramidal geometry; the monomeric Mo(V) complex [MoOCl$_3$(SPPh$_3$)][40] also displays approximate tetragonal pyramidal geometry. Square pyramidal molybdenum environments with apical oxo-groups are also found in the dimeric molecules [Mo$_2$O(S$_2$CNEt$_2$)$_2$(C$_6$H$_5$CON$_2$)$_2$]CH$_2$Cl$_2$,[31]

Table II

Comparison of Dihedral Angles of [MoO(SC$_2$H$_4$PPh$_2$)$_2$] and [Mo$_2$O$_3$(SCH$_2$CH$_2$SCH$_2$CH$_2$S)$_2$]With Ideal Polytopal Forms

Complex	Dihedral Angle[a]	
	Type e[d]	Type a
Ideal TBP, D$_{3h}$	53.1	101.5, 101.5
	53.1, 53.1	101.5, 101.5, 101.5, 101.5
Ideal MoOl'L''L$_2$[b]	58.7	92.2, 97.2
	49.4, 49.4	101.4, 101.4, 108.4, 108.4
Mo$_2$O$_3$(S$_3$)$_2$, (II)	29.4	80.9, 85.2
	62.7, 63.8	109.1, 110.8, 114.4, 114.9
MoO(SC$_2$H$_4$PR$_2$)$_2$, (I)	27.7	88.5, 90.0
	57.2, 57.2	111.5, 111.8, 112.8, 113.8
Ideal MoOL'L''L$_2$[c]	0.0	55.4, 59.6
	65.5, 65.5	131.7, 131.7, 135.4, 135.4
Ideal Sq. Pyr.	0.0	75.6, 75.7
	72.7, 72.7	119.8, 119.8, 119.8, 119.8

[a] Angles are defined in ref. 39.

[b] The calculation assumed a MoO distance of 1.70Å and Mo-L distances of 2.42Å. The oxo-group occupied a position on the equatorial plane.

[c] The distances used for the idealized C$_{2v}$ symmetry were employed. The oxo-group occupied the apical position.

[d] Shape determining angles.

$[Mo_2O(S_2CNEt_2)_2(ClC_6H_4CSN_2)_2]CHCl_3$,[31] $[Mo_2O_4(S_2CNEt_2)_2]$[19], and $[Mo_2O_2S_2(S_2C_2(CN)_2)_2]$.[2-24] Although trigonal bipyramidal geometry is not common for molybdenum complexes, distorted trigonal bipyramidal geometry has been observed for $[Mo_2O_4(cysteine\ ethyl\text{-}ester)_2]$[18] and $[Mo_2O_2S_2(cysteine\ methyl\text{-}ester)_2]$,[21] where the terminal oxo-group, the cysteine sulfur, and a bridging oxo or sulfido group define the equatorial plane.

The distortions from regular polytopal forms exhibited by $MoO[SC_2H_4P(C_6H_5)_2]$ are similar to those previously observed for $ReNCl_2[P(C_6H_5)_3]_2$,[41] where the bulky phosphine ligands are in approximately apical positions and the π-substituent nitrido-group in the equatorial plane. Since equatorial π-bonding has been shown to be stronger than axial for the trigonal bipyramid,[42] the approximately equatorial location of the π-donor ligands in these distorted molecules may be anticipated as the substitutional preference.

Although tetragonal pyramidal geometry is more commonly observed for d^1 and d^2 complexes where the apical interaction between the π-donor oxo-group and the metal d_{xz} and d_{yz} orbitals may be maximized, while the destabilization of the d-electrons is minimized, it is apparent that the conformational preference is due not only to electronic effects but to steric requirements of the coordinated ligands.[43] Thus, the bulky trimethylamine groups in $[VOCl_2(NMe_3)_2]$[44] occupy apical positions of a distorted trigonal bipyramid rather than basal positions in the tetragonal pyramidal geometry more commonly observed for vanadyl complexes.[45/46] Similarly, it appears that nonbonding interactions play an important role in determining the conformational preference of $MoO[SC_2H_4P(C_6H_5)_2]_2$. There are a number of short non-bonding contacts, specifically, S-C(ring) distances of 3.4-3.5Å. Although these distances are quite short and suggest steric interaction, any further deviation from trigonal bipyramidal toward square pyramidal geometry could only make the S-C ring distances shorter and the interactions more unfavorable.

The Mo-O distance of 1.733(9)Å is significantly longer than the distances of 1.664(8) and 1.66(1)Å observed for the Mo(IV) complexes $[MoO(S_2CNPr_2)_2]$[13] and $[MoO(S_2CSR_2)_2]$,[37] respectively. It is not obvious whether this difference is electronic in nature or results from the large steric requirements of the donor ligands. The average Mo-S distance, 2.360(4)Å, is shorter than the distances commonly observed for Mo-S in dithiocarbamate complexes which range from 2.45 to 2.55Å[35]. These longer distances are due in part to structural *trans*-effects produced by *trans* terminal oxo- or bridging oxo-groups, to differences in the hybridization at the sulfur, and to the steric requirements of the four-membered chelate rings in these complexes. Four-membered rings display decreased angles at the metal atom resulting in poorer metal-ligand overlap and consequently longer bond lengths.[47] The expansion of the chelate ring to a five membered system in $MoO[SC_2H_4P(C_6H_5)_2]_2$ increases the chelate angle at the metal from the average values of approximately 73° in the four membered dithiocarbamate rings to 82°, a value significantly closer to the usual metal valence angle of 90°. The Mo-S distance should be compared to the value of 2.38Å observed for $[Mo_2O_4\text{-}(cysteine\ ethyl\text{-}ester)_2]$[18] and $[Mo_2O_2S_2\ (cysteine\ methyl\text{-}ester)_2]$,[21] where the S atom not only displays the same hybridization but is also present in a five-membered chelate ring in an equatorial position of a distorted trigonal bipyramidal coordination.

Tridentate Thiolate Ligands

[Mo₂O₃(SC₂H₄SC₂H₄S)₂]

$[Mo_2O_3(SC_2H_4SC_2H_4S)_2]$

In addition to providing mercapto-sulfur donor groups, ligands of the type 10a and 10b may also yield complexes of five-coordinate Mo(V), providing a geometry with ready access to the sixth coordination site by a potential substrate, without a prior dissociation step.

The reaction of $[MoO_2(acac)_2]$ with either 10a or 10b yields red-green dichroic crystals in about 25 percent yield (See Scheme 1), analyzing for $[Mo_2O_3(SC_2H_4SC_2H_4S)_2]$, (II), and $[Mo_2O_3(SC_2H_4N(CH_3)C_2H_4S)_2]$. Complex (II) proved to react readily with a number of small molecule substrates, such as thiophenol, mercaptoethanol, and pyridine, as anticipated from the formally coordinatively unsaturated nature of the molybdenum (Scheme 1).

The crystal structure of (II) is shown in Figure 2 and is seen to consist of dimeric molecules $[Mo_2O_3(SC_2H_4S\ C_2H_4S)_2]$, with an *anti* terminal oxo-configuration and a rigorously planar O=Mo-O-Mo=O grouping, imposed by the center of symmetry

Scheme 1

Reactions of Tridentate Thiolate Ligands
of the Type HS-X-SH with Molybdenum

$[MoO_2(acac)_2] + HSC_2H_4SC_2H_4SH \rightarrow [Mo_2O_3(SC_2H_4SC_2H_4S)_2]$, (II) (25%)

$[MoO_2(acac)_2] + HSC_2H_4N(CH_3)C_2H_4SH \rightarrow [Mo_2O_3(SC_2H_4N(CH_3)C_2H_4S)_2]$

$[MoCl_5] + HSC_2H_4SC_2H_4SH \rightarrow [Mo(S_2C_2H_4SC_2H_4S)_2]$, (III)

(II) + $HSC_2H_4OH \rightarrow [Mo_2O_2(S_2C_2H_4SC_2H_4S)_2(SC_2H_4O)]$

(II) + $HSC_6H_5 \rightarrow [Mo_2O_3(SC_6H_5)(S_2C_2H_4SC_2H_4S)_2]$

(II) + pyr $\rightarrow [Mo_2O_3(S_2C_2H_4SC_2H_4S)_2(pyr)_2]$

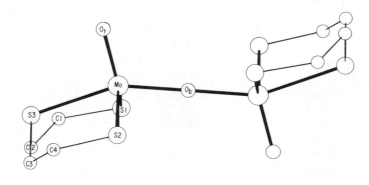

Figure 2. A view of the structure of $[Mo_2O_3(SC_2H_4SC_2H_4S)_2]$.

at the bridging oxo-group.[57] The configuration about each molybdenum is intermediate between square pyramidal and trigonal bipyramidal, as illustrated by the analysis of Table II. The distortions from regular trigonal bipyramidal geometry are evident in the S3-Mo-O_b angle of 155.9(2)° compared to 180° for the regular polyhedron and in the basal plane angles of 129.1(3)°, 115.2(7)°, and 114.4(7)° for S1-Mo-S2, S1-Mo-O_t, and S2-Mo-O_t, respectively (Table III). The coordination geometry is likewise distorted from regular tetragonal pyramidal as illustrated by the significantly different S1-Mo-S2 and S3-Mo-O_b angles and the concomitant deviations of the sulfur and the bridging oxygen atoms from the best least squares plane through these positions, ±0.30Å.

Table III
Selected Bond Lengths and Angles for
$[Mo_2O_3(SCH_2CH_2SCH_2CH_2S)_2]$

Mo-S3	2.484(6)	S1-Mo-S3	82.1(2)
Mo-S1	2.360(7)	S1-Mo-s2	129.1(3)
Mo-S2	2.346(8)	S1-Mo-O_b	115.2(7)
Mo-O_T	1.64(1)	S1-Mo-O_t	86.8(2)
Mo-O_B	1.873(2)	S2-Mo-S3	82.3(2)
S1-C1	1.80(2)	S2-Mo-O_b	88.2(2)
S2-C2	1.81(3)	S2-Mo-O_t	114.4(7)
S3-C3	1.86(2)	S3-Mo-O_b	155.9(2)
S3-C4	1.88(3)	S3-Mo-O_t	96.0(7)
C1-C3	1.49(4)	O_b-Mo-O_t	108.1(7)
C2-4	1.41(6)		

The crystal structure of (I), Figure 1, (Table I) consists of monomeric molecules $[MoO(SC_2H_4PPh_2)_2]$ of geometry intermediate between trigonal bipyramidal and tetragonal pyramidal, exhibiting approximate C_{2v} local symmetry.[56] The distortions from regular trigonal bipyramidal geometry are evident in the P1-Mo-P2 angle of 154.7(1)° compared to 180° for the regular polyhedron and in the angles in the basal plane, 122.3(3)°, 117.3(3)°, and 120.4(3)° for S1-Mo-S2, S1-Mo-O, and S2-Mo,O, respectively, rather than 120°. The significantly different P1-Mo-P2 and S1-Mo-S2 angles (154.7(1)° and 122.3(1)°, respectively) and the concomitant deviations of the sulfur and phosphorus atoms from the best least squares planes through these positions, ±0.30Å, illustrate the distortion from tetragonal pyramidal geometry. Analysis of the molecular geometry in terms of ideal polytopal forms as shown in Table II confirms that the structure is intermediate between the limiting ideal shapes.[39] The table presents not only the ideal dihedral angles calculated by Muetterties and Guggenberger,[39] but also the interplanar angles for idealized C_{2v} polyhedra which take into account the bond length distortion inherent as a result of the strong π-substituent effects of the oxo-ligand.

Table IV

Mean Values of Relevant Structural Parameters for Oxomolybendum(V) Complexes Containing the $Mo_2O_3^{4+}$ Unit and Sulfur-Containing Ligands

Complex	$Mo-O_t$	$Mo-O_b$	$Mo-L_c$	$Mo-L_{t-t}$	$Mo-L_{t-b}$	Mo-O-Mo	O_b-Mo-O_t	Ref.
$[Mo_2O_3(SC_2H_4SC_2H_4S)_2]$	1.64(2)	1.873(2)	Sm,2.353(6)	---	S_t,2.484(6)	180.0	108.1(7)	This work
$[Mo_2O_3(SC_2H_4N(CH_3)C_2H_4S)_2]$	1.667(8)	1.860(19	Sn,2.345(3)	---	N,2.24(1)	180.0	104.8(3)	33
$[Mo_2O_3(S_2COC_2H_5)_4]$	1.65(2)	1.86(2)	S,2.49(1)	S,2.70(2)	S,254(1)	178(4)	104(1)	14
$[Mo_2O_3(S_2P(OC_2H_5)_2)_4]$	1.65(1)	1.86(1)	S,2.47(1)	S,2.801(5)	S,2.547(5)	180.0	103(1)	15
$[Mo_2O_3(S_2CN(C_3H_7)_2)_4]$	1.67(1)	1.87(1)	S,2.48(4)	S,2.68(1)	S,2.53(1)	178(1)	103(1)	13
$[Mo_2O_3(S_2CSC_3H_7)_4]$	1.69(1)	1.84(1)	S,2.489(6)	S,2.694(6)	S,2.565(6)	170.3(8)	104.4(7)	16
$[Mo_2O_3(C_5H_4NS)_4]$	1.673(4)	1.853(1)	S,2.463(2)	N,2.185(5)	N,2.305(5)	180.0	105.9(2)	32

Abbreviations: $O_t \equiv$ terminal oxo-group; $O_b \equiv$ bridging oxo-group; $L_c \equiv$ ligand *cisto* O_t and O_b; $L_{t-t} \equiv$ ligand *trans* to O_t; $L_{t-b} \equiv$ ligand *trans* to O_t; $L_{t-b} \equiv$ ligand *trans* to O_b.

The solution properties of complex (II) are distinct from those observed for the class of complexes $[Mo_2O_3(S_2)_4]$, where S_2 is a dithioacid ligand. The complex does not exhibit the usual disproportionation into Mo(IV) and Mo(VI) species which generally characterizes this type of complex.

$$[Mo_2O_3(S_2)_4] \rightleftharpoons [MoO(S_2)_2] + [MoO_2(S_2)_2]$$

The significant reactivity of this species is thus inherent in the coordination type and does not require a dissociation of the molybdenum centers.

A particularly unusual feature of the complex is the observation of significant

Table V
ESR Parameters

Complex	$<g>'^a$	g_x	g_y	g_z	$<A>'^b$	A_x	A_y	A_z
MoOCl(SNNS)[c]	1.965	1.940	1.949	2.006	37.8	38.0	15.0	60.5
$Mo_2O_3(S_3)_2$[e]	1.979	1.950	1.972	2.014	38	36	20	58
Xanthine oxidase[d]	1.977	1.951	1.956	2.025	31.7	35	22	38

[a] $<g>' = (g_x + g_y + g_z)/3$
[b] $<A>' = (A_x + A_y + A_z)/3$
[c] Ref. 49
[d] Ref. 50
[e] Room temperature spectra, X band, of solid sample

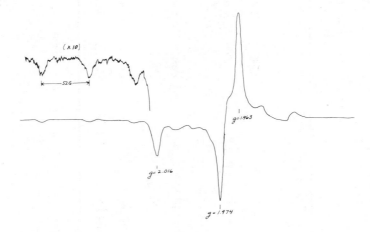

Figure 3. Room temperature epr spectrum of $[Mo_2O_3(SC_2H_4SC_2H_4S)_2]$, taken on a Varian E-4 spectrometer with the following experimental parameters: scan range, 400 G; field setting, 3400 G; time constant, 1 sec; scan time, 8 min; modulation amplitude, .63 G; modulation frequency, 100 KHz; power, 5 mM; microwave frequency, 9.472GHz.

paramagnetism, $\mu_{\text{eff}} = 0.8\text{BM/Mo}$ at room temperature. The observation of and esr signal (Figure 3) rules out temperature independent paramagnetism as the source of the moment and suggests as one possibility the thermal accessibility of a low-lying triplet state. The esr parameters are compared in Table V to those observed for [MoOCl(SNNS)] and xanthine oxidase. The rhombic distortion in the esr spectra of (II) is similar to that observed for the very rapid signal from xanthine oxidase.[50] Some of the anisotropic g and A values and the $<g>$ value are remarkably close to those of the enzyme, but the $<A>$ value is significantly higher than that observed for the xanthine oxidase very rapid signal.

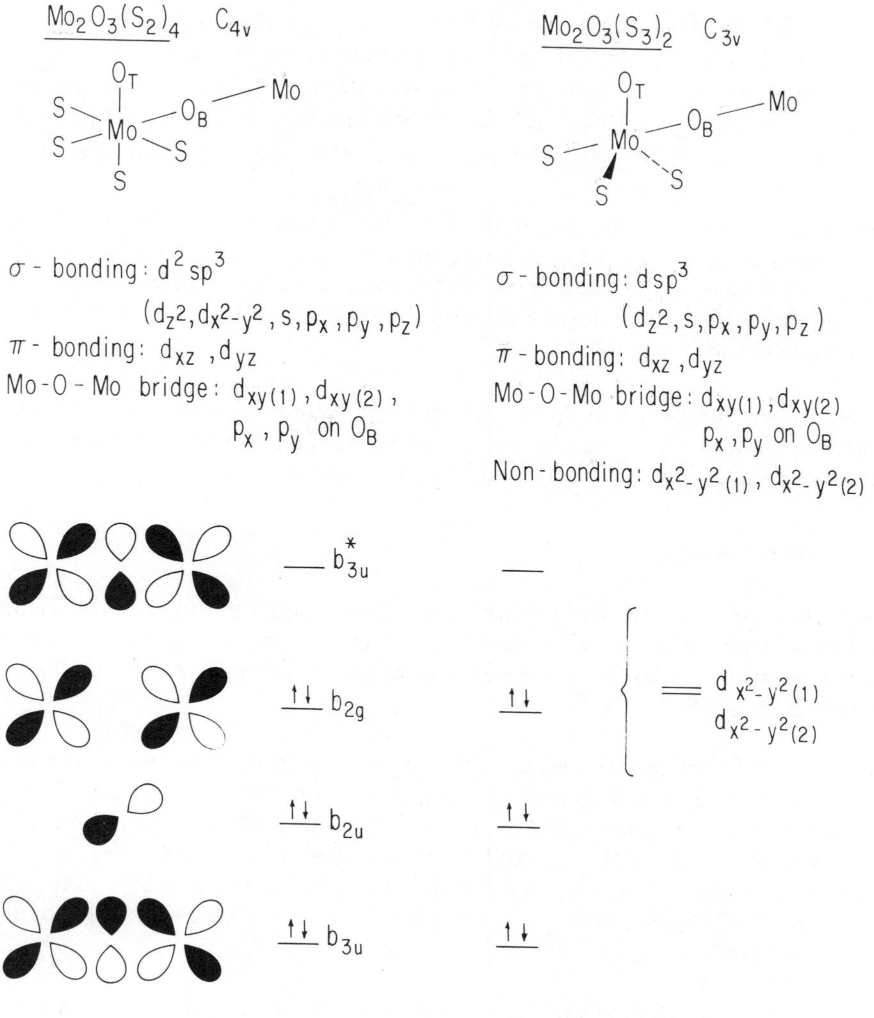

Figure 4. Symmetry-based molecular orbital treatment of the bonding in molecules of the type [Mo$_2$O$_3$L$_8$] and [Mo$_2$O$_3$L$_6$] with a planar Mo$_2$O$_3$ grouping. The shading indicates the phase of the wave function. The electronic population of the Mo-O$_b$-Mo bridging orbitals is indicated. The treatment is adapted from ref. 14.

The observation of paramagnetism in a complex possessing the $Mo_2O_3^{4+}$ core is unusual but may be rationalized in terms of five-coordination about the molybdenum for complex(II), rather than the more usual six-coordination. The usual bonding treatment invoked to rationalize the diamagnetism of complexes of the type $[Mo_2O_3(S_2)_4]$ was first proposed by Blake, Cotton and Wood.[14] Each Mo of the dimer is assumed to possess idealized local symmetry C_{2v} (Figure 4). Each Mo thus utilizes d^2sp^3 hybrid orbitals in σ-bonding to the four sulfur donors, the terminal oxo-group, and the bridging oxo-group. In addition, the metal d_{xy} and d_{yz} orbitals are utilized in π-bonding to the terminal oxo-group.[14]

The Mo-O-Mo bridge system thus employs the d_{xy} orbitals of each metal and the p_x and p_y orbitals of the bridging oxo-group. As long as the $O = Mo-O-Mo = O$ grouping is nearly linear, this results in a bonding interaction, b_{3u}, two essentially nonbonding orbitals, b_{2g} and b_{2u}, and an antibonding b_{3u}^* molecular orbital. The six electrons then fill the b_{3u}, b_{2g} and b_{2u} orbitals, resulting in a diamagnetic ground state.

In the case of complex (II), essentially the same bonding scheme applies except for the availability of two additional, approximately degenerate, non-bonding orbitals. While the scheme suggests a diamagnetic ground state for complexes $[Mo_2O_3(S_3)_2]$, the availability of a set of degenerate non-bonding orbitals may provide a low-lying paramagnetic state.

An alternative explanation is the presence of small amounts of a monomeric, paramagnetic impurity. The absence of a $\Delta M_S = 2$ signal in the epr spectrum of the sample lends some weight to this possibility.[61]

$[Mo(SC_2H_4SC_2H_4S)_2]$

Upon reaction of $[MoCL_5]$ with two equivalents of $HSC_2H_4SC_2H_4SH$ in tetrahydrofuran, a deep green solution developed in several minutes. Careful chromatographic separation yielded an emerald green species, analyzing for $[Mo(SC_2H_4SC_2H_4S)_2]$, (III).

The X-ray structure of (III) (Figure 5) shows an essentially trigonal prismatic structure.[58] The relevant bond lengths and angles are presented in Figure 5. The highly distorted trigonal prismatic geometry is unusual for a d^2 complex, although complexes with d^0 and d^1 electronic configurations display more or less regular trigonal prismatic geometry. The twist angle of 20° may be compared to 28°, that observed for $[Mo(S_2C_2(CN)_2)_3]^{2-}$, another example of Mo(IV) displaying a geometry intermediate between trigonal prismatic and octahedral.

The complex is a unique example of a Mo(IV) species coordinated to a saturated ligand and lacking oxo-coordination. Although dithiolene-type ligands, classified by Jorgenson as "non-innocent,"[51] have been shown to strip molybdenum of its usual oxo-coordination, discrete examples of complexes of molybdenum in the higher oxidation states with thiolate or thioether ligands and lacking oxo- or sulfido-coordination have not been previously described. The Mo(IV) oxidation state ap-

Mo-S (THIOETHER)   2.408(9)
Mo-S (MERCAPTO)   2.365(10)

S1-Mo-S2	77.8(6)	S2-Mo-S6	81.2(7)
S1-Mo-S3	131.7(8)	S3-Mo-S4	77.1(8)
S1-Mo-S4	91.9(9)	S3-Mo-S5	140.9(5)
S1-Mo-S5	84.4(7)	S3-Mo-S6	83.0(7)
S1-Mo-S6	133.9(7)	S4-Mo-S5	88.1(7)
S2-Mo-S3	79.7(9)	S4-Mo-S6	129.7(9)
S2-Mo-S4	137.6(9)	S5-Mo-S6	78.9(7)
S2-Mo-S5	130.6(9)		

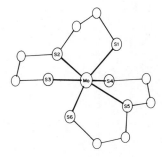

Figure 5. ORTEP rendering of the structure of [Mo(SC₂H₄SC₂H₄S)₂] and the relevant bond lengths and angles.

pears to be stabilized by the "soft" thioether donor groups, such that the usual pronounced tendency toward oxidation to the Mo(VI) state is not exhibited by (III).

The complex is electrochemically active, displaying reversible oxidation and reduction processes (Figure 6). The electrochemical behavior is consistent with the following scheme:

$$[Mo(S_2C_2H_4SC_2H_4S)_2]^{1+} \overset{+0.53V}{\rightleftarrows} [Mo(SC_2H_4SC_2H_4S)_2]$$

$$e^- \uparrow\downarrow -1.07V$$

$$[Mo(SC_2H_4SC_2H_4S)_2]^{1-}$$

$$[Mo(SC_2H_4SC_2H_4)_2]^{1-} \overset{k}{\rightarrow} X \overset{-0.25V}{\rightarrow} X^{1+}$$

The electrochemical behavior is complicated by a number of cathodic processes between -1.25 and -2.00V whose nature has not been elucidated at this time.

Tetradentate Ligands

Tetradentate ligands of the types 11a and 11b react readily with a number of molybdenum starting materials to yield complexes displaying a variety of structural types (Scheme 2 and Table VI).

N,N′-dimethyl-N,N′-bis (2-mercaptoethyl)ethylenediamine, 11a, reacts with

Table VI

Properties of Complexes of Molybdenum with Ligands of the Type HS N N SH

Complex[a]	Color	Analysis[b]			Infra-red Assignments[c]	Electrochemical Results[d]	
		C	H	N		$E_{1/2}$[e]	n[b]
[MoO$_2$(N$_2$S$_2$)]	Yellow	30.0	5.21	8.61	892,(s), 917(s)-ν(MoO$_t$)	-1.34(r)[g]	1.0
		(28.7)	(5.39)	(8.38)			
[MoO$_2$(N$_2$S$_2$')]	Yellow	30.9	5.62	8.26	888(s), 921(s)-ν(MoO$_t$)	-1.09(i)	1.1
		(31.0)	(5.75)	(8.05)			
[Mo$_2$O$_3$(N$_2$S$_2$)$_2$]	Purple	29.4	5.63	8.62	936(s)-ν(MoO$_t$)	+0.15(r)	3.6
		(29.4)	(5.52)	(8.59)	442(m), 739(m)-ν(Mo$_2$O$_b$)		
H$_2$[Mo$_2$O$_4$(N$_2$S$_2$')$_2$]	Orange	30.8	6.00	8.24	952(s)-ν(MoO$_t$)	-1.69(r)	0.5
		(30.9)	(6.02)	(8.02)	495(m), 740(m)-ν(Mo$_2$O$_2$)	-0.72(i)	1.2

[a]N$_2$S$_2$ = $^-$SCH$_2$CH$_2$N(CH$_3$)(CH$_2$)$_2$N(CH$_3$)CH$_2$CH$_2$S$^-$; N$_2$S$_2$' = $^-$SCH$_2$CH$_2$N(CH$_3$)(CH$_2$)$_3$N(CH$_3$)CH$_2$CH$_2$S$^-$.

[b]Calculated value in parentheses.

[c]Recorded as KBr disks; frequencies in cm^{-1}; abbreviations: s, strong; m, medium; t, terminal; b, bridging.

[d]Measured in CH$_2$Cl$_2$, 0.1M in [n-Bu$_4$N]PF$_6$, under argon at room temperature at a platinum wire electrode.

[e]Potentials are in volts vs. the aqueous saturated potassium chloride calomel electrode. $E_{1/2}$ is the half-wave potential recorded by normal pulse voltammetry. These values are in accord with those measured by a.c. and cyclic voltammetry.

[f]n refers to the number of electrons transferred per molybdenum atom. The value was derived by comparison of the current density for the electrode process with that observed for the known one-electron oxidations of [Ni(mnt)$_2$]$^{2-}$ (14). The results were verified by controlled potential electrolysis at potentials 0.2V from the half-peak potential of the process in question.

[g]Abbreviations: r = reversible; i = irreversible. The reversibility of the process was judged from the value of ΔE_p and the rate of i_p^c/i_p, derived from cyclic voltammograms obtained at sweep rates of 50 to 2000 mV/sec. For a reversible process, ΔE_p = 59.5 mV and i_p^c/i_p^a = 1.0. The reversible couples reported above displayed values of 63.0 mV and 1.0 for ΔE_p and i_p^c/i^a. The irreversible processes were characterized by the absence of a peak on the subsequent anodic sweep.

Scheme 2

Reactions of Molybdenum with Tetradentate Ligands

$[MoO_2(acac)_2] + S_4H_2 \rightarrow [MoO_2(S_4)]$

$[MoO_2(acac)_2] + S_4H_2 \text{ (excess)} \rightarrow [Mo_2O_3(S_4)_2]$

$[MoO_2(acac)_2] + N_2S_2H_2 \rightarrow [MoO_2(N_2S_2)]$

$2MoO_2(N_2S_2) + PPh_3 \rightarrow [Mo_2O_3(N_2S_2)_2] + OPPh_3$

$[MoO_2(acac)_2] + N_2S_2'H_2 \rightarrow [MoO_2(N_2S_2')]$

$2MoO_2(N_2S_2') + PPH_3 \rightarrow \text{``}Mo_2O_3(N_2S_2')_2\text{''} \rightarrow [H_2Mo_2O_4(N_2S_2')_2] \text{ (IV)}$

. . . .

$S_4H_2 \equiv HSC_2H_4S(CH_2)_n C_2H_4SH$

$N_2S_2H_2 \equiv HSC_2H_4N(CH_3)C_2H_4N(CH_3)C_2H_4SH$

$N_2S_2'H_2 \equiv HSC_2H_4N(CH_3) C_3H_6N(CH_3)C_2H_4SH$

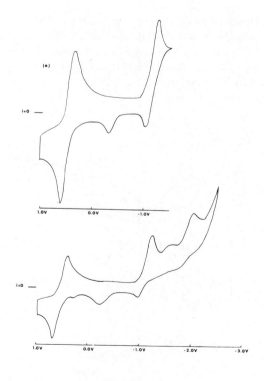

Figure 6. Cyclic voltammograms of $[Mo(SC_2H_4SC_2H_4S)_2]$, 5×10^{-4} M in CH_2Cl_2, $(C_4Hg)_4NPF_6$ supporting electrolyte; sweep rate, 200 mV/sec; pt electrode. (a) Scan range $+1.1V$ to $-1.4V$; (b) Scan range $+1.0V$ to $-3.0V$.

$[MoO_2(acac)_2]$ to give the *cis-* dioxo species $[MoO_2(N_2S_2)]$, which may be treated in turn with PPh_3 to yield a complex with spectroscopic characteristics consistent with the presence of the $Mo_2O_3^{4+}$ structural unit.

The complex $[MoO_2(N_2S_2)]$ is remarkably reactive and may be used as a precursor in the synthesis of a number of hydrazido- and diazenido-complexes (Scheme 3).

Scheme 3

Reactions of Complex $[MoO_2(N_2S_2)]$
with Substituted Hydrazines

$[MoO_2(N_2S_2)] + HSC_6H_5 \rightarrow [Mo_2O_4(SPh)_2(N_2S_2)]$

$[MoO_2(N_2S_2)] + H_2NNAr_2 \rightarrow [MoO(NNAr_2)(N_2S_2)]$

$\quad \downarrow + H_2NNAr_2$

$\quad [Mo(NNAr_2)_2(N_2S_2)]$

$[MoO_2(N_2S_2)] + H_2NNHAr \rightarrow [MoO(NNHAr)(N_2S_2)]$

EtOH, H_2NNAr ↙ ↘ H_2NNHAr

$[Mo(N_2Ar)_2(N_2S_2)] + [Mo(N_2Ar)(N_2S_2)]_2$ $[MoO(NNAr)(N_2S_2)]$

Hydrazido complexes are possible intermediates in the protonation of coordinated dinitrogen. The structure of a related bis- hydrazido complex is shown in Figure 7, demonstrating the linearity of the Mo-N-N linkage. The bond angle at the metal-bound nitrogen and the Mo-N and N-N bond distances suggest extensive delocalization through the Mo-N-N-Ar$_2$ grouping.

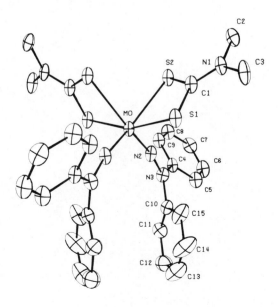

Figure 7. Crystal structure of $Mo(N_2Ph_2)_2(S_2CNR_2)_2$.

The diazenido-complex, [MoO(NNAr)(N$_2$S$_2$)], may be reversibly protonated in methanol solution.

$$[MoO(NNAr)(N_2S_2)] \underset{base}{\overset{acid}{\rightleftarrows}} [MoO(NNHAr)(N_2S_2)]$$

Attempts at further protonation to produce NH$_2$Ar and [MoO(NH)(N$_2$S$_2$)] were unsuccessful as were reactions with hydroxylamino-o-sulfonic acid, a source of NH$_2^+$.

The analogous system with N,N′-bis(2-methyl-2-mercaptopropyl)propylene, 11b, displayed similar reactivity patterns with the exception of the reaction with PPh$_3$. [MoO$_2$(N$_2$S$_2$′)] reacted to give a transient purple color, which bleached in five minutes to give a yellow solution from which precipitated a mass of yellow crystals, whose infrared spectrum is consistent with structural type 5a.

Although this structural type is common to Mo(V) chemistry, it is characterized by the presence of terminal bidentate ligands, leaving some question as to the mode of ligand coordination in the complex H$_2$[Mo$_2$O$_4$(N$_2$S$_2$′)$_2$].

The structural analysis revealed the anticipated Mo$_2$O$_4^{2+}$ core (Figure 8).[51] The most unusual feature of the structure is the bridging, bidentate ligation mode adopted by the ligand, illustrating the unique flexibility of such unsaturated polydentate ligand types. The Mo atoms are in square pyramidal environments, the Mo displaced an average of 0.71 Å from the S$_2$O$_2$ basal plane in the direction of the apical oxygen. The dihedral angle between the planes generated by S1S2O1O2 and S3S4O1O2 is 149.9°, as compared to 153° in the analogous [Mo$_2$O$_4$(histidine)$_2$].[53] (Table VII)

The bridging, bidentate ligation mode adopted by N$_2$S$_2$′ reflects the coordination requirements of the Mo$_2$O$_4^{2+}$ core, the steric flexibility conferred by the additional methylene group in the amine bridge of the ligand, and the protonation of the amine nitrogens. The amine nitrogens of the ligand direct their lone pair orbitals toward a common point, producing the conformation of the protonated diamine portion of the ligand shown in Figure 9. The intraligand N-N distance of 2.74(1)Å, compared to 3.22(1)Å in [(FeN$_2$S$_2$′)$_2$] where N$_2$S$_2$′ functions as a tetradentate ligand[54] and to a van der Waals diameter of approximately 3.0Å for nitrogen,[55] lends further support for

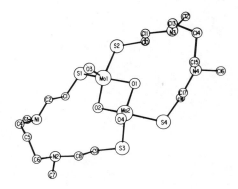

Figure 8. A view of the structure of H$_2$[Mo$_2$O$_4$(SC$_2$H$_4$N(CH$_3$)(CH$_2$)$_3$N(CH$_3$)C$_2$H$_4$S)$_2$].

Table VII

Selected Bond Length and Angles for
$H_2[Mo_2O_4(SCH_2CH_2N(CH_3)(CH_2)_3(CH_3)CH_2CH_2S)_2]$

Bond distances in Å		Bond angles in (°)			
Mo1-Mo2	2.623(1)	Mo2-Mo1-O3	106.9(2)	S3-Mo2-O2	143.2(2)
Mo1-O1	1.954(7)	S1-Mo1-S2	76.9(1)	S3-Mo2-O4	103.9(2)
Mo1-O2	1.946(8)	S1-Mo1-O1	143.6(3)	S4-Mo2-O1	138.1(3)
Mo1-O3	1.68(1)	S1-Mo1-O2	83.7(3)	S4-Mo2-O2	83.6(3)
Mo1-S1	2.447(4)	S1-Mo1-O3	103.8(2)	S4-Mo2-O4	107.7(3)
Mo1-S3	2.434(4)	S2-Mo1-O1	83.9(3)	O1-Mo2-O2	91.2(3)
Mo2-O1	1.933(9)	S2-Mo1-O2	137.9(3)	O1-Mo2-O4	112.8(4)
Mo2-O2	1.950(7)	S2-Mo1-O3	107.6(3)	O2-Mo2-O4	111.6(4)
Mo2-O4	1.671(9)	O1-Mo1-O2	90.7(4)	Mo-S-C(ave)	107.1(5)
Mo2-S3	2.440(3)	O1-Mo1-O3	111.4(4)	S-C-C(ave)	109.6(9)
Mo2-S4	2.434(4)	O2-Mo1-O3	113.2(5)	C-C-N(ave)	109.3(13)
S-C(ave)	1.83(1)	Mo1-Mo2-O4	106.6(2)	C-N-C(ave)	110.9(13)
N-C(ave)	1.50(1)	S3-Mo2-S4	76.6(1)	C-C-C(ave)	117.5(15)
C-C(ave)	1.2(2)	S3-Mo2-O1	83.6(2)		

amine nitrogens as the site for protonation. That N_2S_2 does not form an analogous complex, $[Mo_2O_4(N_2S_2)_2]^{2-}$, reflects the steric contraints imposed by the di-methylene bridge between amine nitrogens, effectively preventing the ligand from achieving the conformation required to bring the amine nitrogens into a suitable orientation for protonation while simultaneously providing an intraligand S-S distance of approximately 6.5Å, necessary to achieve bridging bidentate character.

An unusual feature of the chemistry of this system is the reversibility of the redox processes for $[MoO_2(N_2S_2)]$ and $[Mo_2O_3(N_2S_2)_2]$ (Figure 10), whereas the analogous complexes with delocalized ligand systems such as dithioacid groups[37/12] yield irreversible voltammetric patterns or complex behavior as a result of disproportionation reactions. In particular, the observation of both cathodic and anodic reversible

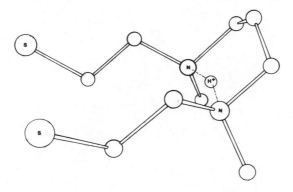

Figure 9. Schematic representation of the ligand geomtry showing the conformation adopted by the protonated diamine segment of $H_2[Mo_2O_4(SC_2H_4N(CH_2)_3C_2H_4S)_2]$.

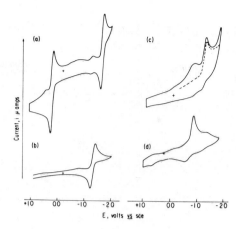

Figure 10. Cyclic voltammograms recorded at 200mV/sec scan speed at a Pt wire electrode vs. the saturated calomel electrode on CH_2Cl_2 solutions 0.1M in [n-Bu$_4$N]PF$_6$ and 0.001 M in (a) [Mo$_2$O$_3$(N$_2$S$_2$)$_2$], (b) [MoO$_2$(N$_2$S$_2$)], (c) [MoO$_2$(N$_2$S$_2$′)], (d) H$_2$[Mo$_2$O$_4$(N$_2$S$_2$′)$_2$].

processes for [Mo$_2$O$_3$(N$_2$S$_2$)$_2$], a complex which retains its integrity in solution as judged by Beer's law plots, suggests that the ligand chelation mode may be unusual. A point of further interest is the irreversibility of the electrode process for [MoO$_2$(N$_2$S$_2$′)] and the shift of the cathodic wave 0.30V more positive when compared to [MoO$_2$(N$_2$S$_2$)], indicating that the introduction of an additional methylene group into the amine bridge may cause drastic reorientation of the donor groups about the molybdenum atom. Crystal structure analyses of [MoO$_2$(N$_2$S$_2$)], [MoO$_2$(N$_2$S$_2$′)], and [Mo$_2$O$_3$(N$_2$S$_2$)$_2$] are in progress to establish the role of ligand flexibility in determining which structural type is adopted.

The coincidence of the cathodic process observed at -0.7V in the voltammogram of [MoO$_2$(N$_2$S$_2$′)] with that observed for the reduction of [Mo$_2$O$_4$(N$_2$S$_2$′)$_2$]$^{2-}$ is attributable to the presence of residual water[60] in the solvent and consistent with the following scheme:

$$[MoO_2(N_2S_2')] + e^- \xrightarrow{-1.09V} [MoO_2(N_2S_2')]^-$$

$$2[MoO_2(N_2S_2')]^- + 2H^+ \rightarrow [H_2Mo_2O_4(N_2S')_2]$$

$$[H_2Mo_2O_4(N_2S_2')_2] + 2e^- \xrightarrow{-0.7V} [H_2MoO_4(N_2S_2')_2]^{2-}$$

The disappearance of the cathodic process at -0.7V upon the suspension of alumina in the solvent confirms the proposed reaction scheme.

A detailed analysis of the electrochemical characteristics of the system, together with the results of the crystallographic investigations of $[MoO_2(N_2S_2)]$ and $[MoO_2(N_2S_2')]$ will be presented in a forthcoming paper.

Conclusions

Although the structural trends established for molybdenum complexes with 1,1 dithioacid ligands are generally maintained for the linear thiolate ligands discussed in this paper, some modifications in structural and chemical behavior have been observed:

1. Ligand flexibility and variation in number of donor groups allows some variation in structural type. In particular, the use of tridentate ligands, such as 10a and 10b, provides dimeric five-coordinate complexes of Mo(V) whose chemical properties are quite distinct from those observed for 1,1-dithioacid complexes of molybdenum. This is a particularly attractive coordination mode for Mo(V) in enzymes as it provides a coordination site for the reacting substrate without necessitating disproportionation or ligand dissociation as a prior step.

2. A number of molybdenum oxidation states may be stabilized by varying the nature of certain donor groups. Thus, in the presence of *class a* donors, as with 11a, the Mo(V) and Mo(VI) oxidation states are preferred, whereas *class b* donors, phosphine in 9c and thioether sulfur in 10a, tend to provide ready access to the Mo(IV) oxidation state. Although not a totally unexpected result, the stability of (I) toward air-oxidation and electrochemical oxidation is exceptional.

3. Complex (III), $[Mo(SC_2H_4SC_2H_2S)_2]$, provides an unusual example of a Mo(IV) coordination complex with "innocent" ligands[51] and lacking oxo-coordination. The reversible oxidative couple demonstrates well-behaved redox characteristics for a complex incorporating exclusively cysteinato- and methionato-type donor groups. In view of recent proposals that the molybdenum in nitrogenase is not oxo-bound, the isolation of this complex demonstrates the feasibility of this type of coordination mode for a simple mononuclear molybdenum complex with thiolate donor ligands.

4. The complex $[Mo_2O_3(SC_2H_4SC_2H_4S)_2]$ (II) appears to provide a unique example of a dimeric Mo(V) species displaying room temperature epr activity. The similarity of the epr parameters to those observed for isolable monomeric Mo(V) species[49] and to the xanthine oxidase very rapid signal suggests that the properties of these complexes are of considerable interest with respect to the possible structure of enzymatic Mo(V) centers and suggests the presence of a monomeric species as the esr active constituent.

5. The electrochemical investigations of the complexes demonstrate that well-behaved redox properties are characteristic of the molybdenum complexes with ligands possessing flexible and saturated backbones.

Acknowledgement

These investigations were supported by a grant from the National Institutes of Health (GM-22566). We thank Dr. J. R. Dilworth of the ARC Unit of Nitrogen Fixation in Sussex, England, for helpful discussions and for permission to include the data on $[MoO(PC_2H_4PPh_2)_2]$ in this article.

References and Footnotes

1. J. T. Spence, *Coord. Chem. Rev., 4*, 475 (1969).
2. E. I. Stiefel, *Prog. Inorg. Chem., 22*, 1 (1977).
3. (a) R. C. Burns and R. W. F. Hardy, *Methods Enzymol., 24*, 480 (1972); (b) W. A. Buelen and J. R. LeComte, ibid., *24*, 456 (1972); (c) V. K. Shek and W. J. Bril, *Biochem. Biophys. Acta 305*, 445 (1972).
4. T. C. Huang, W. G. Zumft, and L. E. Mortenson, *J. Bacteriol. 113*, 884 (1973).
5. (a) R. R. Eady, B. E. Smith, K. A. Cook, and J. R. Postgate, *Biochem. J., 128*, 655 (1972); (b) R. R. Eady, B. E. Smith, R. N. F. Thorneley, D. Ware and J. R. Postgate, *Biochem. Soc. Trans., 1*, 37 (1973); (c) J. R. Postgate, personal communication, 1975.
6. (a) R. C. Bray, P. C. Knowles and L. S. Meriwether, *Wenner-Gren Cent. Int. Sympt. Ser., 9*, 221 (1967); (b) R. C. Bray and J. C. Swann, *Struct. Bonding (Berlin), 11*, 207 (1972).
7. W. E. Newton, J. L. Corbin, D. C. Bravard, J. E. Searles, and J. W. McDonald, *Inorg. Chem., 13*, 1100 (1974).
8. (a) S. P. Cramer, K. O. Hodgson, W. O. Gillum, and L. E. Mortenson, *J. Am. Chem. Soc., 100*, 3398 (1978).
   (b) S. P. Cramer, W. O. Gillum, K. O. Hodgson, L. E. Mortenson, E. I. Stiefel, J. R. Chisnell, W. J. Brill, and V. K. Shah, *J. Am. Chem. Soc., 100*, 3814 (1978).
9. (a) B. J. Weathers, J. H. Grate, and G. N. Schrauzer, *J. Am. Chem. Soc., 101*, 917 (1979); (b) B. J. Weathers, J. H. Grate, N. A. Strampack, and G. N. Schranzer, *J. Am. Chem. Soc., 101*, 925 (1979), and references therein.
10. G. J.-J. Chen, J. W. McDonald, and W. E. Newton, *Inorg. Chem., 15*, 2612 (1976).
11. W. E. Newton, G. J.-J. Chen, and J. W. McDonald, *J. Am. Chem. Soc., 98*, 5388 (1976).
12. L. J. DeHayes, H. C. Faulkner, W. M. Doub, Jr., and D. T. Sawyer, *Inorg. Chem., 14*, 2110 (1975).
13. L. Ricard, J. Estienne, P. Karagiannidis, P. Toledano, J. Fischer, A. Mitschler and R. Weiss, *J. Coord. Chem., 3*, 277 (1974).
14. A. B. Blake, F. A. Cotton, and J. S. Wood, *J. Am. Chem. Soc., 86*, 3024 (1964).
15. J. R. Knox and C. K. Prout, *Acta Crystallogr., Sec. B, 25*, 228-337 (1969).
16. G. B. Maniloff and J. Zubieta, *Inorg. Nucl. Chem. Lett., 12*, 121 (1976).
17. J. R. Knox, and C. K. Prout, *Acta Crystallogr., Sec B, 25*, 1857 (1969).
18. M. G. B. Drew and A. Kay, *J. Chem. Soc., A*, 1846 (1971).
19. L. Ricard, C. Martin, R. West, and R. Weiss, *Inorg. Chem., 14*, 2300 (1975).
20. B. Spivack, Z. Dori, and E. I. Stiefel, *Inorg. Nucl. Chem. Lett., 11*, 501 (1975).
21. M. G. B. Drew and A. Kay, *J. Chem. Soc. A*, 1851 (1971).
22. D. H. Brown and J. A. d. Jeffreys, *J. Chem. Soc., Dalton Trans.*, 732 (1973).
23. L. Ricard, J. Estienne and R. Weiss, *Inorg. Chem., 12*, 2182 (1973).
24. J. I. Gelder and J. H. Enemark, *Inorg. Chem., 15*, 1839 (1976).
25. B. Spivack and Z. Dori, *Coord. Chem. Rev., 17*, 99 (1975), and references therein.
26. F. A. Cotton and S. M. Morehouse, *Inorg. Chem., 4*, 1377 (1965).
27. G. T. J. Delbaere and C. K. Prout, *Chem. Commun.*, 162 (1971).
28. B. Spivack, A. P. Gaughan, and Z. Dori, *J. Am. Chem. Soc., 93*, 5265 (1971).
29. B. Spivack and Z. Dori, *J. Chem. Soc., Dalton Trans.* 1173 (1973).
30. G. Bunzey, W. E. Newton, and N. Parizadath, *J. Less. Common. Met., 54*, 513 (1977).
31. (a) M. W. Bishop, J. Chatt, J. R. Dilworth, G. Kaufman, S. Kim and J. A. Zubieta, *J. Chem. Soc. Chem. Comm.*, 70 (1977).
    (b) M. W. Bishop, J. Chatt, J. R. Dilworth, J. R. Hyde, S. Kim, K. Venkatasubramanian, and J. Zubieta, *Inorg. Chem., 17*, 2917 (1978).
32. F. A. Cotton, P. E. Fanwick, and J. W. Fitch, III, *Inorg. Chem., 17*, 3254 (1978).

33. Y.-Y. P. Tsao, C. J. Fritchie, Jr., and H. A. Levy, *J. Am. Chem. Soc., 100,* 4089 (1978).
34. J. Cragel, Jr., B. V. Pett, M. O. Glick, and R. E. DeSimone, *Inorg. Chem., 17,* 2885 (1978).
35. B. Spivack and Z. Dori, *Coord. Chem. Rev., 17,* 99 (1975), and references therein.
36. (a) G. Bunzey, J. H. Enemark, J. H. Howie, and D. T. Swyer, *J. Am. Chem. Soc., 99,* 4168 (1977).
    (b) G. Bunzey and J. H. Enemark, *Inorg. Chem., 17,* 682 (1978).
    (c) See also J. T. Huneke and J. H. Enemark, *Inorg. Chem., 17,* 3698 (1978).
37. J. Hyde, K. Venkatasubramanian, and J. Zubieta, *Inorg. Chem., 17,* 3698 (1978).
38. E. A. Maatta and R. A. D. Wentworth, *Inorg. Chem., 18,* 524 (1979).
39. E. L. Muetterties and L. J. Guggenberger, *J. Am. Chem. Soc., 95,* 1748 (1974).
40. M. Boorman, C. D. Garner, and F. E. Mabbs, *Chem. Comm.* 663 (1974).
41. R. J. Doedens and J. A. Ibers, *Inorg. Chem. 6,* 204 (1967).
42. A. R. Rossi and R. Hoffman, *Inorg. Chem., 14,* 365 (1975).
43. J. S. Wood, *Progr. Inorg. Chem., 16,* 227 (1972).
44. J. E. Drake, J. Vekris, and J. S. Wood, *J. Chem. Soc. (A),* 1000 (1968).
45. R. P. Dodge, D. H. Templeton, and A. Zalkin, *J. Chem. Phys., 35,* 55 (1961).
46. K. Dickman, G. Hamer, S. C. Nyburg, and W. F. Reynolds, *Chem. Commun.,* 1295 (1970).
47. D. Coucouvanis, S. J. Lippard, and J. A. Zubieta, *Inorg. Chem., 9,* 2775 (1970).
48. R. E. DeSimone and M. D. Glick, *Inorg. Chem., 17,* 3574 (1978).
49. (a) J. T. Spence, M. Minelli, P. Kroneck, M. I. Scullane, and N. D. Chasteen, *J. Am. Chem. Soc., 100,* 8002 (1978).
    (b) R. D. Taylor, J. P. Street, M. Minelli, and J. T. Spence, *Inorg. Chem., 17,* 3207 (1978).
50. R. C. Bray in "The Enzymes," Vol. 12, 34th ed., P. O. Boyer, Ed, Academic Press, New York, p. 299.
51. C. K. Jorgenson, *Inorg. Chim. Acta Rev., 2,* 65 (1968).
52. N. Kim, S. Kim, P. Vella, and J. Zubieta, *Inorg. Nucl. Chem. Lett., 14,* 457 (1978).
53. L. T. J. Delbeare and C. K. Prout, *Chem. Commun.,* 162 (1971).
54. W.-J. Hu and S. J. Lippard, *J. Am. Chem. Soc., 96,* 2366 (1974).
55. L. Pauling, "The Nature of the Chemical Bond," 3rd Ed., Cornell Univ. Press, New York, 1960.
56. Summary of crystal data: a $=10.263(3)\text{\r{A}}$, b $+$ $10.346(1)\text{\r{A}}$, c $= 26.657(5)\text{\r{A}}$, $\alpha = \beta = \gamma = 90°$, orthorhombic crystal system, p2,2,2, with Z $= 4$; $p_{calc} = 1.43\text{g/cm}^3$, $\varrho_{found} =1.44\text{g/cm}^3$, F(000) $= 1232$; CuKα radiation, $\lambda = 1.5418\text{\r{A}}$; 2719 symmetry independent reflections with $0 < \theta < 128$.
57. Summary of crystal data a $= 14.027(5)\text{\r{A}}$, b $= 13.815(5)\text{\r{A}}$, c $= 8.970(4)\text{\r{A}}$, $\alpha = \beta =\gamma = 90.0°$; orthorhombic system, P_{bca}, with Z $= 4$; $\varrho_{calc} = 2.08\text{g/cm}^3$, $\varrho_{found} = 2.05\text{g/cm}^2$; F(000) $= 1072$; MoKα radiation, $\lambda = 0.71073\text{\r{A}}$; 701 symmetry independent reflections $2 < 2\theta < 50°$.
58. Summary of crystal data: a $= 8.281(2)$, b $= 14.800(3)$, c $= 12.180(2)\text{\r{A}}$, $\alpha = \gamma = 90.0$, $\beta = 103.68(2)°$; monoclinic crystal system, P2$_1$ or P2$_1$/m with Z $= 4$; $\varrho_{calc} = 1.83\text{g/cm}^3$, $\varrho_{found} =1.81\text{g/cm}^3$; F(000) $= 808$; MoKα radiation, $\lambda = 0.71073$; 1581 symmetry independent reflections $2 < 2\theta < 50°$.
59. See ref. 52 for summary of crystal data.
60. Methylene chloride "dried" by conventional methods was found to be 20-25 mM in H_2O, which addition of alumina directly into the electrochemical cell reduced H_2O concentration to 1-5 mM.
61. Subsequent studies have established that the source of the esr spectrum is a monomeric impurity. The magnetic susceptibility of the complex $Mo_2O_3(S_3)_2$ was found to be *independent* of temperature in the range 77°K to 350°K. The observed paramagnetism is consistent with the presence of 2% of paramagnetic monomer in the bulk sample. Preparations using [MoOCl$_3$•2THF] as starting material yield small amounts of the complex [MoO(S$_3$)Cl•THF], as Mo(V) monomer with unexceptional magnetic properties.

Part IV
Systems of Higher Order Nucleic Acid Statics and Dynamics

The Diverse Spatial Configurations of DNA
Evidence for a Vertically Stabilized Double Helix

Ramaswamy H. Sarma and M. M. Dhingra[1]
Institute of Biomolecular Stereodynamics
Department of Chemistry
State University of New York at Albany
Albany, New York 12222

and

Richard J. Feldmann
Division of Computer Research and Technology
National Institutes of Health
Bethesda, Maryland 20014

Introduction

Albert Szent-Gyorgye[2] in his opus on the electronic theory of life states "Life is a miracle, but even miracles must have their underlying mechanism. The archives in which the basic blueprints of this mechanism are preserved and xeroxed are the nucleic acids, while the business of life is carried on by the proteins." This paper and several of them which follow are concerned solely with nucleic acids and in these papers we explore their spatial configuration and dynamics.

Extensive NMR studies have revealed several important aspects of nucleic acid conformation.[3-12] Thus (i) in aqueous solution nucleic acid structures are conformationally pluralistic. The composition of the conformational blend and the preferred intramolecular order are largely determined by the constitution and sequence which control the torsional and flexural movements about the internucleotide phosphodiester bonds; (ii) transmission of conformation in nucleic acids takes place through a series of stereochemical domino effects due to the presence of a synchronous coupled set of base sequence dependent spatial configuration parameters and interdependent structural changes linked to base base stacking interactions (iii) they have 3'5', and not 2'5', internucleotide bonds because the intrinsic molecular stereodynamics of 2'5' systems are such that they cannot support stable helical structures with base stacking interactions at the single or double stranded level. One of the imperatives of molecular biology is that polynucleotides must be present either completely or significantly in stable helical configurations in order to participate in the fundamental processes of information storage, transfer and retrieval, and the ability of 3'5' systems to express themselves in helical configurations probably provided them an evolutionary advantage over their 2'5' analogs.

251

Intimate details of the spatial configurations of nucleic acid structures have come from single crystal x-ray diffraction studies.[13-19] An important achievement of this decade is the determination of the detailed geometry of yeast phenylalanine tRNA.[15-18] An artistic rendition of this structure by the celebrated Irving Geis is depicted in Plate 1. Fiber diffraction studies have revealed that DNA may assume different shapes depending upon the mode of preparation and the extent of humidity content. The well known are the A, B, C, and D forms. In Plates 2 through 5 are depicted the computer generated stereo perspectives of these forms. These are generated using the coordinates of Arnott, et al.[20-22] and Marvin et al..[23] An artistic rendition of B and A-DNA are shown in Plates 6 and 7. In Table I are summarized the various torsion angles in the different forms of DNA. Inspection of Plates 1 through 5 and examination of the data in Table I clearly reveal that changes in sugar pucker and/or small variation in the backbone torsion angles indeed affect the dimensions as well as the geometric relationships between the base pairs. However, there is considerable similarity in the gross morphology in all the four forms. Thus the helical organization of the sugar phosphate backbone is right handed and except for a tilt of 19° in A-DNA, the base planes are almost perpendicular to the helix axis. Since nucleic acid structures in aqueous solution are flexible there is considerable interest to determine the effect of changes in individual torsion angles on their overall spatial configuration. The question is whether nucleic acids could adopt biologically functional forms which vastly differ in morphology and whether such forms can be created by torsion about the various bonds. For example, in all the forms in Table I the sugar base torsion angle (χ, see Dhingra and Sarma[24] for nomenclature) is in the *anti* domain. It is known that in aqueous solution, the magnitude of the sugar-base torsion angle has a decisive effect on the mode of sugar pucker and the C4'-C5' (ϵ/ψ) torsion[3-6] of nucleic acid structures. It has further been shown[25] experimentally that when χ is fixed at a high anti domain ($\simeq 120°$) the *sense of the base stack* is left handed. However, the effect of a high anti arrangement on the *sense of the helical organization* of the backbone of a polynucleotide is not known with any certainty, and two diametrically opposed views have been reported about its helix handedness.

The Controversy

The observation that the cotton effect in the spectra of high anti polynucleotides was opposite to that of normal nucleic acid counterparts[26/27] suggested the possibility

Table I

Backbone and Glycosidic Torsion Angles for Various Forms of DNA

Molecule	Sugar Pucker	Torsion Angles In Degrees						Ref.
		$\alpha(\phi')$	$\beta(\omega')$	$\gamma(\omega)$	$\delta(\phi)$	$\epsilon(\psi)$	χ	
A-DNA	³E	178	313	285	208	45	26	20
B-DNA	²E	155	264	314	214	36	82	21
C-DNA	²E	211	212	315	143	48	73	23
D-DNA	²E	141	260	298	208	69	84	22

that their helical organization may be left handed. Fujii and Tomita[28] carried out conformation energy calculations and concluded that energetically these molecules prefer a left handed helical backbone. Recently, Yathindra and Sundaralingam[29/30] using the "rigid nucleotide concept," a concept whose validity has been challenged,[8/31] reported the helical parameters n and h where n is the number of nucleotide residues per turn and h is the residue height along the helix axis) for high anti polynucleotides and concluded that the helical sense of the backbone in both single and double stranded systems is left handed and that the sense of base stack is also left handed. They have shown a model of the left handed duplex in reference 30.

However, Olson and Dasika[32] have advanced an alternative model for high anti polynucleotides. From extensive theoretical studies and model building these authors demontrated that the right handed helical backbone conformation with the bases in a left handed stack and with the base planes oriented parallel to the helix axis are energetically more favored than any of the alternative structures. They argued that the left handed base stack would explain the inverse cotton effect and maintained that the helical sense of the backbone remained right handed. At the double stranded level such a structure generates a right handed vertical double helix[33] in which the base planes are parallel to the helix axis and the duplex is stabiliz- ed by vertical base stacking and familiar Watson-Crick hydrogen bonding. This is absolutely an out of the ordinary and a novel double helix and vastly different in gross morphology from the one advocated by Sundaralingam and Yathindra.[30] A computer generated stereo space filling perspective of this vertical double helix is shown in Plate 8. A striking feature of the vertical double helix is the large vacant in- ner core ($\simeq$ 36) which is absent in B-DNA and is about 6-7Å in A-DNA (Plate 9). This paper describes our experimental attempts to settle the controversy between Olson[32/33] on the one hand and Sundaralingam and Yathindra[29/30] on the other hand about the helical organization of the single and double stranded high anti polynucleotides.

The Resolution of the Controversy

In the absence of single crystal x-ray data on short mini helices of high anti nucleic acid structures we have attempted to resolve the controversy by NMR studies of such systems. The high lights of our findings are summarized below. Details will be presented elsewhere.[34]

Experimental

The most difficult part of this investigation was the procurement of a nucleic acid constituent (i) in which the sugar base torsion can be chemically engineered to adopt a fixed value of $\simeq$ 120° (ii) in which a potentiality exists for the formation of double helices and (iii) whose NMR signals, particularly those from the base protons can be unambiguously assigned. All the above requirements are easily met by the cyclodinucleoside monophosphate A^{s}pU^o whose structure is shown in Figue 1. The A^{s}pU^o and the constituent mononucleotides A^{s}p and pU^o were synthesized in the laboratory of Ikehara and Uesugi.[34] r-ApU and d-pGpC and the constituent mononucleotides were purchased commercially. Pulsed ¹H NMR measurements in

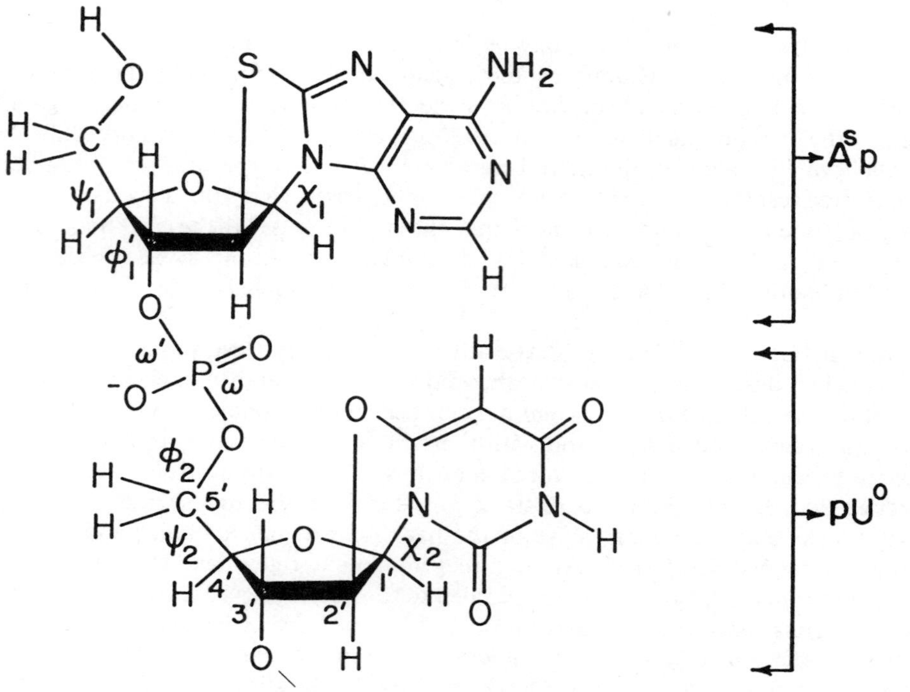

Figure 1. The structure of A^{s}pU^o.

the time domain were performed using the Bruker 270 MHz FT system at the Southern New England High Field Facility, New Haven and the data were Fourier transformed to the frequency domain using a BNC 12 data system located at the laboratory in Albany. ^{31}P FT NMR measurements were made at 109.3 MHz at National Magnet Laboratory, MIT.

Evidence For the Formation of a Self-Complementary Miniature Double Helix by A^{s}pU^o

The chemical shift data at 5°C and 20 mM indicated that the base proton H2 of the A^{s}p- segment of A^{s}pU^o has moved markedly upfield (0.32 ppm) relative to the corresponding proton in the mononucleotide A^{s}p, while the chemical shift of the uridine H5 was little affected (Figure 2). This kind of a shift pattern for PUPY (PU = purine, PY = pyrimidine) dinucleoside monophosphates can be described as bizarre because such a situation has never been encountered in our detailed study of all the naturally occurring systems, i.e., ApC, ApU, GpC, GpU and the four corresponding deoxyribose systems.[3-10] In the naturally occurring PYPU and PUPY dinucleoside monophosphates one invariably observes the base protons of the PY residue shifting significantly to higher fields with very small upfield shifts for the base protons of th PU residue. These observations have been rationalized on the ground that PYPU and PUPY molecules show a general preference to exist in intramolecularly stacked arrays and in such an array the strong ring current shielding abilities of the purine causes significant upfield shifts of the pyrimidine protons and

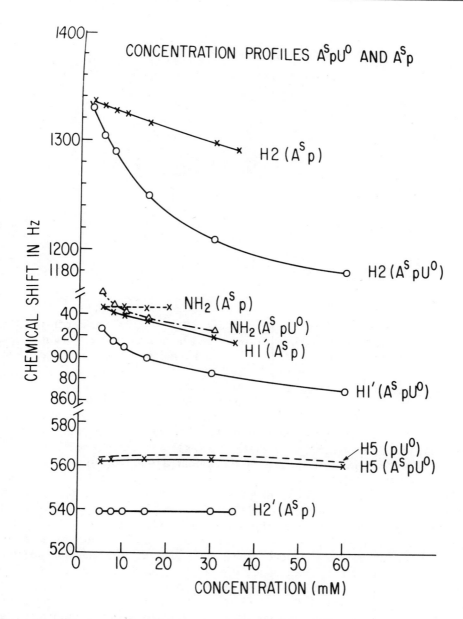

Figure 2. Concentration dependence of the chemical shifts of A^{s}pU^o and the mononucleotide A^{s}p.

that the purine protons do move to higher fields only to a small extent because of the poor ring current effects of the pyrimidine systems. For example, in ApU, the H5 and H6 of -pU are shifted to higher fields by 0.374 and 0.263 ppm respectively whereas the adenine H2 and H8 of Ap- are affected in the range of 0.005 and 0.021 ppm respectively[3/10] but the situation has reversed in A^{s}pU^o.

The only way to rationalize the observed upfield shifts of H2 of A^{s}p- of A^{s}pU^o is to invoke adenine-adenine interactions. There are two ways in which this could hap-

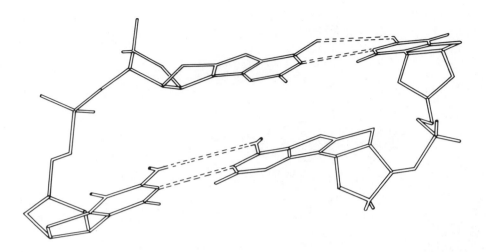

Figure 3. Self-complementary miniature double helix of A^{s}pU^o.

pen: (i) aggregation of A^{s}pU^o molecules like all the nucleic acid constituents resulting in base stacking and column formation, (ii) formation of a Watson- Crick hydrogen bonded miniature double helix in which adenine of molecule 1 shields the adenine of molecule 2 and vice versa (Figure 3).

It has been shown[35] that in the adenine nucleotide systems intermolecular aggregations become dominant and cause marked upfield shifts only beond 50 mM concentrations. The concentration dependence of the chemical shifts of A^{s}pU^o and the mononucleotide A^{s}p (Figure 2) show that at even as low as 15 mM, the H2 of A^{s}pU^o is shifted to higher fields by as much as 66 Hz compared to the corresponding proton in the mononucleotide A^{s}p. In the concentration range of 5-30 mM at 5°C the H2 and H1′ of the mononucleotide show a linear dependence and undergo an upfield shift of 30-35 Hz. In the case of A^{s}pU^o, the H2 and H1′ of A^{s}p- part show a nonlinear dependence and undergo shifts to higher fields by 120 Hz (!) and 41 Hz respectively. The H5 of -pU^o of A^{s}pU^o is unaffected in the same concentration range. The concentration study indeed shows that weak and strong adenine- adenine interactions exist in the mononucleotide A^{s}p and the dinucleoside monophosphate A^{s}pU^o respectively and molecular topologies that cause these interactions are most likely different in the two systems. We interpret the upfield shifts with increasing concentration from 5 to 30 mM for the mononucleotide essentially reflect increasing aggregations, whereas the marked high field shift for A^{s}pU^o comes mostly from the increasing formation of adenine-adenine overlapping Watson-Crick base paired miniature double helices with increasing concentration from 5 to 30 mM. Krugh, *et al.*[36] and Young and Krugh,[37] by monitoring the shift value of the Watson-Crick hydrogen bonding protons, have demonstrated that at low temperatures self-complementary systems such as GpC, CpG, etc. form increasing amounts of double helices as the concentration is increased from 5 to 30 mM. Cross and Crothers[38] and Patel and coworkers[39-41] have shown that in systems which readily form double helical arrays, such as tetra-, penta-, and hexanucleotides, formation of double helical structures always results in large upfield shifts for the base protons as we have observed for A^{s}pU^o. The only difference is that our system is just a dinucleoside monophosphate. It appears likely that the rigid high anti χ_{CN} of $\simeq 120°$ in A^{s}pU^o renders the easy formation of a double helix stabilized by adenine-adenine interactions.

The direct evidence for the formation of miniature double helices in A^{s}pU^o comes from the actual experimenal observation of the helix-coil melting curve for A^{s}pU^o.

Helix Coil Transition in A^{s}pU^o

NMR spectroscopy has an advantage over such methods as circular dichroism, ultraviolet absorption and infrared spectroscopy to monitor the helix coil transition in polynucleotides because more than one chemical shift can be followed as a function of temperature. Patel and coworkers[39-41] Cross and Crothers[38] and Arter, *et al.*[42] have used proton chemical shifts to monitor the helix coil transition in tetra-,

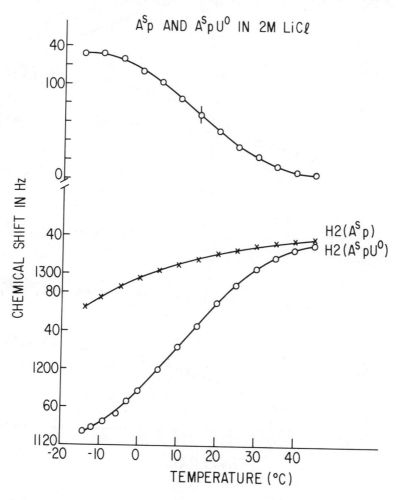

Figure 4. Temperature profiles of the chemical shifts of H2 of A^{s}pU^o and A^{s}p in 2 M LiCl. The one on the top is the difference curve. Concentration 20 mM.

penta-, and hexanucleotides. From the sigmoidal curve, the melting temperature $T_{1/2}$ has been determined. The transition from coil to helix has been shown[38/39/42] to be accompanied by large chemical shift changes for some of the protons.

We have carried out our melting studies in 2.0 M LiCl (plus 0.01 M sodium cacodylate) the data are plotted in Figures 4 and 5 which show that the H2, H1' and H2' of A^{s}p- part and H1' of -pU^o part show the sigmoidal behavior but H5 and H2' of -pU^o does not. It is important to emphasize that all protons do not have to display

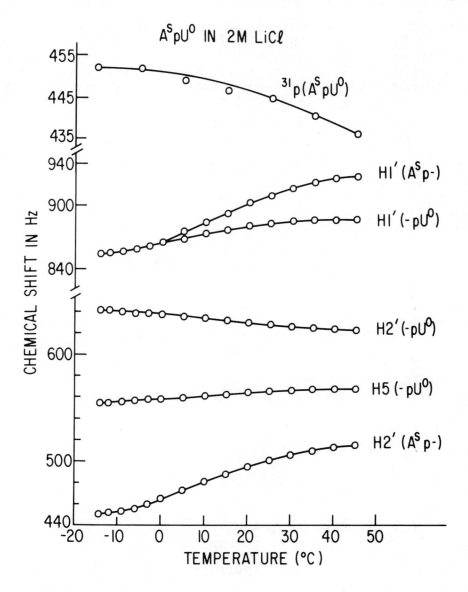

Figure 5. Temperature profiles of the chemical shifts of ^{31}P and protons other than H2 in A^{s}pU^o in 2 M LiCl, Concentration 20 mM.

the melting pattern. Chemical shift of a given proton is local geometry and environment dependent and formation of a helix from coil seriously perturbs the chemical shift of certain protons, some protons undergo little perturbations and in other cases a large number of factors such as ring current fields, electric field polarizations, atomic diamagnetic anisotropy, etc. interplay and compensate the shifts. A very important observation is that the ^{31}P chemical shifts (Figure 5) measured at a frequency of 109.3 MHz do not show any sigmoidal behavior and the total change in chemical shifts as a result of helix coil transition is only about 15 hertz. This indicates that during helix- coil transition there is little change in the O-P-O bond angle.[43/44]

One can derive the thermodynamic parameters for the helix-coil transition from the melting curve. In order to obtain good values, it is necessary to correct the melting curve from contributions of aggregation effects. This is particularly so when conditions such as low temperature and high salt concentrations which favor aggregations are employed. We have attempted to correct for this by determining the temperature dependence of the shift of adenine H2 in the mononucleotide A^{s}p on the assumption that aggregation and column formation in A^{s}p and A^{s}pU^o may affect the shift values in the same way. This is the best one could do with NMR methodology. The data for A^{s}p in 2 M LiCl is presented in Figure 4. From the data for A^{s}p and A^{s}pU^o, the difference curve is plotted and is shown in Figure 4 (top) and the curve is a clean melting curve with a transition temperature of 15°C in 2 M LiCl and the first observed melting curve for a dinucleoside monophosphate. Calculations based on Grella-Crothers[45] expression indicate that the helix coil transition of self-complementry A^{s}pU^o is associated with a reaction enthalpy of 15.7 Kcal/mole. The equilibrium constant calculated by the methods in Patel[39] for the helix coil transition of 20 mM solution in 2 M LiCl at T$_{1/2}$ = 15° is 50 (i.e., K$_{formation}$). The magnitude of reaction enthalpy and the formation constant appears to be reasonable when compared to those reported for tetra-, pentanucleotides, etc.[38-42] In the absence of any such data for dinucleoside monophosphates, it is necessary to state that the reported thermodynamic quantities be treated qualitatively.

Solution of the Controversy and Evidence in Favor of a
Right-Handed Vertical Double Helix
for High Anti Polynucleotides

Evidence presented above unmistakably shows that the high anti self- complementary A^{s}pU^o forms miniature double helices and in doing so the adenine H2 undergoes overwhelming shielding with little shielding of the H5 of -pU^o, i.e., the structure of the miniature double helix is characterized by strong interstrand adenine-adenine interaction with little intramolecular stacking between the adenine and uracil of the same strand, i.e., *significant interstrand base stacking and little intrastrand base stacking* (Figure 3).

Our experimental observation that under conditions in which χ_{CN} is fixed at $\simeq$ 120°, A^{s}pU^o forms miniature double helices with *significant interstrand base stacking* and *little intrastrand base stacking* enables to settle the question about the helix handedness of the duplexes of high anti polynucleotides. Examination of the Sundaralingam-Yathindra model (Fig. 5ab in Reference 30) clearly reveals that in their their model there is strong intrastrand and little interstrand base stacking. This is exactly the reverse of what is observed experimentally in our laboratory, unmistakably leading to the conclusion that the theoretical helix parameters that Sundaralingam and Yathindra[30] have derived from the rigid nucleotide concept is untenable, and consequently their proposed molecular mechanics of untwisting a helix and switching the sense of the helix[30] are of questionable validity. The observed significant interstrand and little intrastrand base stacking in the miniature duplex of A^spU^o are also not a feature of A-DNA, B-DNA, C-DNA and D-DNA as is evident from Plates 10-13. In plates 14 and 15 are shown two exploded views of the Olson's right handed vertical double helix for poly(rA)•poly(rU), and one immediately notices that the conspicuous feature of this double helix is the significant interaction between interstrand bases ad lttle intrastrand base-base interactions. Thus our experimental observations that when χ_{CN} is fixed at $\simeq 120°$ the self-complementary A^spU^o forms miniature double helices with significant interstrand and little intrastrand base-base interactions provide the first direct experimental support for the novel vertical double helix for high anti polynucleotides.

The arguments that are presented in favor of a vertical double helix are qualitative based on the geometric relationships of the base pairs in a poly(rA)•poly(rU) duplex. However, our system is A^spU^o and the corresponding polymer is made of the self- complementary poly rArU system. Hence Olson's coordinates for a self-complementary → ApU UpA ← system was derived and these coordinates were utilized to generate the miniature double helix conforming to Olson's torsion angles. This is shown in Figure 6 where the interstrand base stacking can be seen very clearly. For the structure shown in Figure 6 we have made detailed ring current and diamanetic and paramagnetic susceptibility anisotropy calculations. The experimentally observed and theoretically predicted shielding for a miniature double helix conforming to Olson's torsion angles are given in Table II. Except for the H1' the agreement between the two is remarkable. The lack of agreement of H1' is a direct consequence of the presence of a sulphur or oxygen atom situated very near the H1' in the experimental system of A^spU^o. From these studies it appears reasonable to conclude for a nucleic acid system if the sugar base torsion angle can adopt a high anti value, it most likely may take up the shape of the vertical double helix proposed by Olson.[33]

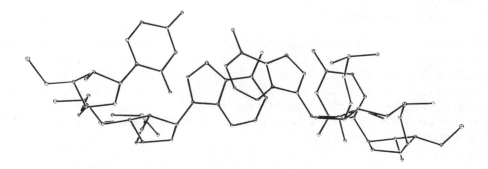

Figure 6. Self complementary miniature double helix of ApU conforming to the torsion angles of vertical double helix. Torsion angles employed were: χ = 120°, $\alpha(\phi') = 198°$, $\beta(\omega') = 268°$, $\gamma(\omega) = 295°$, $\delta(\phi) = 178°$, $\epsilon(\psi) = 58°$, sugar pucker ^3E.

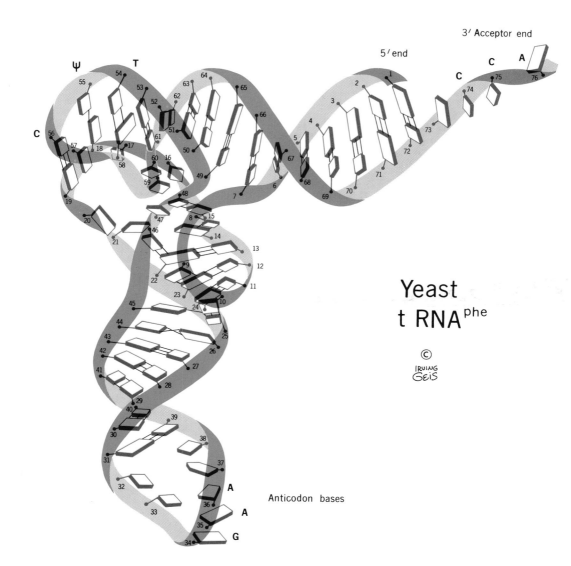

Plate 1. An illustration of the crystal structure of yeast phenylalanine transfer RNA. The phosphate ribose backbone is represented as a continuous blue ribbon with the bases attached to the light side of the ribbon. Bases are shown in red with purines as longer boards, pyrimidines as shorter boards and the highly modified base 37 as a pentagonal board. Hydrogen bonds between bases are shown with thin black lines. All the base pairs are of the Watson-Crick type with the following exceptions: 4-69, 8-14, 9-23, 10-45, 15-48, 18-55, 22-46, 26-44, and 54-58. Note especially the three way base pairing of 46-22-13, 9-23-12; the "propellor" twist of 44-25, and the "wobble" pair 4-69. Illustration by Irving Geis in collaboration with Sung-Hou Kim. Copyright 1979 by Irving Geis, 4700 Broadway, New York, New York 10040.

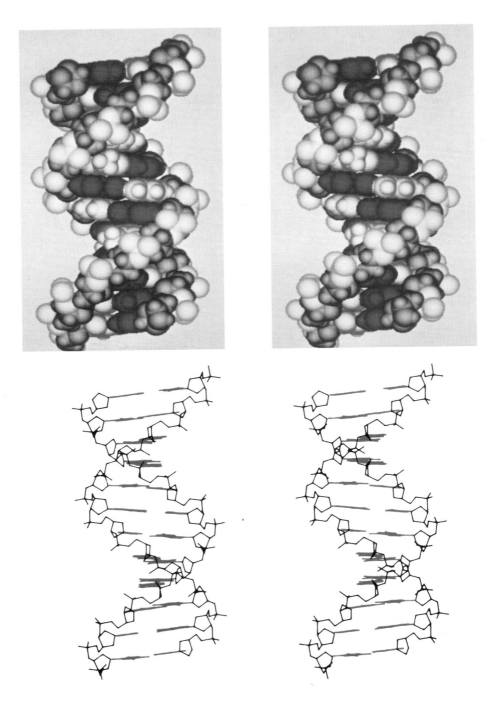

Plate 2. Computer generated space filling and the corresponding line drawing of B-DNA in stereo.

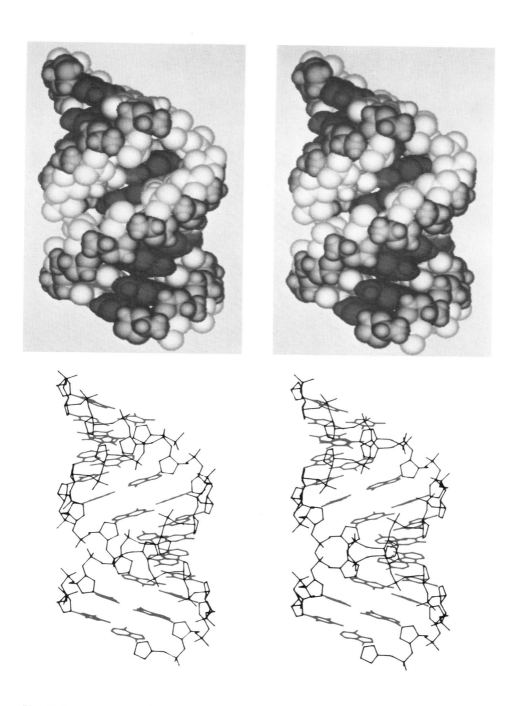

Plate 3. Computer generated space filling and the corresponding line drawing of A-DNA in stereo.

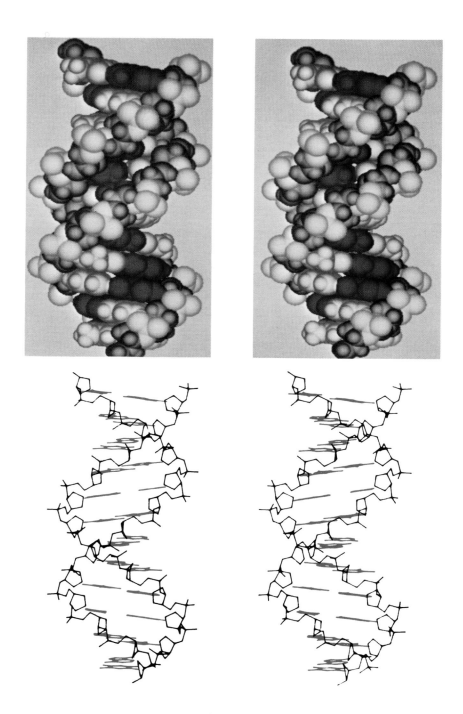

Plate 4. Computer generated space filling and the corresponding line drawing of C-DNA in stereo.

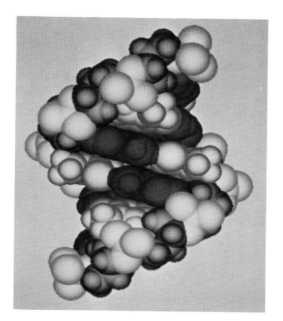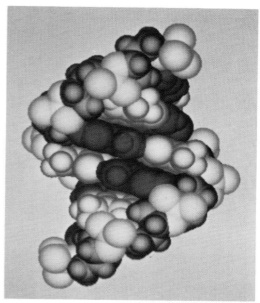

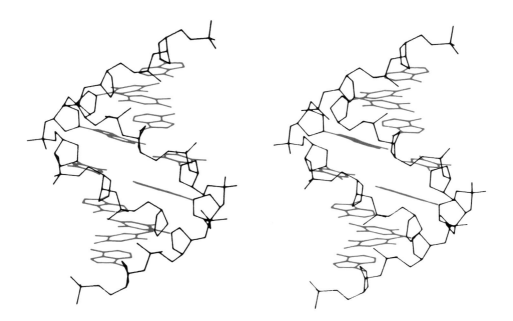

Plate 5. Computer generated space filling and the corresponding line drawing of D-DNA in stereo.

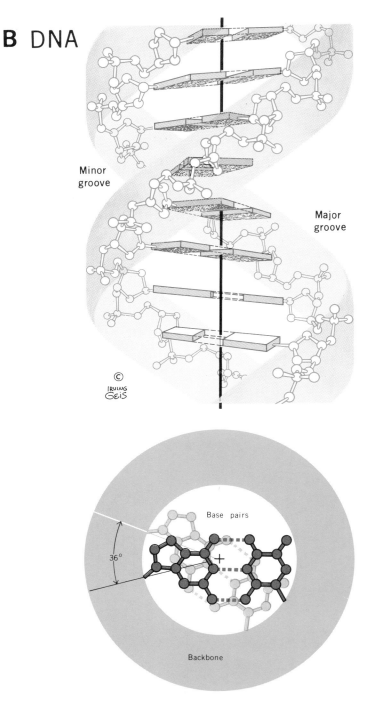

B DNA

Minor groove

Major groove

© IRVING GEIS

Base pairs

36°

+

Backbone

Plate 6. Double helical model of B-DNA, viewed perpendicular to the helical axis (top illustration). Base pairs are drawn in red and they are almost perpendicular to the helical axis. Ten base pairs make one complete turn. The phosphate deoxyribose backbone is enclosed in a blue ribbon. View along the helical axis (bottom illustration). Only two consecutive base pairs are shown. Note that the planes of base pairs intersect the helical axis.

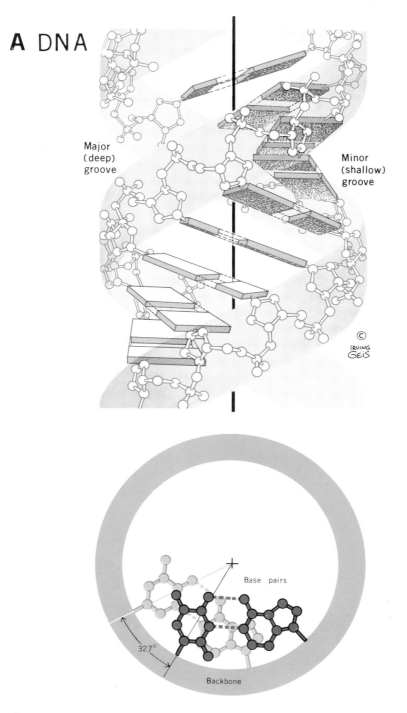

A DNA

Major
(deep)
groove

Minor
(shallow)
groove

© IRVING GEIS

Base pairs

32.7°

Backbone

Plate 7. Double helical model of A-DNA, viewed perpendicular to the helical axis. (top illustration) Base pairs are shown in red and they are inclined to the helical axis. Eleven base pairs make one complete turn. As in the B-form, the phosphate deoxyribose backbone is enclosed in a blue ribbon. View along the helical axis (bottom illustration). Note that the base pairs go around the helical axis leaving the vertical axis open. Double helical A-RNA is very similar in conformation to A-DNA. Illustrations in these two pages, copyright by Irving Geis.

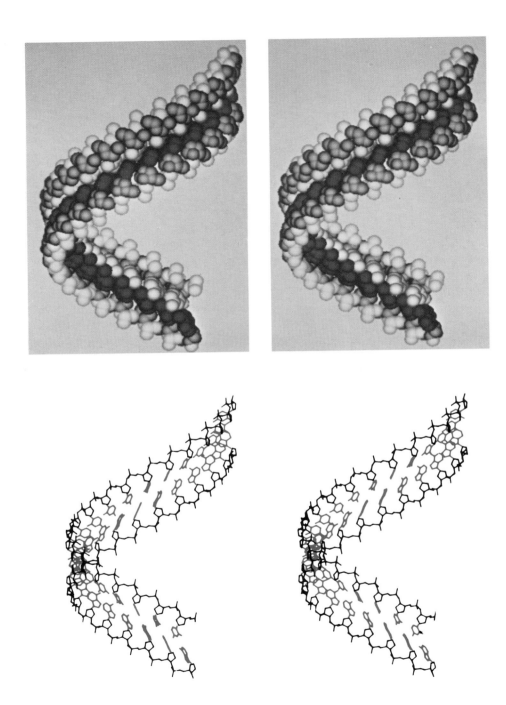

Plate 8. Computer generated space filling and the corresponding line drawing of the vertical double helix.

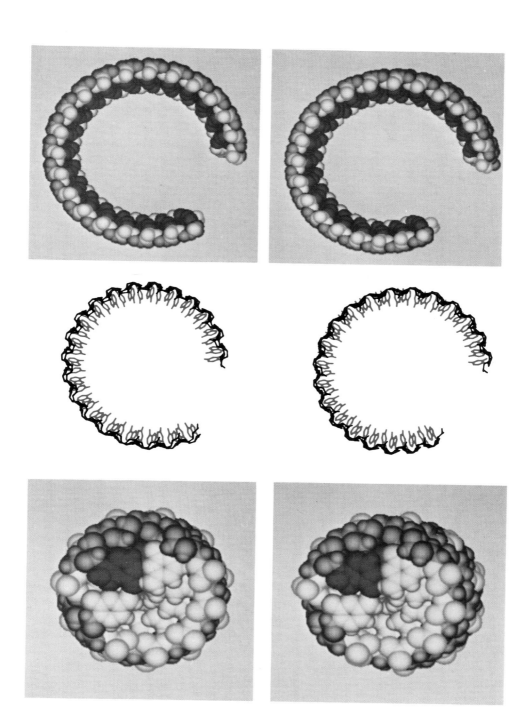

Plate 9. Computer generated top view of the vertical double helix (top two stereos) and that of B-DNA (bottom). Notice the large vacant inner core in the former and none in the latter.

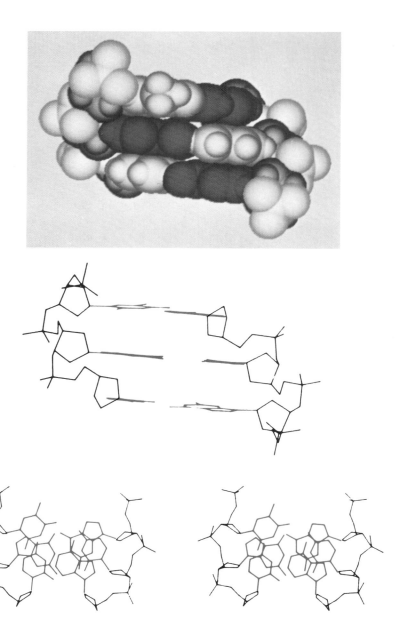

Plate 10. Computer generated geometrical relationships among the base pairs in a trimer duplex of B-DNA. The top two views are along the base planes and the bottom one is top view in stereo.

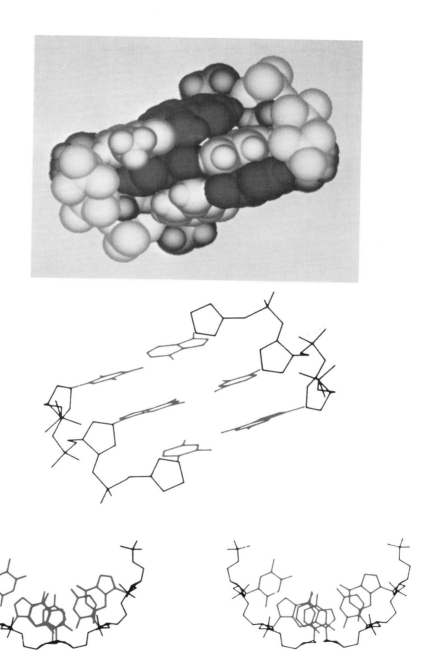

Plate 11. Computer generated geometrical relationships among the base pairs in a trimer duplex of A-DNA. Details same as in Plate 10.

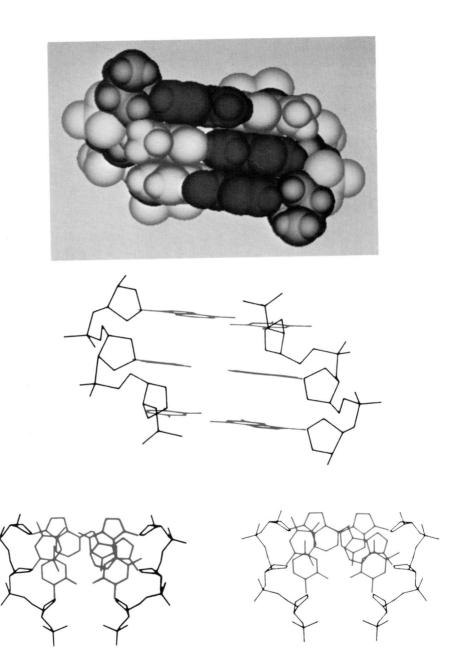

Plate 12. Computer generated geometrical relationships among the base pairs in a trimer duplex of C-DNA. Details same as in Plate 10.

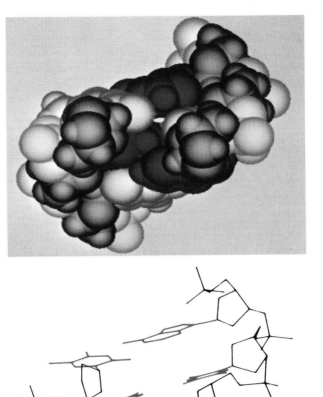

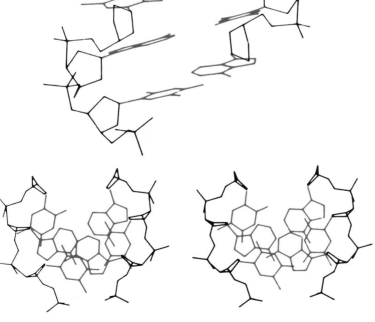

Plate 13. Computer generated geometrical relationships among the base pairs in a trimer duplex of D-DNA. Details same as in Plate 10.

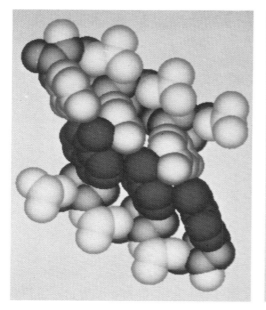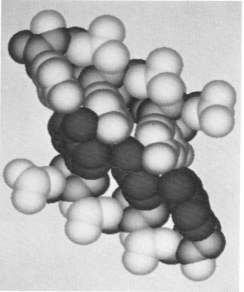

Plate 14. Computer generated duplex segment of Olson's vertical double helix.

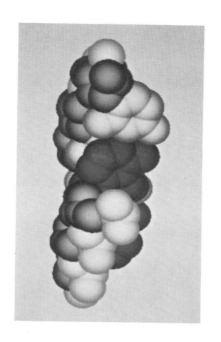
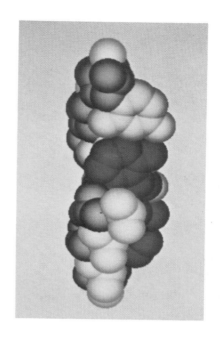

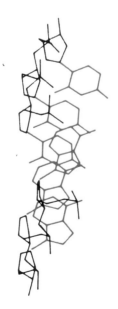
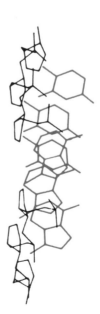

Plate 15. Same segment as in Plate 14, but the view more clearly reveals geometric relationships among the bases.

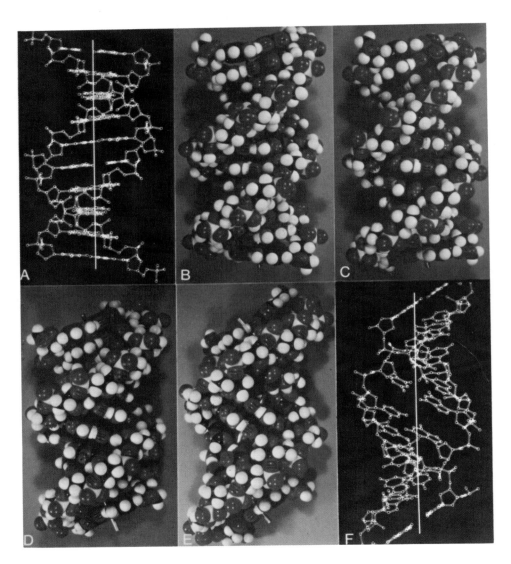

Plate 16. Illustration to show the normal mode oscillation in DNA structure that gives rise to the formation of β kinked DNA. A. B DNA, drawn by computer graphics. B. B DNA, as visualized with the Corey-Pauling-Koltun (CPK) space filling molecular models. C. Low amplitude normal mode oscillation. D. Higher amplitude normal mode oscillation. E. β kinked DNA structure. F. β kinked DNA, drawn by computer graphics. See text by Lozansky, Sobell and Lessen for detailed discussion.

Table II

(A1pU1)(A2pU2) Calculated and Observed Shifts
for Miniature Double Helix Effect of Adenine (A1)

Moiety	Proton	Calculated Shift in Parts Per Million (ppm)				Observed Shift (ppm)	
		Ring Current	Diamagnetic anisotropy	Paramagnetic anisotropy	Total ∂_{cal}	∂_{obs}	
A2	H-8	0.125	-0.005	0.050	0.170	--
	H-2	0.292	-0.005	0.099	0.386	0.47
Sugar (A2)	H-1'	0.305	0.003	0.091	0.399	0.14
	H-2'	0.096	0.001	0.049	0.146	0.19
Sugar (U2)	H-1'	-0.085	0.005	-0.017	-0.101	0.12
	H-2'	-0.048	0.001	-0.018	-0.064	-0.06
U2	H-5	-0.079	0.003	-0.047	-0.123	0.02
	H-6	-0.060	0.002	-0.002	-0.081	--
U1	H-5	-0.044	0.002	-0.014	-0.056	0.00
	H-6	-0.032	0.002	-0.004	-0.035	--

Vertical Double Helices for Naturally Occurring Polynucleotides?

In her opus on the novel vertical double helix[33] Olson advances it as a potential alter-native ordered structure available to naturally occurring nucleic acid systems. If naturally occurring ApU, GpC, etc. have potentialities to become part of self- com-plementary vertical double helices, they should show proclivities to adopt a high anti sugar base torsion. Detailed NMR studies of all the naturally occurring dinucleoside monophosphates[3-10] indicate that they are flexible in aqueous solution and this flex-ibility gives rise to conformational pluralism,[5-8] i.e., they exist as a conformational blend in which certain ones predominate over the others. NMR methods have not been able to determine accurately the magnitude of χ_{CN} except that they occupy the anti domain in general. If ApU in aqueous solution has tendencies to adopt high anti χ_{CN}, one indeed would expect ApU to go into high anti double helical arrays like A^spU° and display shift patterns similar to A^spU°. In Figure 7, we have plotted the temperature dependence of H2, H8 and H1' of Ap- and H5, H6 and H1'of -pU fragments of ApU under conditions identical to Figures 4 and 5 for A^spU°. A com-parison of the data for A^spU° and ApU indicate that they adopt different spatial configurations and ApU has no proclivity whatsoever to adopt a χ_{CN} of $\simeq 120^\circ$ and enter into a double helical array similar to A^spU°. Our argument is that if any real potential existed for ApU to adopt a $\chi_{CN} \simeq 120^\circ$ the imperatives of molecular stereodynamics demand that it adopts such values under conditions which favor double helices formation because such values of χ_{CN} generate a double helix stabiliz-ed by adenine-adenine stacking interactions, so stabilizing is the interaction that one is able to observe the helix coil melting curve for the dinucleoside monophosphate. We have done identical experimens with d-pGpC and there was no indication of the formation of miniature double helices of the type in Figure 6. Our conclusion is that self-complementary ApU and d-pGpC in 2 M LiCl have no potentiality to form

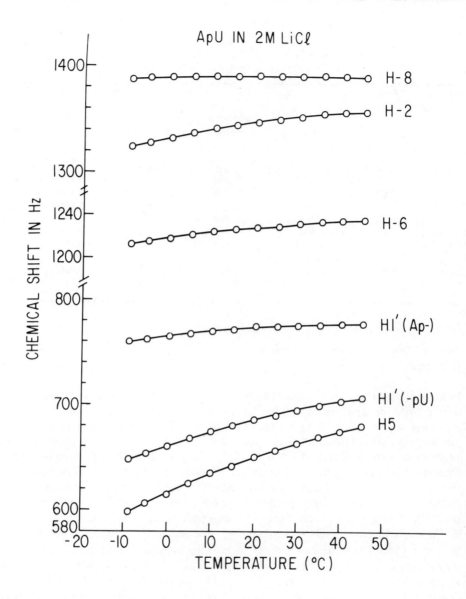

Figure 7. Temperature profiles of the chemical shifts of the various protons in ApU in 2 M LiCl. Concentration 20 mM.

miniature double helices which can be considered part of an ordered vertical double helix for the corresponding polynucleotide.

This is not surprising because we have shown[3-10] that one of the important contributing conformers in the conformational blend of single stranded naturally occurring dinucleoside monophosphates in aqueous solution is a right handed stacked one (03'-P and P-05' torsions $\simeq$ 290° and $\simeq$ 290°) with low values for χ_1 and χ_2. If the double helix of such a structure is made a repeating unit of a polynucleoside duplex, this will result in traditional double helical arrays. Our solution results are in eminent agreement with the single crystal X-ray results on dinucleotide monophosphates.[13-14] Obviously, in the absence of experimental data to the contrary, which are so difficult to obtain, one cannot dismiss the claims in which naturally occurring polynucleotides take up reasonable spatial configurations such as vertical, side by side, kinky, propellor twisted and so on under conditions in which they interact in vivo with other biological structures. And it will be long before one would know the truth.

Acknowledgment

The authors thank National Cancer Institute DHEW (CA12462) and National Science Foundation (PCM-7822531) for the support of this investigation. The authors are deeply indebted to Irving Geis for the artistic rendition of A-DNA, B-DNA, and tRNA. We thank Wilma Olson for supplying the coordinates and Bernard Pullman and Giessner-Prettre for programs to some of the complex calculations. We are indebted to M. Ikehara for a sample of A^{s}pU°.

References and Footnotes

1. Permanent address: Tata Institute of Fundamental Research, Bombay, India.
2. Szent-Gyorgyi, A., *International J. Quant. Chem. QBS 4*, 179 (1977).
3. Lee, C. H., Ezra, F. S., Kondo, N. S., Sarma, R. H., and Danyluk, S. S., *Biochemistry 15*, 3627 (1976).
4. Ezra, F. S., Lee, C. H., Kondo, N. S., Danyluk, S. S., and Sarma, R. H., *Biochemistry, 16*, 1977 (1977).
5. Sarma, R. H., and Danyluk, S. S., *International J. Quant. Chem. QBS 4*, 269 (1977).
6. Dhingra, M. M., and Sarma, R. H., "Proceedings, International Symposium on Biomolecular Structure, Conformation, Function and Evolution"R. Srinivasan (Ed.), Pergamon Press, Inc. (in press).
7. Dhingra, M. M., and Sarma, R. H., *International J. Quant. Chem. QBS 6* (in press).
8. Evans, F. E., and Sarma, R. H., *Nature 263*, 567 (1976).
9. Cheng, D. M., and Sarma, R. H., *J. Am. Chem. Soc., 99*, 7333 (1977).
10. Dhingra, M. M., and Sarma, R. H., Giessner- Prettre, C., and Pullman, B., *Biochemistry, 17*, 5815 (1978).
11. Dhingra, M. M., and Sarma, R. H., *Nature, 272*, 798 (1978).
12. Sarma, R. H., and Dhingra, M. M., *Nature, 278*, 582 (1979).
13. Rosenberg, J. M., Seeman, N. C., Day, R, O., and Rich, A., *J. Mol. Biol. 104*, 145 (1976).
14. Seeman, N. C., Rosenberg, T. M., Suddath, F. L., Kim, J. J., and Rich, A., *J. Mol. Biol. 104*, 109 (1976).
15. Quigley, G. I., Seeman, N. C., Wang, A. H. J., Suddath, F. L. and Rich, A., *Nucleic Acids Research 2*, 2329 (1975).
16. Sussman, J. L., Kim, S. H., *Biochem. Biophys. Res. Commun., 68*, 89 (1976).

17. Ladner, J. E., Jack, A., Robertus, J. D., Brown, R. S., Rhodes, D., Clark, B. F. C., and Klug, A., *Nucleic Acids Research 2*, 1623 (1975).
18. Stout, C. D., Mizuno, H., Rubin, J. Brennen, T., Rao, S. T., and Sundaralingam, M., *Nucleic Acids Research 3*, 1111 (1976).
19. Schevitz, R. W., Podjarny, A. D., Krishnamachari, N., Hughes, J. J., Sigler, P. B. and Sussman, J. L. *Nature 278*, 188 (1979).
20. Arnott, S., Dover, S. D., and Wonacott, A. J., *Acta Cryst., 25*, 2192 (1969).
21. Arnott, S., Hukins, D. W. L., Dover, S. D., Fuller, W., Hodgson, A. R. *J. Mol. Biol. 81*, 107 (1973).
22. Arnott, S., Chandrasekaran, R., Hukins, D. W. J., Smith, R. S. C., and Watts, L., *J. Mol. Biol., 88*, 523 (1974).
23. Marvin, D. A., Spencer, M., Wilkins, M. F. H. and Hamilton, L. D. *J. Mol. Biol. 3*, 547 (1961).
24. Dhingra, M. M., and Sarma, R. H. in "Stereodynamics of Molecular Systems", R. H. Sarma (Ed.), Pergamon Press, Inc., (1979).
25. Dhingra, M. M., and Sarma, R. H., Uesugi, S., and Ikehara, M., *J. Am. Chem. Soc., 100*, 4669 (1978).
26. Ikehara, M., and Uesugi, S., *J. Am. Chem. Soc., 94*, 9189 (1972).
27. Ikehara, M., and Tezuka, T., *J. Am. Chem. Soc., 95*, 4054 (1973).
28. Fujii, S., and Tomita, K., *Nucleic Acids Research, 3*, 1973 (1976).
29. Yathindra, N., and Sundaralingam, M., *Nucleic Acids Research, 3*, 729 (1976).
30. Sundaralingam, M., and Yathindra, N., *International J. Quant. Chem. QBS 4*, 285 (1977).
31. Neidle, S., Taylor, G., Sanderson, M., Huey- Sheng, S., and Berman, H. M., *Nucleic Acids Research, 5*, 4417 (1978).
32. Olson, W. K., and Dasika, R. D., *J. Am. Chem. Soc., 98*, 5371 (1976).
33. Olson, W. K., *Proc. Natl. Acad. Sci. USA, 74*, 1775 (1977).
34. Dhingra, M. M., Sarma, R. H., Uesugi, S., and Ikehara, M. (submitted).
35. Evans, F. E., and Sarma, R. H., *Biopolymers, 13*, 2117 (1974).
36. Krugh, T. R., Laing, J. W., and Young, M. A., *Biochemistry, 15*, 1224 (1976).
37. Young, M. A., and Krugh, T. R., *Biochemistry, 14*, 4841 (1975).
38. Cross, A. D., and Crothers, D. M., *Biochemistry, 10*, 4015 (1971).
39. Patel, D. J., *Biochemistry, 14*, 3984 (1975).
40. Patel, D. J., and Canuel, L., *Proc. Nat. Acad. Sci. USA, 73*, 647 (1976).
41. Patel, D. J., and Tonelli, A. E., *Biochemistry, 14*, 3990 (1975).
42. Arter, D. B., Walker, G. C., Uhelenbeck, O. C., and Schmidt, P. G., *Biochem. Biophys. Res. Commun., 61*, 1089 (1974).
43. Ribas Prado, F., Giessner-Prettre, C., Pullman, B., and Daudey, J-P., *J. Am. Chem. Soc.* (in press).
44. Gorenstein, D. G., "Nuclear Magnetic Resonance Spectroscopy in Molecular Biology,"B. Pullman (Ed.) D. Reidel Publishing Company, pp. 1-15 (1978).
45. Grella, J., and Crothers, D. M., *J. Mol. Biol., 78*, 301 (1973).

Does DNA Have Two Structures in Solution That Coexist At Equilibrium?

Edward D. Lozansky[1/2], Henry M. Sobell[2/3] and Martin Lessen[4]

Department of Physics[1] and Chemistry[2]
College of Arts and Sciences
University of Rochester
Rochester, New York 14627

[3]Department of Radiation Biology and Biophysics
The University of Rochester School of Medicine and Dentistry
Rochester, New York 14642

[4]Department of Mechanical and Aerospace Sciences
College of Engineering
The University of Rochester
Rochester, New York 14627

What is the structure of DNA in solution at equilibrium?

Is it only B DNA?

In this paper, we propose that different regions of DNA could have *two* discrete structures that coexist at equilibrium at a given temperature: B DNA and β kinked DNA. β kinked DNA corresponds to a second order phase transition in the polymer (different regions of DNA undergoing this transition at different temperatures) and arises from a specific normal mode oscillation in DNA structure that is excited by Brownian motion of solvent molecules.

Thus, DNA in solution at lower temperatures could exist only in the B form. At higher temperatures, however, permanently premelted β kinked DNA regions could appear. Due to their enhanced flexibility, these regions could be particularly prone to undergo DNA breathing motions (i.e., further base unstacking and hydrogen-bond breakage) due to thermal fluctuations. These breathing motions arise from both local fluctuations and, in addition, from fluctuations originating in neighboring regions that propagate energy in the form of mechanical waves (i.e., acoustic phonons) along DNA.

Our theory allows us to understand a variety of action at a distance phenomena observed for DNA.[1-3] It provides a model for RNA polymerase-promoter recognition and, possibly, for other protein-DNA interactions as well.[4-7] It allows us to understand how transient fluctuations elsewhere on the polymer can give rise to DNA breathing and to (mono- and bis- functional) drug intercalation.[8-12] Finally, it provides an insight into the transition state intermediate that connects B with A DNA in the B ⇌ A transition.[13]

A preliminary description of our theory has already appeared.[14] This paper further describes the theory and its biological implications.

Nature of Normal Mode Oscillation in DNA Structure that Gives Rise to the Formation of β Kinked DNA

We envision DNA in solution to be continuously bombarded by solvent molecules along its length. Although the vast majority of these solvent collisions transfer energy into the polymer to give rise to anharmonic motions in its structure that eventually dissipate energy back into solution due to viscous damping, occasionally a solvent molecule with the appropriate momentum (i.e., having its direction oriented along the dyad axes in DNA) collides with DNA and this gives rise to harmonic motion in the polymer.

By harmonic motion, we mean a specific normal mode oscillation in DNA structure that either remains localized to a site or travels along the helix in the form of a normal mode wave. At low amplitudes, these harmonic motions cause DNA to oscillate between B DNA and a right-handed superhelical variant of B DNA that contains approximately 10 base-pairs per turn and has a pitch of about 34 Å (compare Plate 16B and C). Such structures balance the unwinding in the helix with right-handed superhelical writhe to keep the linkage invariant,[15] a feature that creates minimal perturbation in DNA structure. Higher amplitude harmonic motions give rise to similar structures except that these contain somewhat more than 10 base-pairs per turn and have base-pairs inclined more acutely to the helix axis (See Plate 16D). Their presence could contribute to the apparent net unwinding of DNA and the altered tilt of base-pairs relative to the helix axis, features observed for DNA in solution (see discussion).

At small amplitudes of oscillation, DNA behaves as an elastic body that accumulates strain energy in its structure through small changes in torsional angles that define the geometry of the sugar-phosphate backbone. These changes are localized primarily in the furanose rings of alternate deoxyribose sugar residues (normally, the puckering of the furanose ring in B DNA is C2'-endo; however, the effect of introducing strain energy into the helix is to alter the magnitude and direction of this puckering). At larger amplitudes of oscillation, the enhanced strain energy in the sugar-phosphate chains begins to flatten out the furanose ring. Finally, at some critical oscillation amplitude, alternate sugars "snap into" a C3' endo sugar conformation with a concommitant partial unstacking of base-pairs (see Plate 16E). This structure (denoted β kinked DNA) corresponds to an inelastic distortion in DNA structure and arises from a transient high energy normal mode oscillation in the helix localized at a specific site.

We are attempting to compute the detailed energy surface that defines this pathway of conformational change more precisely, and will report on this at a later time.

Structure of DNA In Solution At Equilibrium

The structural considerations described above lead us to postulate the existence of two discrete structures for DNA at equilibrium. That is, at any given temperature, DNA may consist of Watson-Crick B DNA regions and other regions that are permanently β kinked. We emphasize that these β kinked regions are *not* fluctuational conformational changes that occur transiently along the polymer—rather, they are permanent conformational changes associated with the absorption of heat from solution as the system approaches equilibrium.

We identify these multiply-kinked premelted regions as promoters. This makes the strong prediction that promoter regions (when activated) should be permanently accessible for attachment by the RNA polymerase enzyme. An important component in RNA polymerase-promoter recognition, therefore, is the interaction between RNA polymerase and these freely breathing DNA regions (by partial intercalation of aromatic side chains into DNA) and through recognition of other structural elements (possibly a specific class of nucleotide sequences) to form the tight binding complex.[5-8] Our model describes the process of RNA polymerase-promoter recognition in a simple and attractive way in comparison to alternative models that require large (and therefore less likely) fluctuations leading to a similar recognition process.

In addition to these permanently premelted β kinked promoter regions, short regions of β kinked DNA can form transiently elsewhere along the helix due to thermal fluctuations that exist at equilibrium (see below). These structures can undergo breathing motions during their lifetime to allow drug intercalation—structural aspects of this process have already been described elsewhere.[14/16-17]

Wave Propagation (Phonons) In DNA

We now ask—can acoustic waves exist in DNA, continuously propagating back and forth along the helix?

By an acoustic wave we mean a normal mode wave in the DNA helix that can travel at the speed of sound along the polymer. Such waves could arise due to thermal fluctuations at equilibrium (i.e., as discussed previously, when an appropriate "hot" solvent molecule strikes DNA this gives rise to a normal mode oscillation in DNA structure that can either remain at that site or travel along the molecule in the form of a normal mode wave). Of course, any travelling wave in DNA will experience damping due to viscous drag by the solvent; however, we think it possible that waves of the type we describe could propagate significant distances along the helix before being completely damped.

If this is true, then an important consequence of phonon propagation in DNA is the altered nature of fluctuations that exist along the helix at equilibrium. This is because—when calculating the magnitude of fluctuations at specific places on

DNA—one has to take into account not only fluctuations that arise locally at this site but, in addition, fluctuations that occur at a distance and travel to this site in the form of phonons. Since DNA is a heterogeneous polymer (having variable flexibility due to different stacking energies between base-pairs in different sequences), it follows (see appendix) that the probability a specific region experiences a fluctuation with a given energy depends on the elasticity of this region and of neighboring regions. Generally, more flexible DNA regions will experience larger fluctuations. The magnitude of these enhanced fluctuations increase the probability that transient DNA conformational change can occur (i.e., B DNA → β kinked DNA). Subsequent additional fluctuations in this region could then give rise to breathing motions to permit intercalation by a drug molecule into DNA.

The same considerations apply when one discusses quasi-equilibrium processes. For example, if one raises the temperature of a DNA solution very slowly one can expect fluctuational changes in conformation to arise along the polymer. As energy continues to enter the polymer, the probability that a fluctuation exceeds the threshold energies needed for inelastic deformation increases at specific places. Eventually, at a given temperature, a new equilibrium structure for DNA will be formed. It is for this reason that we have postulated two discrete structures for DNA to coexist at equilibrium.

Discussion

How does our model for DNA structure and energetics relate to the data? First of all, it clarifies structural aspects of DNA breathing and drug intercalation—two phenomena that reflect the dynamic structure of DNA in solution. Second, it allows understanding of premelting phenomena in DNA and relates this to the process of protein-DNA recognition—in particular, to RNA polymerase-promoter recognition. Third, it provides a reasonable explanation for action at a distance phenomena in DNA.

One of the clearest examples of action at a distance along DNA is the arabinose operon.[3] Here, it has been demonstrated that the araC promoter interacts with the araBAD promoter, even though these DNA regions are separated by about 150 base-pairs. Under conditions where one promoter is active—the other is not. This limitation in promoter function is releaved by cleavage between the promoters at a site close to the araBAD promoter by the BamI restriction nuclease. This now allows the araC promoter to operate in the presence of araC protein and L-arabinose, conditions under which the araC promoter is normally inactive.

Additional physico-chemical evidence for action at a distance along DNA has been presented by Wells and his colleagues.[1/2] The properties of the duplex block polymer $d(C_{15}A_{15})d(T_{15}G_{15})$ were examined by thermal denaturation and nuclease sensitivity in the absence and presence of drugs (actinomycin and netrospin) which bind specifically to only one end of the block polymer. These studies indicate that the properties of one region of a DNA molecule can be influenced by another remote region.

Another question frequently raised is: How many base-pairs per turn does DNA have in solution? Wang[18] has recently provided evidence that DNA has 10.4 base-pairs per turn at room temperature. Levitt[19] using semiempirical energy calculations proposes DNA to have an equilibrium structure centered at about 10.5 base-pairs per turn. An interesting feature of Levitt's model is the predicted base tilt of 78° relative to the helix axis and a propellor-like twist between base-pairs of about 28°. Crothers and his colleagues,[20] using transient electric dichroism to study the structure of monodisperse rod-like DNA molecules in solution estimate that base-pairs are inclined 73° or less to the helix axis, a value consistent with Levitt's proposal.

Our model, however, provides an alternative interpretation to these data. The presence of phonons in DNA distorts the helix for *two* reasons. First, there may be regions of DNA that remain permanently β kinked and therefore partially unwound. Second, travelling wave structures of the type we have described appear continuously due to fluctuations and their presence further acts to distort the DNA double helix. For these reasons, DNA in solution may have an average structure possessing slightly more than 10 base-pairs per turn and having a base tilt somewhat less than 84° relative to the helix axis.

Of course, one can argue that significant wave propagation in DNA is not possible due to viscous damping by the solvent. Clearly, this question must be resolved by experiment. Such an experiment has been done recently by Maret *et al.*,[21] and to our knowledge represents the first experiment of its kind. They have used inelastic laser light (Brillouin) scattering on partially crystalline fibers of calf thymus DNA to detect acoustic phonons travelling at about 1800 m/sec along DNA. These data appear to confirm the existence of wave propagation along the DNA molecule.

Acknowledgements

This work has been supported by grants from the National Institutes of Health, the American Cancer Society, and the Department of Energy (DOE). This paper has been assigned report no. UR 3490-1569 at the DOE, the University of Rochester.

Appendix

The importance of phonon propagation in DNA can be easily demonstrated if one considers a wave arising from a single fluctuation and propagating along the polymer. Since different regions of DNA have different elasticities, transmitted and reflected waves will arise at the boundaries. From Fresnel's formulae, the coefficients of transmission and reflection are

$$T = \frac{2\sqrt{E_1}}{(\sqrt{E_1} + \sqrt{E_2})} \qquad R = \frac{(\sqrt{E_1} - \sqrt{E_2})}{(\sqrt{E_1} + \sqrt{E_2})}$$

where E_1 and E_2 correspond to different values of Young's modulus in two neighboring regions of DNA (we have assumed the density of DNA to be constant along its length).

From this one can easily obtain a relationship between the transmitted and incident energy density. This is

$$\frac{W_t}{W_i} = |T|^2 = \frac{4E_1}{(\sqrt{E_1} + \sqrt{E_2})^2}$$

If $E_1 << E_2$ and the wave originates in area 1 then only a small part of the energy will be transmitted into area 2. On the other hand, if $E_1 >> E_2$ the transmitted energy density is: $W_t \approx 4W_i$. Notice that it makes no difference where the fluctuation originates—the energy of this fluctuation will be distributed with higher density in the region having lower elasticity (i.e., higher flexibility).

Of course, DNA has many regions with different elasticities and therefore multiple reflections and transmissions from a single fluctuation have to be taken into account. In addition, one must consider not only the effect of one fluctuation but more generally the influence of multiple fluctuations having different energies and arising at different times. This problem becomes very complicated and we will describe it elsewhere.

References and Footnotes

1. Burd, J. F., Wartell, R. M., Dodgson, J. B. and Wells, R. D. *J. Biol. Chem. 250,* 5109-5113 (1975).
2. Burd, J. F., Larson, J. E. and Wells, R. D. *J. Biol. Chem. 250,* 6002-6007 (1975).
3. Hirsh, J. and Schleif, R. *Cell, 11,* 545-550 (1977).
4. Travers, A. *Cell, 3,* 97-100 (1974).
5. Dickson, R. C., Abelson, J., Barnes, W. M. and Reznikoff, W. S. *Science, 187,* 27-30 (1975).
6. Stahl, S. J. and Chamberlin, M. J. *J. Mol. Biol., 112,* 577-601 (1977).
7. Von Hippel, P. H. (1978) in *Biological Regulation and Development,* Volume 1, ed. R. F. Goldberger, Plenum Publishing Company, New York, New York, pp. 279-347.
8. Teitelbaum, H. and Englander, S. W. *J. Mol. Biol., 92,* 55-78; 79-82 (1975).
9. McGhee, J. D. and von Hippel, P. H. *Biochemistry, 14,* 1281-1296; 1297-1303 (1975).
10. McGhee, J. D. and von Hippel, P. H. *Biochemistry, 16,* 3267-3275; 3276-3293 (1977).
11. Li, H. J. and Crothers, D. M. *J. Mol. Biol. 39,* 461-477 (1969).
12. Wakelin, L. P. G. and Waring, M. J. *Biochem. J., 157,* 721-740 (1976).
13. Franklin, R. E. and Gosling, R. G. *Acta Cryst., 6,* 673-678 (1953).
14. Sobell, H. M., Lozansky, E. D. and Lessen, M. *Cold Spring Harb. Symp. Quant. Biol., 43,* 11-15 (1978).
15. Fuller, B. F. *Proc. Nat. Acad. Sci. USA, 86,* 815-818 (1971).
16. Sobell, H. M., Tsai, C.-C., Jain, S. C. and Gilbert, S. G. *J. Mol. Biol., 114,* 333-365 (1977).
17. Sobell, H. M., Reddy, B. S., Bhandary, K. K., Jain, S. C., Sakore, T. D. and Seshadri, T. P. *Cold Spring Harb. Symp. Quant. Biol., 42,* 87-101 (1977).
18. Wang, J. C. *Proc. Nat. Acad. Sci. USA* in press.
19. Levitt, M. *Proc. Nat. Acad. Sci. USA, 75,* 640-644 (1978).
20. Hogan, M., Dattagupta, N. and Crothers, D. M. *Proc. Nat. Acad. Sci. USA, 75,* 195-199 (1978).
21. Maret, G., Oldenbourg, R., Winterling, G. and Dransfeld, K. (1978), *Abstracts VI Int. Bioph. Congress,* Kyoto, Japan.

Structure, Fluctuations and Interactions
of the Double Helix

N. R. Kallenbach[1], C. Mandal[2], and S. W. Englander[2]
Departments of Biology[1], Biochemistry, and Biophysics[2]
University of Pennsylvania
Philadelphia, PA 19104

The rate of exchange of nucleic acid protons with solvent protons is much slower in the double helix than in unstructured molecules such as the free bases or single stranded polynucleotides.[1-4] Figure 1 illustrates the behavior of exchangeable base protons in the synthetic duplex polynucleotide, poly(rA)•poly(rU).[5] This measurement provides an approach to the dynamics of an important structural opening process in this helix, some characteristics of which can be inferred from analysis of the salt, pH and temperature dependence of the H- exchange rates. In addition, measurement of H- exchange rate can yield information about the extent of base pairing in partially ordered molecules such as tRNA or rRNA, and potentially can even furnish details about nucleic acid structure in the presence of large amounts of protein as in ribosome, chromatin, and viruses. The experiment in Figure 1 was performed by monitoring the long wavelength UV transmittance changes upon rapidly diluting into H_2O a solution of poly(rA)•poly(rU) in D_2O using a stopped-flow spectrophotometer. Exchange of deuterons for protons is accompanied by a detectible difference spectrum near 290 nm.

Two kinetic processes are revealed in poly (rA)•poly(rU). The slower proves to represent the H-D exchange of the two exocyclic amino protons of A in the duplex. These two hydrogens exchange at identical rates despite the obvious structural difference between them: one is a member of an A-NH to U-C=O hydrogen bond while the second is freely exposed to solvent. The identity in rates results from the exchange of both from a conformational fluctuation of the double helix which severs the Watson-Crick H-bonds and makes both $A-NH_2$ protons equally vulnerable to exchange. This fluctuation is referred to as the H- exchange open state. The faster process in Figure 1 registers the exchange of the N_3-H ring proton of U. Despite its apparent inaccessibility due to the intramolecular $U-N_3$-H to $A-N_1$ hydrogen bond and the bases stacked above and below it, this hydrogen exchanges at a rate equal to the rate of the base pair separation reaction just described.

The basis for these assertions is presented below, with a discussion of the possible nature of the H- exchange open state in poly(rA)•poly(rU) and in other nucleic acid duplexes. Finally, a new application of H- exchange is presented in which details of

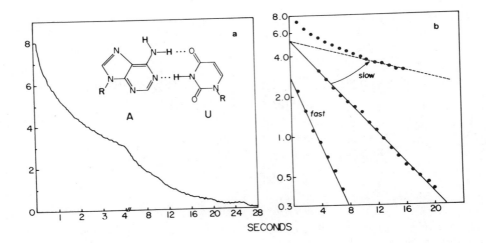

Figure 1. HD exchange of poly(rA)•poly(rU) detected by stopped-flow mixing at λ = 285nm. In this experiment, a sample of poly(rA)•poly(rU) in 0.1M phosphate buffer, D_2O at pD7, 20°C has been mixed with a ten-fold excess of H_2O containing salt and buffer at the same concentration. Final concentration of polymer was 0.22mM in phosphorus. The fast and slow exchange processes are assigned to U-N_3H and A-NH_2 respectively as described in the text. (a) Kinetic trace of transmittance recorded on two time scales. (b) Semi-logarithmic plots of the trace: curve (1) shows the slow reaction, and curve (2) the early time data on a two-fold expanded scale.

the intercalation of the phenanthridine dye ethidium bromide (EB) into DNA can be studied owing to the slowing of the free exchange rate of at least one of the amino protons of the dye molecule. The H- exchange slowing seems to indicate an intermolecular EB-DNA hydrogen bond in the complex.

Exchange Chemistry

Any interpretation of the exchange data must rest on an understanding of the exchange behavior of the bases in the absence of duplex structure. Figure 2 summarizes the exchange of the U N_3-H proton as a function of pH.[5] The two branches of the profile correspond to an H^+ catalyzed process below pH 4 and an OH^--catalyzed domain above pH 4. In the alkaline region the U-N_3-H proton is catalyzed by OH^- with a rate constant near the diffusion limited value, 1.5×10^{10} M^{-1} sec^{-1}. The chemical step in this case appears straightforward near neutral pH:

$$HO^{\ominus} + D-N \longrightarrow HOD + {}^{\ominus}N \qquad (1)$$

Catalysis of the exchange of the U-N_3-H proton by buffers such as imidazole confirms this, since the measured rate in the presence of imidazole base as acceptor is closely predicted by the difference in pK between imidazole (7.1) and uracil (9.5), i.e., $10^{10} \times 10^{-2.4}$ or near 10^7 M^{-1} sec^{-1}.

By contrast, the exchange behavior of adenosine is more complex, as seen in Figure 3. Near neutral pH, the amino proton of adenine exchanges by a pH independent mechanism involving two proton transfer steps-protonation of N_1 on the ring which reduces a highly unfavorable pK for deprotonating the amino group, facilitating removal of H from this group by OH^- or other bases;[2/6/7]

$$(2)$$

The pK for the first step is 4; the reduction in pK of the amino group following protonation is seen in the analogous case of N_1-methyl adenosine, for which the effective pK is near 8, rather than an estimated value near 20 in the case of adenosine itself.[2] The pathway in Equation 2 involves the product of (H^+) and (OH^-), which is just the constant Kw. Hence the independence of the rate with pH. Added general base catalysts can greatly increase the A-NH_2 exchange rate.[6] The pH independent exchange rate in neutral poly(rA) is found to be slower by a factor of three than in AMP.[5]

Evidence for An Open State in Poly (A•U) And Identity of the Exchanging Protons

From previous tritium exchange experiments, it is found that all three nitrogen protons of the bases exchange in A•T and A•U duplexes (and all five in the case of G•C duplexes).[1-3] A number of observations demonstrate that a transient breaking of base pairs in the duplex A•U structure acts as an intermediate state in the exchange reactions. i) Despite the relatively greater protection of ring protons from solvent, especially in the case of the central G-N_1-H to C-N_3 hydrogen bond of G•C pairs, these exchange readily under conditions in which duplex structure is stable and unperturbed by denaturaton. In fact, they exchange faster than the more exposed amino protons. ii) Both A-NH_2 protons exchange at the same rate in poly(rA•rU), poly(rAU)•poly(rUA) and in poly(dAT)•poly(dTA),[2] despite the involvement of only one in hydrogen bonded structure. iii) Exchange of A-NH_2 protons in A•U or A•T duplexes is pH-independent over the same pH range as for free A, implying

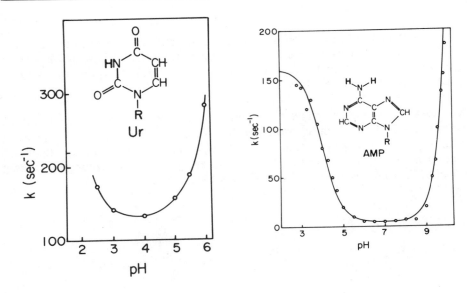

Figure 2. *Left* Exchange of the ring proton N_3-H in uridine at different pH determined by proton NMR in H_2O 0.1M solution in the absence of buffer. The rate is determined by the excess line width of the ring proton (near 11 ppm downfield from DSS), at 20°C.

Figure 3. *Right* Exchange of the amino protons of AMP measured by stopped-flow H-D exchange in the ultraviolet.[6] Near neutral pH, hydroxide is less effective than the pathway summarized in reaction (2) in which the ring N_1 position first protonates followed by removal of the now favorable amino protons by OH^- (or any buffer base).

that the pathway in Equation 2, requiring prior protonation at N_1 of the adenine ring, is operative. This site is blocked by an internal hydrogen bond in the duplex. iv) The similar pathway involving proton removal by buffer base, which also requires a pre-protonation at the normally protected N_1 of adenine, goes just as well in the double helix as in free AMP. We conclude that the exchange from stable hydrogen bonded duplexes takes place via a transient fluctuational base unpairing within the double helix.

A minimal description of an opening-dependent pathway, involving three states of an exchangeable proton, can be written as follows (see, for example, 2):

$$\text{closed} \underset{k_{cl}}{\overset{k_{op}}{\rightleftharpoons}} \text{open} \overset{k_{tr}}{\rightarrow} \text{exchanged} \tag{3}$$

The rate constants k_{op}, k_{cl} pertain to the conformational process, for which the equilibrium constant $K_{eq} = k_{op}/k_{cl}$, and k_{tr} represents the chemical transfer rates discussed in the previous section. The overall exchange rate corresponding to scheme (3) is given by:

$$k_{ex} = \frac{k_{op}k_{tr}}{k_{op} + k_{cl} + k_{tr}} \tag{4}$$

As has often been noted before, Equation 4 admits of two simple limiting situations: (1) if $k_{tr} \geq\geq k_{cl}$, so that the transfer is rapid, exchange becomes opening-limited and:

$$k_{ex} = k_{op} \tag{4a}$$

(2) conversely if $k_{tr} \leq\leq k_{cl}$,

$$k_{ex} = K_{eq}/(1 + K_{eq}) k_{tr} \sim K_{eq}k_{tr} \tag{4b}$$

the preequilibrium limit. Equation 4 also has a simple Lineweaver-Burke type of reciprocal form:

$$1/k_{ex} = 1/k_{op} + [(1 + K_{eq})/K_{eq}] 1/k_{tr} \tag{5}$$

Assignment of the slower rate class in poly (rA•rU) to the two A-NH$_2$ protons is on the basis of their pH independence over the same range as in AMP, and on their identical response to buffer catalysts as the NH$_2$ protons in AMP.[5] Most critically for the mechanism we have postulated as well as for the identity of the faster rate class, the experiment summarized in Figure 4 makes it clear that the faster process is in fact insensitive to general base catalysis as predicted by Equation 4a. The base used in this experiment, trifluoroethylamine (TFEA) would accelerate the free U-N$_3$-H exchange rate by orders of magnitude over the concentration range shown. But U-N$_3$-H, already exchanging at the opening-limited rate (Equation 4a), is unaffected. Only the slower process, representing the A-NH$_2$ protons, exchanging via a

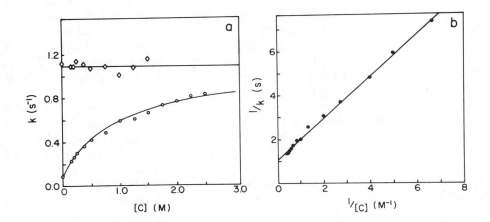

Figure 4. Test of open limiting exchange in poly(rA)•poly(rU) by response of exchange rates for the fast and slow processes in the polymer to a general base catalyst. The base used in this case was trifluoroethylamine (TFEA) (a). It is seen that the faster rate does not respond to buffer base over the range indicated, under which conditions free U-N$_3$-H would experience an approximate 10^5 increase in rate. On the other hand, the slower process asymptotically approaches the faster rate if catalyst is added. In (b) the reciprocal-plot of Equation 5 is illustrated, from which both k_{ex} and k_{op} can be evaluated. The data are those for the slower process in panel (a).

pre-equilibrium opening pathway (Equation 4b), is accelerated by TFEA. The slower rate class asymptotically approaches the same value of k_{op} inferred from Equation 4a, just as required by Equation 5.

The importance of this experiment is hard to exaggerate. Not only does it furnish a quantitative test of the concepts which lead to Equation 5, but it also establishes for the first time that both classes of hydrogens in poly(rA•rU) exchange from the same open state. If we accept the notion that the unperturbed closed form of the double helix is in equilibrium with a variety of fluctuational open states varying in energy from minimal openings through gross denaturation, this result implies that the H-exchange open state is the lowest in energy of these states and also displays the fastest rate at which the duplex can open. As such, the structure of this opening is of some interest.

The facts which any model for the H- exchange open state must reconcile are these: i) $\Delta G°$ for opening is about $+2$ kcal/mole at $25°C$, with $\Delta H°$ near $+5$ kcal/mole and $\Delta S°$ 10 e.u. ii) The opening rate at $20°C$ is 1 sec^{-1}, the closing rate 20 sec^{-1}, with an energy of activation of $+15$ kcal/mole for opening. iii) The open state does not exhibit significant features of a bulk denaturation process of the helix: in particular, neither salt nor divalent ion concentrations, which affect Tm appreciably, have much effect on exchange rates. The structure we presently lean toward represents a minimal opening in which an internal A•U pair ruptures and the U swings out, leaving the A more or less stacked. The extreme slowness of this process may be due to two important factors: i) The high activation barrier required to rupture the internal hydrogen bonds without simultaneously transferring the U-N_3-H and A-N_1 groups to water, as occurs in breaking base pairs at the ends of a duplex. ii) The difficulty of swinging out one member of a base pair from an internal bonded position because of steric contacts with adjacent base pairs. Most flexibility in nucleotides is associated with the O-P torsional angles, rotation about which is highly restricted in this situation. The $\Delta G°$ for unpairing a U base and moving it out of the helix seems within range of $+2$ kcal/mole, based on model A•U duplexes incorporating non-complementing U residues (see 8). It is more difficult to rationalize the large activation energy corresponding to the open state.

In this connection, it should be noted that the enthalpy corresponding to A•U hydrogen bonding in anhydrous chloroform is about -6 kcal/mole,[9] while that for transfer of U from its crystal to water is about as large.[10] The observed activation energy for opening might include both kinds of contribution. Finally we should indicate that the possibility of coupling the opening to internal fluctuations of the double helix cannot be eliminated, and experiments with short oligonucleotide duplexes as well as solvents of much higher viscosity are planned to investigate this.

The existence of this open state raises a number of questions concerning its relevance to protein-nucleic acid recognition and the dynamics of the DNA duplex in chromatin. If G•C base pairs are refractory both to melting and opening in the sense we have described,[3] one can imagine that A•T sequences provide a more rapid path to melting of DNA by proteins such as the RNA polymerase holoenzyme for example. At sites defined by a sequence including the Pribnow box[11/12] RNA polymerase forms an open complex with DNA in which a short region of DNA melts.[13/14] The highest efficiency promoters in *E. coli* and a number of phages correspond to a version of this sequence with six adjacent A•T pairs, while lower efficiency promoters include one or more G•C pairs at this site. Is this due to more rapid access of enzyme to the open state(s) of the A•T pairs? To answer this we need to define the open state in DNA carefully, in order to determine whether there are neighbor base pair effects for instance that influence opening of individual pairs in a heterogeneous environment.

A second line of enquiry concerns the dynamics of the DNA molecule within the nucleosome core structure in chromatin.[15] In this case, we can measure the open state and its kinetic properties for isolated core-particles released from chromatin by nuclease digestion. The particle from calf thymus consists of about 145 b.p. of DNA with a protein complement involving 26 tyrosines but no tryptophanes.[16] Hence the signal near 290nm should register the nucleic acid exchange as in free DNA. In this case, it is conceivable that changes in the dynamics might prove interesting, regardless of the magnitude of difference in equilibrium opening present. The point is that the open state defines both an equilibrium and dynamic aspect of nucleic acid behavior, and should thus be significant.

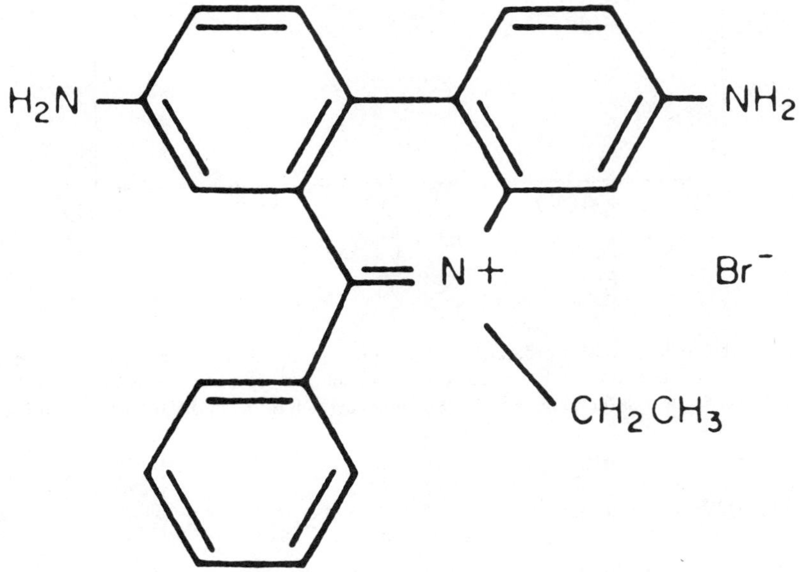

Figure 5. Structure of the phenanthridim dye ethidium bromide.

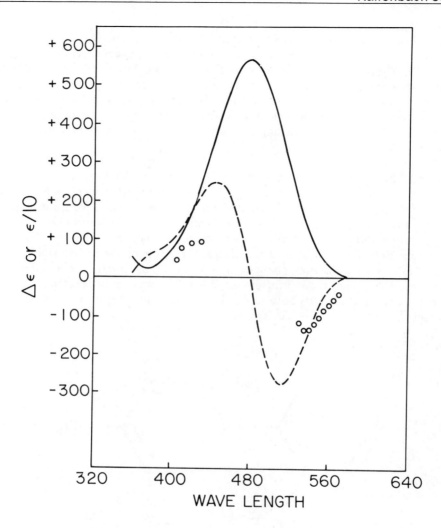

Figure 6. Absorbance and differential absorbance of ethidium bromide in H_2O and D_2O. The figure shows the visible absorbance spectrum in H_2O reduced twenty-fold as a solid line, the equilibrium difference spectrum (H_2O-D_2O) as a dashed line, and the exchange derived (kinetic) difference spectrum as individual points.

New Applications of Exchange Methods

The advantage of H-D exchange measurements using stopped-flow spectrophotometry over tritium exchange methods lies both in the rapidity and selectivity that can be achieved. While tritium-exchange experiments cannot directly resolve protein from nucleic acid exchange in protein-nucleic acid complexes, the isotope effect on absorption spectra of the base chromophores may well make this possible. As pointed out above, the requirements are simply that the protein not possess sufficient aromatic amino acids to dominate the contribution of nucleic acid around 280-290 nm. We believe that the exchange of nucleic acids in the presence of considerable

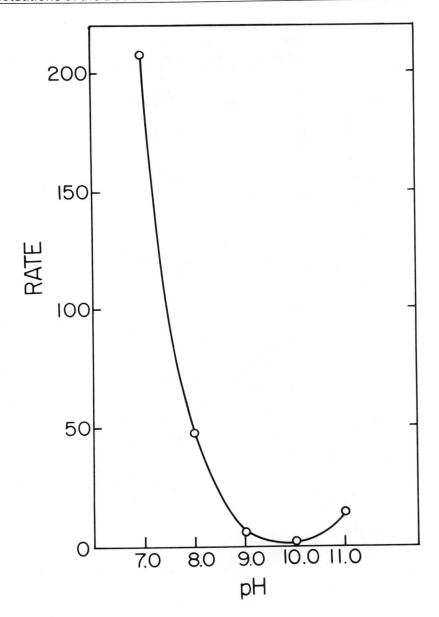

Figure 7. Dependence of exchange rate of amino protons of ethidium bromide on pH. The rates were determined using the difference spectrum of $\lambda = 540$nm, and a stopped-flow spectrophotometer, diluting dye in D_2O into H_2O. Each solution contained 1mM phosphate to stabilize pH.

amounts of protein can be followed by this technique. Englander *et al.*[17] have recently shown that exchange of the peptide group itself can be monitored in the vicinity of 230nm, and it may therefore be possible to detect exchange of proteins separately from nucleic acid in this region. Obviously, individual protons of protein and nucleic acid can be readily resolved using proton NMR, but it is frequently difficult to obtain high resolution spectra on high molecular weight systems, and chemical exchange is not necessarily the most significant factor in line-broadening or saturation transfer experiments.[18]

We have recently carried out an exchange study of protons in the intercalating dye ethidium bromide (EB), both free in solution and complexed with excess DNA. There are two sets of -NH_2 protons in this molecule (see Figure 5), which has a

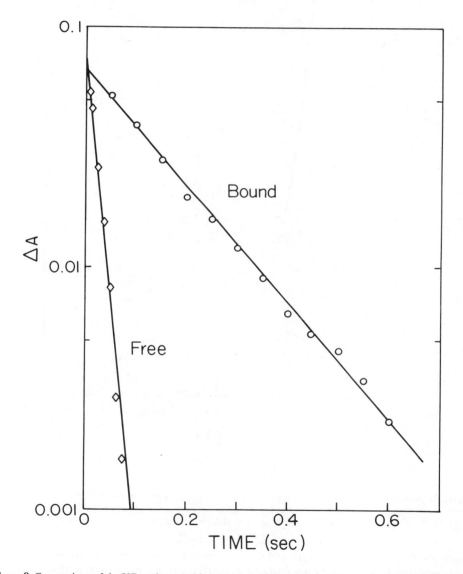

Figure 8. Comparison of the HD exchange of free ethidium bromide (0.5mM) with that of the dye complexed in the presence of 3mM calf thymus DNA. The retardation is a factor of seven, leading to an estimated dissociation constant for dye from complex of 5×10^{-4}M. So far we cannot discriminate which bound state the immobilized amino protons are retarded by; it seems likely to be the fully intercalated state.

strong absorption band in the visible. Spectral and difference spectral properties of EB are shown in Figure 6. Exchange of the free EB in solution as a function of pH (Figure 7) shows that in this case exchange proceeds by a diffusion-limited acid-catalyzed pathway below pH 9, according to the scheme:

$$\text{>}-ND_2 + H_3O^{\oplus} \longrightarrow \text{>}-\overset{\oplus}{N}HD_2$$

$$+ H_2O \quad (6)$$

Exchange of EB from the intercalated state proves to be slowed down significantly with respect to the free state at the same pH (Figure 8). This behavior can be understood in terms of participation of the $-NH_2$ drug protons in the complex:

$$\text{EB}(ND_2) + \text{DNA} \underset{k_{off}}{\overset{k_{on}}{\rightleftharpoons}} \text{DNA - EB}(-ND_2)$$

$$\downarrow k_{ex}^{free} \qquad\qquad\qquad \downarrow k_{ex}^{bound}$$

$$\text{EB}(NH_2) \qquad\qquad\qquad \text{DNA - EB}(-NH_2) \qquad\qquad (7)$$

The complexing of dye to DNA proceeds by a minimal two stage kinetic process;[19] the rate constants k_{on} and k_{off} apply to that step involved in restricting exchange of dye protons. If $k_{off} \geq\geq k_{ex}^{free}$ exchange of dye in the complex will approximate a preequilibrium pathway as in Equation 4b, with the equilibrium constant now corresponding to dissociation of EB from DNA. The calculated value of the dissociation constant for drug at pH 9 is 5×10^{-4} M, in fair agreement with the value determined from binding isotherms measured using fluorescence. Since the rate k_{ex}^{free} can be accelerated by acid catalysis at fixed pH or by shifting pH below 7, a situation in which $k_{ex}^{free} \geq k_{off}$ can also be achieved permitting direct determination of the off rate for EB from the complex. Whether or not the slowing in exchange of dye amino protons is due specifically to their participation in hydrogen bonds to DNA groups or to some indirect effect of intercalation remains to be established.

Acknowledgements

This research was supported by grants HL21757, AM11295 and 1RO1-CA24101-01 from the National Institutes of Health and PCM 77-26740 from the National Science Foundation.

References and Footnotes

1. Englander, J. J., Kallenbach, N. R. and Englander, S. W. *J. Mol. Biol., 63*, 153-169 (1972).
2. Teitelbaum, H. and Englander, S. W., *J. Mol. Biol., 92*, 55-78 (1975a).
3. Teitelbaum, H. and Englander, S. W. *J. Mol. Biol., 92*, 79-92 (1975b).
4. Nakanishi, M. and Tsuboi, M., *J. Mol. Biol., 124*, 61-71 (1978a).
5. Mandal, C., Kallenbach, N. R. and Englander, S. W., submitted for publication (1979).
6. Cross, D. G., *Biochemistry, 14*, 357-362 (1975).
7. McConnell, B., *Biochemistry, 13*, 4516-4523 (1974).
8. Lomant, A. J. and Fresco., J. R., *Prog. Nuc. Acid Research and Mol. Biol., 15*, 185-216 (1975).
9. Binford, J. S. and Holloway, D. M., *J. Mol. Biol., 31*, 91-99 (1963).
10. Scruggs, R. L., Achter, E. K. and Ross, P. D., *Biopolymers, 11*, 1961-1972 (1972).
11. Pribnow, D., *Proc. Natl. Acad. Sci. U.S.A., 72*, 784 (1975a).
12. Pribnow, D., *J. Mol. Biol. 99*, 419 (1975b).
13. Hinkle, D. C. and Chamberlin, M. J., *J. Mol. Biol. 70*, 187-195 (1972).
14. Chamberlin, M., in *DNA Polymerase* (Eds., Losick, R. and Chamberlin, M.), Cold Spring Harbor, 1976, pp. 159-192.
15. Felsenfeld, G., *Nature, 271*, 115-121 (1978).
16. Bostock, C. J. and Summer, A. T., *The Eukaryotic Chromosome*, North-Holland Publishing Company, Amsterdam, New York, 1978, Chapter 4.
17. Englander, J. J., Calhoun, D. B. and Englander, S. W., *Anal. Biochem. 92*, 517-524 (1979).
18. Johnston, P. D., Figueroa, N. and Redfield, A. G., *Proc. Natl. Acad. Sci. U. S. A.* (in press), (1979).
19. Bresloff, J. and Crothers, D. M., *J. Mol. Biol. 95*, 103-123 (1975).

Internal Twisting of Short DNA Segments

B. H. Robinson, Ian Hurley, Charles P. Scholes

and

L. S. Lerman
Center for Biological Macromolecules
State University of New York at Albany
Albany, New York 12222

Introduction

In order to understand the thermal motion of the unconstrained double helix, we have studied the electron paramagnetic resonance (EPR) of a series of intercalating probes carrying a stable nitroxyl radical. Theoretical advances in the simulation of EPR spectra permit the geometry of probe attachment and the complex dynamics of probe motion to be analyzed. A recently developed technique of saturation transfer electron paramagnetic resonance promises to extend our capacity to detect slower motion whose time scale extends well into the realm of biologically imposed movement.

In order that a probe follow faithfully the motions of the DNA to which it is attached, three compatible conditions should be fulfilled:

1. Attachment of the probe should not significantly alter the elastic properties of the helix.

2. A probe should not be capable of any independent movement.

3. The geometry of probe attachment should be uniform, with at least helical symmetry.

It is especially important that this third condition be met if the direction of motion is to be inferred. We have supposed that the third condition might best be met by an intercalating probe, since for no other type of bonding has any structural relation yet been well established. Unfortunately, intercalation must alter the elasticity at least locally; this was clear in the earliest configurational study for flexural considerations.[1] However, we expect the collective motional behavior will be determined by lengths of helix much larger than the intercalation site, and the stiffening of two or four base pairs may not be consequential. Oscillation of the probe relative to DNA in the plane of the intercalation site is a possibility; we shall be concerned here to determine whether it is a serious problem. It will be convenient first to review the

283

work of Robinson, *et al.*[2] and then consider our recent measurements with variously sized DNA molecules.

Fiber Studies

Fully hydrated DNA fibers were made by extruding DNA solutions through the fine orifices of a rayon spinneret into an aqueous ethanol solution and manually stretching them. The alcohol was displaced by an aqueous solvent containing a substantial amount of poly(ethylene oxide), the polymer being necessary to maintain the condensed state of the DNA. On the basis of x-ray diffraction studies under similar conditions,[3] we estimate the typical distance between helix axes in these fibers to be 44Å, allowing the solvent a space of about 20 Å between outermost atoms of neighboring helices. The approach to equilibrium binding when we then added a spin labeled intercalator to this system was indicated by the color of the fibers. Spectra were taken from the blotted fibers enclosed in a capillary to prevent drying.

Acridine spin probes were prepared by reacting 2,2,5,5-tetramethyl-3-pyrrolin-1-oxyl-3-carboxylic acid N=hydroxysuccinimide ester with a 6-substituent or a 9-substituent on an acridine ring. Structural formulas of these probes appear in Figure 1. These compounds have good resistance to hydrolysis and a strong affinity for DNA near neutral pH.

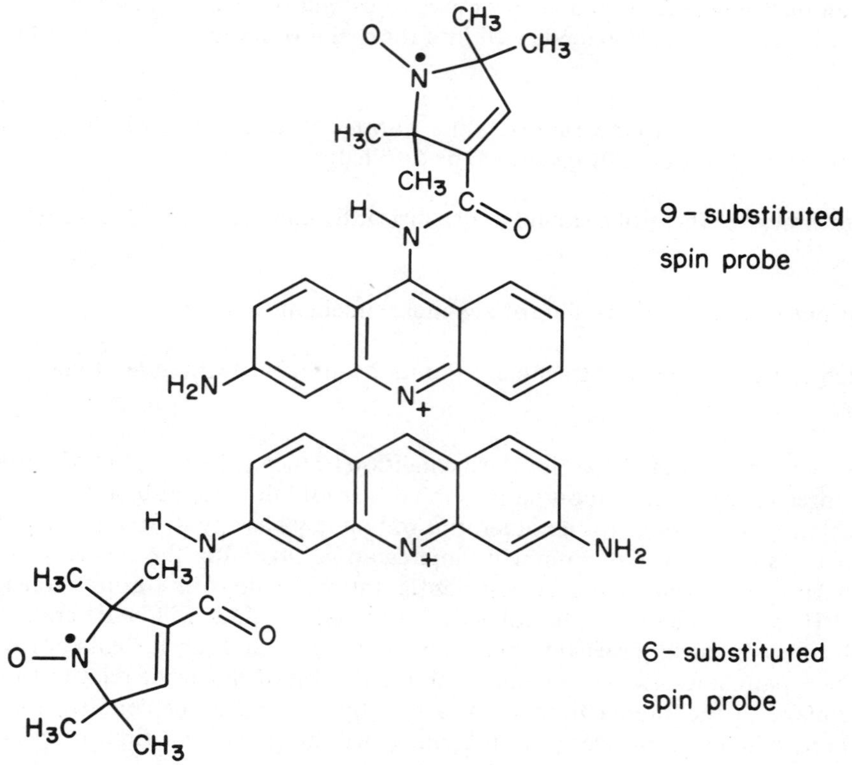

Figure 1. Structural diagram of the acridine spin probes. (a) 9-substituted spin probe: 6-amino-9-N(2,2,5,5-tetramethyl-3-pyrrolin-1-oxyl-3-carboxamido) acridinium cation. (b) 6-substituted spin probe: 3-amino-6-N(2,2,5,5-tetramethyl-3-pyrrolin-1-oxyl-3-carboxamido) acridinium cation.

The probes were found to be ordered with respect to the helix axis and in a fashion characteristic of probe structure. It was found that the spectrum of the 6- spin label-ed acridine bound to a fiber perpendicular to the magnetic field corresponded close-ly to the spectrum of the 9-substituted probe bound to a fiber oriented parallel to the magnetic field and vice versa (Figure 2). Detailed simulations of these spectra, which took into account the fact that a small amount of the DNA will not be in well-ordered strands and that the DNA strands may deviate a little statistically from the fiber axis, suggest that the spin orbital of the 6-substituted probe lies at about 40° from the helix axis and that the spin orbital of the 9-substituted probe lies at about 90°. The basis of this strong tilt away from the plane of intercalation is shown in the photographs of a space-filling model of the 6-substituted probe cation in Figure 3. It reveals that van der Waals contact between the oxygen of the amide group and the hydrogens attached to the 5- or 7- position of the acridine prevents coplanarity. Both the tilt based upon conventional van der Waals radii and the crystal structure of a simple analogous compound, acetanilide, suggest that the spin orbital is oriented at 40° to the acridine ring. This in turn implies that the plane of the acridine ring must be nearly perpendicular to the helix axis, as is indeed expected in intercalation. The larger tilt observed in the instance of the 9-substituted acridines must result from the even closer proximity of the 1- and 8- hydrogens to the amide oxygen bound to the 9-position of the ring.

Solution Studies

The spectrum of the 6-substituted aminoacridine spin probe in a solution of sonical-ly fragmented DNA was found to be identical to the spectrum of the same probe

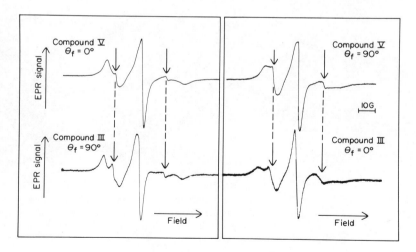

Figure 2. EPR spectra of spin probes bound to DNA fibers as a function of fiber orientation and the posi-tion of the nitroxyl-containing substituent on the acridine ring. Compound V is the 6-labeled acridine; Compound III is the 9-labeled acridine. DNA fibers were equilibrated with these substances in a buffer containing 100 mg/ml poly(ethylene oxide), 0.2 M sodium chloride, 100 mM sodium acetate, pH 5. θ_f is the angle between the fiber axis and the magnetic field. Note that similar orientations are diagonally ar-ranged in the Figure. Arrows mark the cross-over points of the freely tumbling (unbound) intercalator.[2]

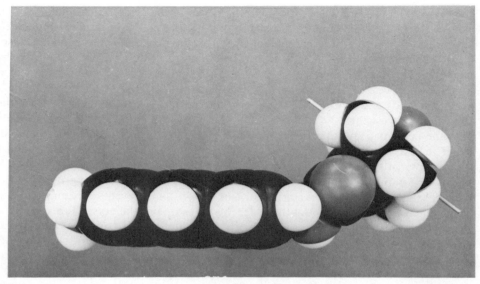

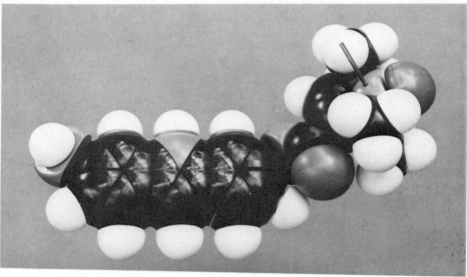

Figure 3. Photographs of a space-filling model of 6-substituted spin probe viewed a) in the plane of the acridine ring; (b) perpendicular to the plane of the acridine ring. The metal rod through the nitrogen atom of the pyrroline ring shows the orientation of the spin orbital. Contact between the amide oxygen and the 5-hydrogen of acridine prevents the pyrroline ring from becoming coplanar with acridine.

bound to DNA fibers, suggesting that the mode of binding was the same in two cases. The rotational correlation times for these spin probes together with an additional spin label based upon propidium (Figure 4) are listed in Table I. The propidium spin probe gives by far the largest correlation time. It seems in reasonable accord with the differences in structure that the motion of the acridines is less well coupled to the motion of the adjacent base pairs than is the motion of propidium.

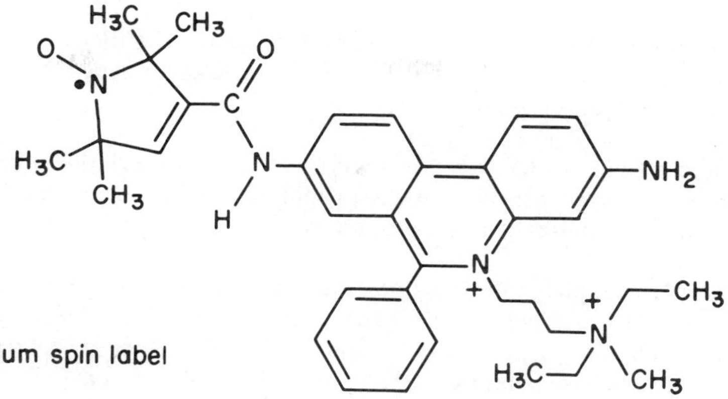

propidium spin label

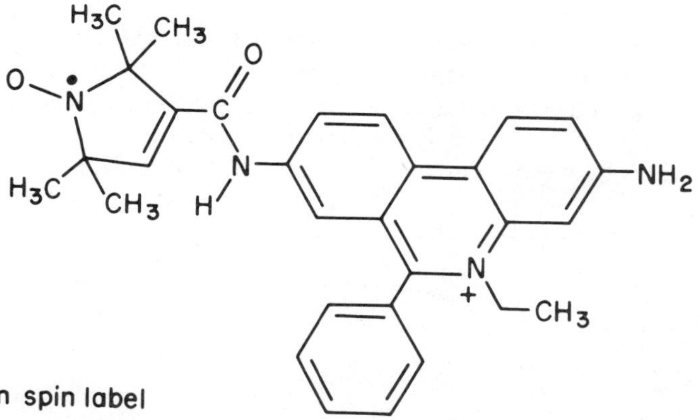

ethidium spin label

Figure 4. The phenathradinium spin probes. (a) Propidium spin probe: 3-amino-5 [3-(diethylmethylammonio)propyl]-6-phenyl-8-N(2,2,5,5-tetramethyl-3-pyrrolin-1-oxyl-3-carboxamido) phenanthridinium cation. (b) ethidium spin probe: 3-amino-5-ethyl-6-phenyl-8-N(2,2,5,5- tetramethyl-3-pyrrolin-1-oxyl-3-carboxamido)phenanthridinium cation.

Table I

Geometrical and Motional Parameters for Some Probes

Compound	Tilt Angles	Correlation Time*	DNA Form†
6-spin 3-amino acridine	40	0.9	solution and fiber
9-spin 6-amino acridine	90	2.2	solution
9-spin 6-amino acridine	90	9.0	fiber
spin propidium	40¶	30.0	solution

*Correlation time is reported in nanoseconds.

†The results are reported for probes bound either to sonicated DNA (as solution) or condensed DNA (as fibers), at 20°C.

¶An assumed value based on the results of the studies on the acridine derivatives.

While the intercalation socket constrains movement so that very little rotation about an axis perpendicular to the helix axis is expected, independent rotation about the helix axis remains possible.

Four dynamic models were considered in order to examine whether the observed rotational correlation time of the propidium spin probe is plausible and consistent with other estimates of DNA properties.

Model 1. The spin probe is proposed to be rigidly attached to the double helix, which rotates uniformly as if it were a speedometer cable. The correlation time will be approximately proportional to the length of the cable. The model requires that a DNA molecule 550 base pairs long (the average size of the sonicated DNA used in these experiments as estimated from viscosity measurements) has a correlation time about seven times greater than our experimental value. This indicates either that the rigid rod model is unsuitable or that the model must be modified to allow the observed correlation time to include the effect of independent motion of the probe.

Model 2. The spin probe is rigidly attached to relatively short uniformly rotating segments of double helix connected by swivels in Model 2. This model is compatible with our measured correlation time if the segment length is about 100 base pairs and if no coupling between adjacent segments through the swivels is permitted. While such junctions between segments might exist if the double helix were near its melting

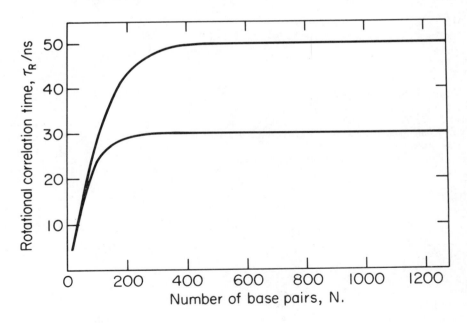

Figure 5. Calculated rotational correlation time, τ_r, of Model 4 as a function of DNA length measured in base pairs at 20°C. The upper curve corresponds to an rms oscillation amplitude, A, of 6°; the lower curve to an rms oscillation amplitude of 4°. The curves were obtained by applying equations 1-4 in the text. Equation 2 was evaluated by numerical integration over the time interval 0 to 10^{-7} seconds. Contributions to the value of the integral from longer time intervals were negligible.

temperature, it is hard to imagine any basis for such discontinuities in coupling at room temperature.

Model 3. The spin probe is hypothesized to jump rapidly between two equilibrium positions within the double helix, the positions being separated by a large potential barrier. This model requires there to be an Arrhenius temperature dependence of the rotational correlation time with the pre-exponential factor of the Arrhenius expression corresponding to the jump frequency. The pre-exponential factor we found to be several orders of magnitude too large for this requirement to be met.[2]

Model 4. Model 4 conceives of the double helix as consisting of a stack of coaxially arranged discs linked by simple torsional springs. This is equivalent to oscillation of nucleotide pairs about the helix by means of complex motion within the phosphate-deoxyribose bridges. The intercalator is assumed to act as if it were a typical disc of the stack at least to a first approximation. Our experimental correlation time can be accommodated to this model if the root mean square amplitude of rotational oscillation between adjacent base pairs is approximately $5°$.

Models 1 and 4 appear to be not inconsistent with the initial set of solution experiments. Before we describe the experiments we did to distinguish between the models, it is appropriate to go somewhat deeper into the theory we developed.

Theoretical analysis of the dynamics of linearly coupled rotational oscillators in a viscous medium[2/4] led to the following equations for the rotational correlation time, τ_R, for a double helix of N base pairs in the case of our Model 4:

$$\tau_R = \tau_F^{-1} - T_e^{-1} \tag{1}$$

$$\tau_F = {}_0\!\int^\infty [e^{-t/T} e^{-\Omega[g,N]t/\tau_0}]dt \tag{2}$$

$$\Omega[g,N] = \frac{1}{N} \sum_{k=1}^{N} \frac{1-e^{-g\sin^2[\pi k/N]}}{g\sin^2[\pi k/N]} \tag{3}$$

$$g = 2t/[A^2\tau_0] \tag{4}$$

where g is a composite of t, time (the variable of integration); τ_0, the correlation time for a single disc; and A, the rms oscillation amplitude of each disc. T_e is the appropriate relaxation time for the magnetization of the electron. The DNA is assumed circular and no end effects are considered. The rotational correlation time, τ_R, is shown as a function of N for two values of A in Figure 5. The lower curve corresponds to an rms amplitude of oscillation of $6°$ and the upper curve to an amplitude of $4°$. The other values used in the calculation were $T_e = 25$ nanoseconds and $\tau_0 = 0.35$ nanoseconds. The former is an estimate of the spin-spin dephasing time and the latter was arrived at by a method to be described later. Notice that τ_R reaches a limiting value below 400 base pairs and is independent of N for larger double helices. Dynamics equivalent to these have been developed independently for the analysis of the time-dependent anisotropic fluorescence of eithidium bound to DNA.

The corresponding equation in the case of Model 1 relates the rotational correlation time to the radius of a rigid cylinder, R, rotating about its unique (long) axis in a medium of viscosity, η, at a temperature, T:

$$\tau_R = \frac{\pi R^2 \eta \; 3.4N}{kT} \tag{5}$$

where π and k have their usual values as constants. (This equation was used to estimate τ_0 at 293°K by setting N, the number of base pairs, equal to 1.) This equation, 5, predicts that the correlation time ought to increase uniformly with the length of the DNA. It seemed reasonable, therefore, to determine the length dependence of the rotational correlation time in order to distinguish between modified Model 1 and Model 4.

Sonicated chicken erythrocyte DNA was fractionated by controlled[6] precipitation from aqueous ethanol at 301°K. We used the high molecular weight DNA and three of the five fractions obtained to prepare samples for taking spectra. Figure 6 is a photograph of the DNA distribution after electrophoresis in 50 mg/ml polyacrylamide gel of the DNA samples. The mass mode in each sample estimated visually was compared with the positions of the Hae III restriction fragments of ϕX-174 RF DNA at each side.

Spin labeled ethidium was prepared by S. Archer and P. Osei-Gyimah made by reacting ethidium bromide with 2,2,5,5-tetramethyl-3-pyrrolin-1-oxyl-3-carboxylchloride. Figure 4b is a structural diagram of their presumed product. Notice that it is very similar to the propidium spin probe.

EPR spectra of our solutions A (dotted line) and C (solid line) appear in Figure 7. The mass mode of the DNA molecules in sample A was estimated to be approximately 200 base pairs, while the mass mode of the DNA in solution C was approximately 1200 base pairs.

The difference between these spectra reflect the motion of the DNA to which they are bound. Free or partly free rotation of the probe within its intercalation pocket would not be expected to vary with length of the molecule when the length is much larger than the intercalation site.

One important difference between these spectra of partially immobilized spin probes is that the outermost features, which we have labeled *a* and *b* in Figure 7, are narrower and more widely separated in C than in A. Empirical methods have been

Sample in Channels:	Volumes of 95 percent ethanol added/ml	
	Before Precipitation	Precipitation Complete
4 and 5	7.57	10.63
6 and 7	6.79	7.21
8 and 9	6.30	6.45

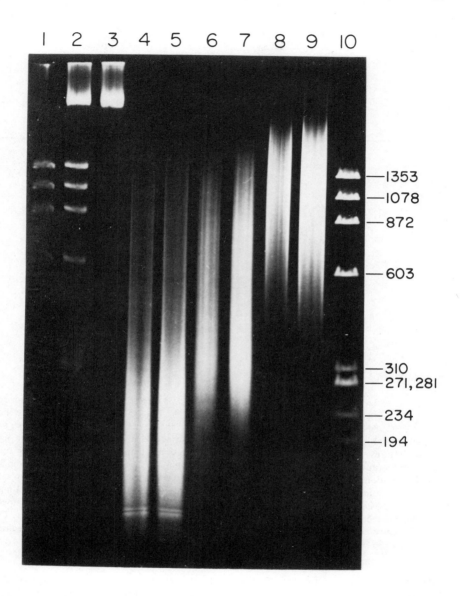

Figure 6. Electrophoresis was carried out in a 50 mg/ml polyacrylamide gel for 1½ hours at about 8 v/cm. Ethidium stain was photographed with 254 mm excitation. The outside channels (1 and 10) were loaded with 0.5 /m/g of a Hae III digest of ϕX174 RF DNA. The size of each fragment is noted on the figures. Lower molecular weight fragments were visible on the original which are not reproduced in this print. Channels 2 and 3 contained 0.5 μg and 1.0 μg high molecular weight DNA respectively. A 10 mg portion of this DNA was sheared by sonication, diluted to 6.77 ml in 0.73 mM sodium phosphate, 2.5 mM sodium acetate, 2.5 mM sodium cacodylate and then fractionated by repeatedly adding small portions of 95 percent ethanol at 28°C. The even-numbered channels were filled with 0.5 μg DNA and the odd-number channels with 1.0 μg of samples of the fractions used in the measurements shown in Table II. The table left summarizes the composition of the mixtures in which precipitation occurred.

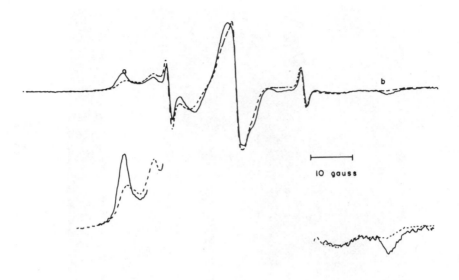

10 gauss

Figure 7. Room temperature electron paramagnetic resonance (EPR) spectra of ethidium spin probe bonded to chicken erythrocyte DNA. (_____Solution A; DNA model length ∿ 200 base pairs). (----Solution C; DNA model length ∿ 1200 base pairs). At the top of the figure, the complete spectra are superimposed. At the bottom, the outermost features (labeled *a* and *b* above) appear at higher gain. These features are markedly narrower in the case of solution C than in solution A and show significantly greater separation. These differences indicate that the spin label is reorienting more slowly in solution C than in solution A. Our test solutions were made by dissolving a weighed sample of ethidium spin label in a 1:10 (v/v) dimethylformamide methanol mixture to make a spin label solution containing 1 mg of solute per ml of solvent. A 25 μl portion of this solution was combined with the volume of DNA stock solution which contained 300 μg of DNA and diluted to 500 μl with 1 mM sodium phosphate buffer. The spectra were taken using a Bruker ER 420-X band spectrometer at 10 milliwatts power and 0.63 gauss modulation. Up to 80 50-second scans were averaged using a Tracor Northern NS-570 digital signal averager.

developed to relate these line widths and splittings to the rotational correlation time of the spin probe.[7/8] Application of these methods gives the results listed in Table II. They show that the correlation time increase with helix length and appears to approach an assumptive limit, as expected for Model 1. The close agreement between these values and that for spin propidium in the previous study again argues against independent motion of the probe since, if independent probe motion contributed significantly, it would be expected that the cationic side chain of propidium, which increases the binding constant relative to ethidium, would provide additional anchoring and its correlation time would be longer.

The consequences of neglecting end effects in a similar dynamic model has been considered by Forgacs, *et al.*,[4] who conclude that the mean correlation times seen for open-ended molecules will correspond to the values for endless molecules about half

Table II*

Rotational Correlation Time in nanoseconds of DNA
as a Function of Number of Base Pairs

Solution	DNA length base pairs (mass mode)	Rotational correlation time, τ_R			Calculated length of DNA segment[++] base pairs
		From splitting	From high field linewidth	From low field linewidth	
A	200	12.	6.	6.	40
B	300	16.	10.	9.	62
C	200	28.	28.	40.	> 150
D	high mw	26.			> 150

*A value of 68 gauss for the separation between outermost features in totally immobilized ethidium spin label was used in this determination. This value was obtained by determining the splitting of ethidium spin label bound to DNA in sucrose solution of various concentrations and then extrapolating a plot of viscosity^{-1} versus splitting to the limit of infinitive viscosity. A value of ∂ of 1.0 gauss used in the calculation was determined by finding the best fit to the low field and high field line width as recommended by Mason and Freed.[7]

[++] The length of a DNA circular segment which would have a rotational correlation time equal to that obtained from the measured splitting; calculated using Model 4 with $T_e = 25$ nanoseconds. $\tau_o = 0.35$ nanoseconds and A = 0.109 radians.

as long. The departure of the measured correlation times listed in Table II from the theoretical values with parameters giving the same infinite length value is somewhat in excess of this adjustment—our points correspond to endless fragments 1/5 as long, rather than 1/2 as long. The basis of the deviation is not now clear, and it may be necessary to include flexural mobility at the ends.

Torsional Rigidity of DNA

We can calculate the harmonic force constant, C, for internal twisting of a DNA helix from Model 4 and our experimentally determined rotational correlation times. The force constant is related to the rms oscillation amplitude of a single disc, A, by:

$$C = \frac{lkT}{A^2} \tag{6}$$

where l is the separation of adjacent discs, 0.34 nm, and k is Boltzmann's constant. (A is related to the rotational correlation time, τ_r, by equations 1-4 cited above.) We find that the rotational correlation times of ethidium spin probe bound to relatively long (> 400 base pairs) DNA are fitted by an rms oscillation amplitude of 6° at 20°C and a corresponding torsional force constant of 1.2×10^{-19} erg-cm/radian^2. The propidium spin probe results require 4° and 3×10^{-19} erg-cm/radian^2 for a best fit.

It is of interest to compare our results with other estimates of torsional force constant and rms oscillation amplitude, which can be inferred from two types of experiments; measurement of fluorescence depolarization, and supercoil linking number distributions, as well as from theoretical simulations of DNA structure.

Assuming that the depolarization of fluorescence of ethidium bound to DNA is attributable to the rate of internal twisting, Barkley and Zimm[5] calculate a torsional force constant of 4.1×10^{-19} erg-cm/radian^2. This number corresponds to an rms oscillation amplitude of $3.3°$ at $20°C$.

Two sets of investigators, Pulleyblank, et al.[9] and Depew and Wang[10] obtained estimates of the Gibbs free energy involved in forming supercoiled DNA by observing the variation in linking number of nicked circular DNA when it is restored to double stranded continuity upon incubation with lignase at various temperatures. If we assume that all this free energy is due to torsion and none to flexure we can obtain the following estimates of torsional force constant and oscillation amplitude:

Pulleyblank, et al.[9] $C = 1.5 \times 10^{-19}$ erg-cm/radian^2 and
                                  $A = 5.6°$ at $20°C$;

Depew and Wang[10] $C = 0.64 - 1.1 \times 10^{-19}$ erg-cm/radian^2 and
                                  $A = 6.5 - 8.4°$.

Since some of the free energy may be due to bending these estimates from supercoiling must represent lower limits to the force constant and upper limits to the oscillation amplitude. Barkley and Zimm[5] have also calculated a torsional force constant from hydrodynamic and light scattering estimates of persistence length (Hays, et al.,[11] Jolly and Eisenberg[12]) under the assumption that DNA behaved as a homogeneous isotropic elastic rod. They obtain $C = 1.8 \times 10^{-19}$ erg-cm/radian^2. This corresponds to $A = 5.2°$.

The modeling of DNA structure by Levitt[13] and by Miller[14] lead to independent estimates of the force constant from the variation in total free energy in the system upon introducing a small twist into the helix. Levitt estimated the total energy of a 20 base pair segment of straight DNA as a function of the pitch of the double helix. He used empirical energy functions which allowed for bond stretching, twisting, bending, van der Waals interactions and hydrogen bonding. However, he did constrain the terminal base pairs to remain coaxial, perpendicular to the helix axis, and at a specified net rotation. His relation between energy and pitch shows a parabolic minimum near 10.5 base pairs. If we assume the potential function fits a harmonic form for small displacement as it appears to do for his Figure 4 we obtain a torsional force constant of 20×10^{-19} erg-cm/radian^2 and an rms oscillation amplitude of $1.5°$. Miller[14] in his calculation specified the atomic positions of two Watson-Crick base pairs in accordance with experimental equilibrium bond lengths and angles and permitted only the conformational angles to change in order to minimize the total energy of the system. He reported a dependence of the potential energy of a (dC)p(dG) dinucleoside upon twist angle which implies a harmonic torsional force constant of 36×10^{-19} erg-cm/radian^2 and an rms oscillation amplitude of $1.1°$.

The inference of the oscillation amplitude from each type of experimental observation depends on a particular set of special assumptions. It is clear that the EPR approach has not yet reached its full development in terms of rotational discrimination and accuracy. With these qualifications, it appears that the estimates from the several experimental approaches are in reasonable agreement. The estimates based on theoretical calculations from bonding parameters all imply a stiffer helix and a smaller amplitude than the experimental values. The difference suggests that the calculations impose excessive constraints or the consequences of interaction with the solvent have not been accounted for adequately.

Acknowledgements

We are grateful for the programming assistance of C. S. Carmack. This work was supported by a grant PCM-7725583 from the National Science Foundation.

References and Footnotes

1. Lerman, L. S., J. Mol. Biol. *3* 18 (1961).
2. Robinson, B. H., Lerman, L. S., Beth, A. H., Frisch, H. L., Dalton, L. R., and Auer, C. (submitted for publication).
3. Maniatis, T. P., Venable, J. H., Jr., and Lerman, L. S., *J. Mol. Biol. 83*, 28 (1974).
4. Forgacs, G., Robinson, B. H. and Frisch, H. L. (in preparation).
5. Barkley, M. D., and Zimm, B. H., *J. Chem. Phys.* (in press).
6. Lerman, L. S., Wilkerson, L. S., Venable, J. H., Jr., and Robinson, B. H., *J. Mol. Biol. 108*, 271 (1976).
7. Mason, R. P., and Freed, J. H., *J. Phys. Chem. 78*, 1321 (1974).
8. McCalley, R. C., Shimshick, E. J., and McConnell, *Chem. Phys. Lett. 13*, 115 (1972).
9. Pulleyblank, D. E., Shure, M., Tang, D., Vinograd, J., Vosberg, H. *Proc. Nat. Acad. Sci. 72*, 4280 (1975).
10. Depew, R. E., and Wang, J. C., *Proc. Nat. Acad. Sci. USA 72*, 4275 (1975).
11. Hays, J. B., Mogar, M. E., and Zimm, B. H., *Biopolymers 8*, 531 (1969).
12. Jolly, D., and Eisenberg, H., *Biopolymers 15*, 61 (1976).
13. Levitt, M., *Proc. Nat. Acad. Sci. USA 75*, 640 (1978).
14. Miller, K. J., *Biopolymers 18*, 959 (1979).

The Flexible DNA Double Helix

Wilma K. Olson
Department of Chemistry
Rutgers University
New Brunswick, New Jersey 08903

The macroscopic properties of the DNA double helix in dilute solution depend dramatically upon chain length.[1,2] In the range of low molecular weight the behavior of this molecule closely resembles that anticipated for a rigid rodlike helix found in the solid state. In contrast, extremely long DNA chains of molecular weight 10^7 or more adopt on the average very compact conformations in comparison to the extended regular helix. At this size range the molecule further qualifies as a random coil in the sense that the distribution of chain conformations is completely random or Gaussian. In between these extremes of chain lengths the DNA exhibits the limited motions of a gradually and smoothly bending wormlike model.[3-5] No single mathematical model yet proposed can adequately represent the physical properties of DNA chain in this intermediate category.

The detailed local molecular motions that give rise to the macroscopic flexibility of DNA appear on the basis of available experimental evidence to be highly restricted. According to various physical measurements (that include Raman spectroscopy,[6] CD,[7,8] NMR,[9,10] electron microscopy,[11,12] infrared dichroism,[13,14] electric birefringence,[15] wide-angle X-ray scattering,[16,17] and others) DNA in dilute aqueous salt solutions at neutral pH exhibits the characteristic B-type backbone observed directly by X-ray crystallograhic analysis of low molecular weight DNA analogs[18] and deduced indirectly from X-ray fiber diffraction studies.[19] These solution measurements, however, cannot detect the minor variations of backbone conformation seen within the B- family of structures in the solid state. Fluctuations in polynucleotide conformation are undoubtedly responsible for the well known "breathing" of the double helix in solution.[20-24] At temperatures far below the normal thermal transition point some base pairs of DNA occur in non-hydrogen bonded states open to interactions with the aqueous environment. Protons normally involved in Watson-Crick base pairing exchange[20-22] rapidly with deuterium or tritium in the solvent while exocyclic amino and endocyclic imino groups on the purines and pyrimidines react easily with formaldehyde probes.[23,24] Since the bases attached to the DNA maintain strong stacking interactions[22] under conditions of breathing, these molecular motions presumably are limited in scope.[25]

The computations of polymer chain dimensions outlined below demonstrate how the rigidity of the local conformation of the polynucleotide backbone is magnified into the enormous degree of flexibility of long DNA double helices. The molecular model we adopt is consistent with limited local breathing motions of DNA. In addition, the model accounts satisfactorily for the gross dimensions of the polymer over a broad range of molecular weights. The chain bends smoothly without the occurrence of drastic turns or kinks. The flexibility of the chain as described by the spatial density distribution functions is further consistent with the macroscopic descriptions previously attributed to DNA of various chain lengths (e.g., rigid rod, wormlike coil, Gaussian). Despite the local stiffness in the polynucleotide backbone, longer molecules exhibit a tendency to bend backwards into hairpin loops and to cyclize into "circular" structures. The predicted probabilities of occurrence of such structures appear to be related to the sizes of loops and rings observed in experimental studies.

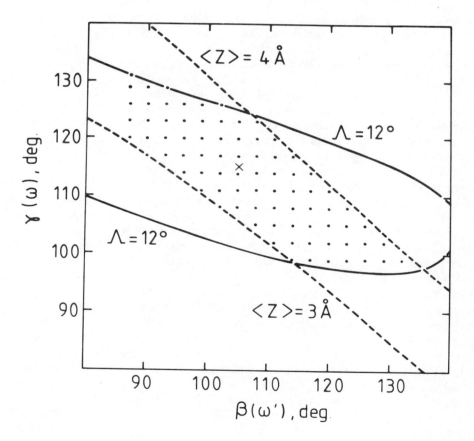

Figure 1. Contour diagram of the $\Lambda = 12°$ base stacking angle (solid line) and the $<Z> = 3Å$ and $4Å$ base stacking distances as a function of the phosphodiester angles for a flexible B-DNA helix. Points denoted on the diagram, where base overlap is also observed, are the 81 discrete conformational states included in the model. The × denotes the conformation repeated in the reference helix of the model. Rotation angles are defined with respect to *trans* = 0°. Rotation angle nomenclature $\omega'\omega$ used in previous publications is noted in parentheses.

Residue Flexibility

For the purpose of simplicity the molecular motions attributed in this study to the DNA double helix are limited to minor angular variations in the phosphodiester rotation angles (β and γ) of each chain repeating unit. These two angles are chosen on the basis of X-ray,[18/26] NMR,[9/27/28] and theoretical studies[29/30] that implicate them as the major source of flexibility in the polynucleotide backbone. Upon assignment of fixed values to the remaining two C-C and two C-O bonds of each nucleotide repeating unit the distance between successive phosphorus atoms is fixed and imaginary virtual bonds[31] may be drawn to span each chemical residue. As a consequence of this mathematical device, statistical mechanical treatment of polymer chain averages is greatly simplified.[31/32]

The extent of motion associated with the flexible B-DNA helix of this study is indicated by the contour surface in Figure 1. In the absence of reliable estimates of conformational energy, the β and γ angles are restricted with equal probability to the 81 points denoted on the surface. In each of these states adjacent bases may be regarded as stacked in that (1) the angle Λ between base planes is 12° or less, (2) the mean distance $<Z>$ between base planes falls in the range 3-4 Å, and (3) the bases exhibit partial overlap. Only one point on the surface (denoted in Figure 1 by $\times$), however, may be considered the building block of an ideal double helix with both base stacking and Watson-Crick base pairing (cf. seq.). The limited motions set by the restrictions in Λ, $<Z>$, and overlap reproduce the radius of gyration $<s^2>$ of DNA spanning a range of chain lengths between 2^8 and 2^{13} nucleotide units.[32]

The relative motions of adjacent bases in the above model are also illustrated by the adenine dinucleoside triphosphate structures of Figure 2. The dimer represented in Figure 2(a) is a reference conformation, that when repeated throughout two complementary DNA strands, produces a theoretical 13-fold double helix. This duplex (denoted in Figure 1 by $\times$) is characterized by structurally ideal linear hydrogen bonds of the Watson-Crick variety.[32/33] Minor (i.e., $\sim$20°) variation of β and/or γ from the above reference conformation is sufficient to distort and in some cases to break interstrand hydrogen bonds. The dimers represented in Figures 2(b) and 2(c) are illustrative of the extrema of flexibility allowed in the present model. Upon changes in the phosphodiester conformation the bases are found to slide back and forth in a manner reminiscent of the oscillating single-stranded base stacking suggested a number of years ago by Davis and Tinoco[34] to account for the observed CD of dinucleoside monophosphates. As evident from the high base overlap in Figure 2(b), this conformation is more tightly wound (by $\sim$20°) than the reference state. The bases in Figure 2(c), in contrast, are more loosely wound (by $\sim$-12°) than those in the reference structure. As detailed below, the bases in the latter unwound dimer are more exposed to the outside of the helix and hence more susceptible to interactions with the chemical probes of breathing and intercalation studies.

The alternate views in Figure 3 parallel to the local helical axes of the above three dimers illustrate the orientational motions in adjacent bases in our flexible model. The bases attached to the reference backbone in Figure 3(a) adopt the classical parallel self-alignment and horizontal tilt of B-DNA in the solid state. In contrast, the bases in the highly wound dimer in Figure 3(b) form a wedge of 11° that opens toward the core of the helix. This orientational feature appears to disfavor interactions between the base and the aqueous environment. As evident from Figure 3(b) the two bases of the tight dimer also tilt by -9° from the horizontal alignment found in the reference dimer. The bases attached to the loose dimer in Figure 3(c) describe a wedge of 12° that opens toward the outside of the helix. Furthermore, these bases upon tilting 52° from the normal horizontal arrangement move close to the outer surface of the structure. Such conformational changes appear to favor base-solvent associations.

The relative motions of the bases and the backbone about the reference conformation of our flexible model are apparent from the overlapping structures in Figures 4(a) and 4(b). Superimposed upon each of the two extreme conformations (darkened bonds) at the O(3′)-P-O(5′) linkage is the reference dimer structure (light bonds). Terminal phosphates and bases are displaced in opposing directions by the two types of motion (winding vs. unwinding). The wedges introduced between the bases are also apparent in the figure.

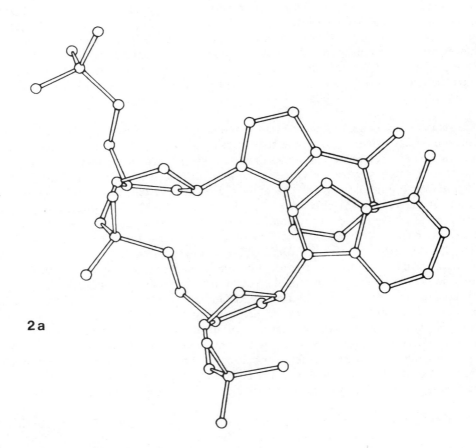

2a

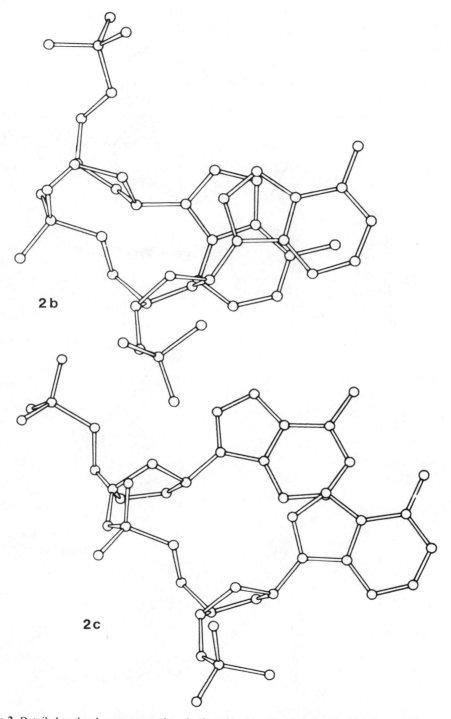

Figure 2. Detailed molecular representations in the pdApdAp dinucleoside triphosphate of the internal motions of the flexible B-DNA helix. All views are drawn perpendicular to the plane of the 5′-adenine moiety to illustrate base stacking. (a) The $\beta,\gamma = 105°, 115°$ reference conformation. (b) The tightly wound $\beta,\gamma = 132°, 99°$ state of extreme flexibility. (c) The loosely wound $\beta,\gamma = 87°, 132°$ state of extreme flexibility.

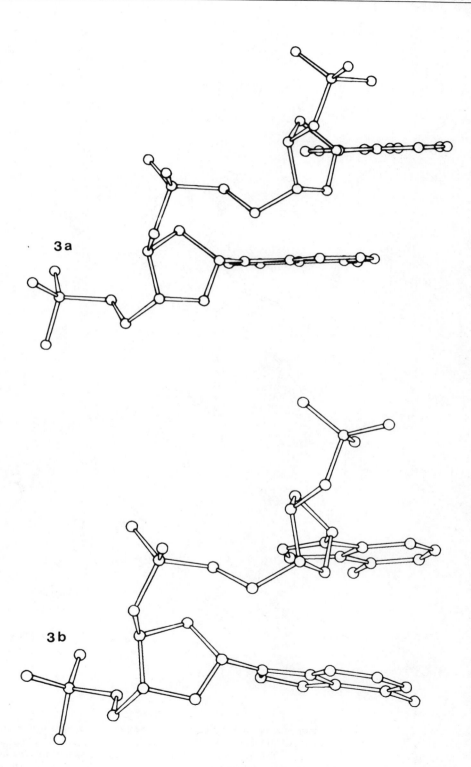

3a

3b

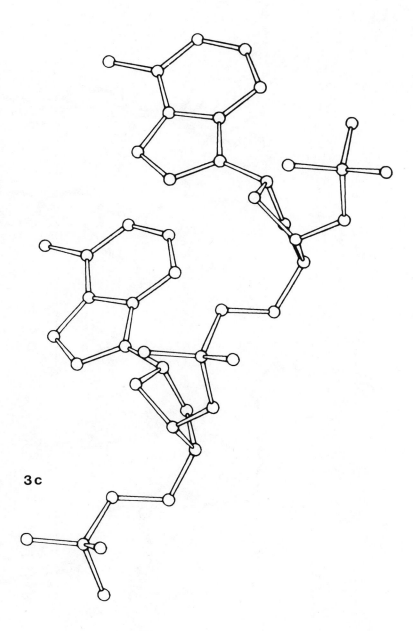

3 c

Figure 3. Additional views drawn perpendicular to the local helical axes of the three dimers described in Figure 2.

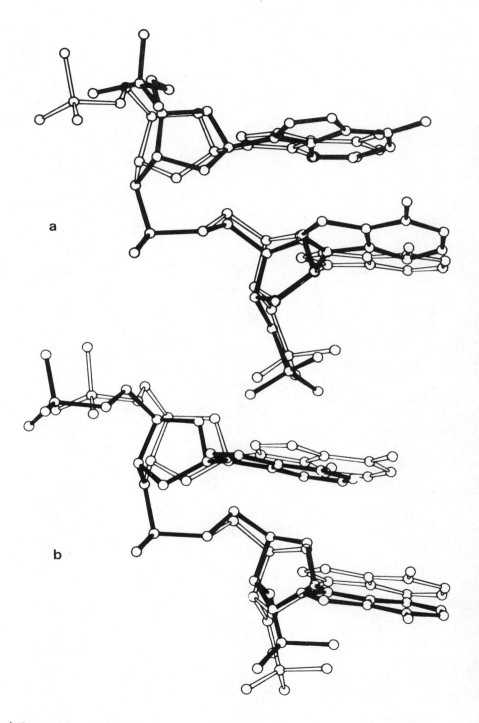

Figure 4. Comparative molecular representations of flexibility in the B-DNA model. The extreme conformations (dark bonds) are superimposed at the central phosphodiester linkage upon the reference state (light bonds). (a) $\beta,\gamma = 132°, 99°$. (b) $\beta,\gamma = 87°, 132°$.

Polymer Flexibility

As illustrated in the perspective drawings of Figure 5, the above limitations upon local flexibility in the DNA backbone produce a smoothly bending structure that approximately resembles a regular helix of the same chain length. The flexible molecule illustrated in Figure 5(a) is a 128-residue single-stranded B-DNA chain chosen at random by Monte Carlo methods.[32/35] This structure represents a single "snapshot" of the chain as it passes through the multitude of conformations (81^{128} or approximately 10^{171}) accessible in the present scheme. For comparison, the regular helix generated by 128 consecutive occurrences of the reference dimer conformation appears in Figure 5(b). For clarity, the sugar and base moieties are omitted in the two figures. Segments of the chains are represented instead by virtual bonds connecting successive phosphorus atoms. The flexible, wormlike chain possesses a pseudohelical backbone and pseudohelical axis that changes direction continually. The individual turns of the pseudohelix also vary considerably in size as a consequence of local winding and unwinding of the dimer chain units. The gradual bending of the pseudohelical backbone describes a trajectory with a large radius of curvature. The end-to-end separation of the flexible chain is shorter than that of the regular helix.

The flexible DNA backbone is best described by the complete array of conformations it can assume. A comprehensive picture of this array is provided by the three-dimensional spatial probability density function $W_0(\mathbf{r})$ of all possible end-to-end vectors.[32/36] This function represents the probability per unit volume in space that the flexible chain terminates at vector position $\mathbf{r}$ relative to the chain origin $\mathbf{O}$ as reference. For short chains $W_0(\mathbf{r})$ may be estimated with reasonable accuracy by direct Monte Carlo calculations and for long chains in terms of a three-dimensional

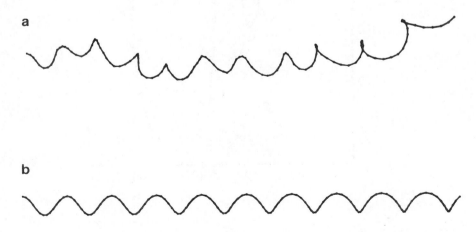

Figure 5. Computer generated perspective representations of a 128-residue single-stranded B-DNA chain. The chain backbone is represented by the sequence of virtual bonds that connect successive phosphorus atoms in the chain. (a) A representative flexible helix generated by Monte Carlo methods. (b) The reference helix of the model.

Hermite series expansion of the Gaussian centered at the so-called persistence vector
(cf. seq.).[36/37] Distribution functions obtained using the Hermite approximation
with correction terms through fourth order for flexible DNA chains of length $2^5 =$
32, $2^6 = 64$, and $2^7 = 128$ are presented in Figure 6. The three-dimensional distribution functions are represented by three two-dimensional contour slices in the xz
helical plane of the reference dimer and hence in the xz plane of the regular helix
depicted in Figure 5(b). The regions enclosed by the contours account for approximately 90 percent of all spatial arrangements accessible to the chains of specified
length. Superimposed upon the contours in Figure 6 are the backbone trajectories of
three flexible Monte Carlo chain of length 128. This group of random molecules is
seen to describe approximately the three different distribution functions. The trajectories of the three molecules are defined with respect to the coordinate system of the
reference helix. The first few segmensts of the chains are roughly superimposable

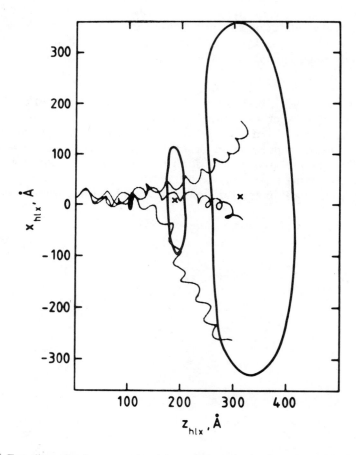

Figure 6. Two-dimensional contours that delineate from left to right the array of conformations described
by a flexible B-DNA helix of lengths 2^5, 2^6, and 2^7. Contours are based upon the three-dimensional spatial
probability density function $W_0(\mathbf{r})$ and are drawn in the xz helical plane of the reference structure illustrated in Figure 5b. persistence vectors **a** for the three chain lengths are denoted by ×. Superimposed
upon the contours are the xz projections of three of the 81^{128} possible random helices of length 128. The
flexible helices are also drawn in the internal coordinate system of the reference helix.

upon the corresponding positions of the reference helix. Segments more removed from the chain origin, however, are found to deviate appreciably from the regular structure.

The position, size, and shape of the density distribution functions of flexible DNA depend markedly upon chain length. The three distribution functions presented in Figure 6 are described with respect to the persistence vector **a** characteristic of the chain length. These positions denoted by the ×'s in Figure 6 are the mean coordinates ($<x>$, $<y>$, $<z>$) of all the possible spatial configurations of the chains in the fixed coordinate system. Because the DNA helix is subject to structural constraints of fixed bond lengths and valence angles as well as to limitations in internal rotations, **a** is a non-null vector at all chain lengths. The limiting asymptotic coordinates of $\mathbf{a}_\infty$ are attained by chain length $2^{11} = 2048$ after which all distribution functions are centered about position (20.0Å, 17.6Å, 553.4Å) in the reference helix coordinate system. As evident from Figure 6 at short chain lengths there is little deviation of the flexible chain from the average structure defined either by **a** or by the reference helix and the distribution is confined to a very limited domain. Since the distribution functions exhibit approximately cylindrical symmetry about the helical z-axis, the volume of space available to a chain of length 32 is best described as a small ellipsoid. With the perpetuation of smooth bending to DNA of longer chain lengths the random helices are found to deviate significantly from both **a** and the reference helix. The volume encompassing chain termini for DNA of length 128 is approximately 14,000 times the volume available to a chain of length 32 and about 50 times that found at length 64. The spatial distribution function at x = 128 is similar to the cap of a rather flat mushroom. By the time the DNA contains 2^8 or 256 residues, the chain backbone is able to bend back upon itself. The distribution functions associated with chains of length 2^7 (enclosed by dashed lines) and 2^8 (shaded area) are compared in Figure 7. The volume described by the termini of the longer chains in the figure is similar to that described by a huge bouquet of flowers. As evident from the shaded contour region in Figure 7, it is still difficult (although not impossible) for a flexible helix of 256 residues to terminate in the "stem" region of the distribution. At chain lengths beyond $2^{10} = 1024$ residues the distribution function once again becomes ellipsoidal in shape. At this chain length the DNA is equally likely to describe a pseudohelix longer or shorter than the persistence vector. A chain of 1024 units is, however, somewhat more flexible in its motions along the x- and y-axis than along the z-axis of the reference helix. At length $2^{13} = 8096$ the distribution volume may be described as a sphere and the probability function is an ideal Gaussian centered at **a**. Beyond this size, the distribution functions are simply larger spheres centered at a common point. At very large chain lengths the magnitude of $\mathbf{a}_\infty$ is small compared to the size of the distribution. At this limit the probability distribution may be represented in the classical fashion as a Gaussian centered at the origin.

Circle and Loop Formation

The long more tortuous DNA chains described above are potentially capable of folding into the cyclic and looped structure long observed in experimental systems. Recent biochemical data indicate that the sizes of so-calld foldback duplexes formed

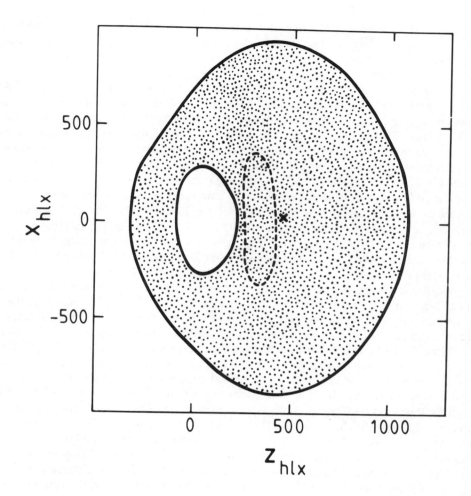

Figure 7. Two-dimensional contours of probability density for flexible B-DNA helices of lengths 2^7 (dashed line) and 2^8 (shaded area). The persistence vector at 2^8 is denoted by ×.

upon extremely fast renaturation of DNA are nonrandom.[38/39] These hairpin structures, whether looped (i.e., single-stranded with bases mismatched) or unlooped (i.e., double-stranded with bases paired), span a narrow distribution of lengths between 70-600 base pairs. Ring structures formed by annealing a variety of eukaryotic DNA fragments exhibit a strikingly similar size preference (300-600 residues).[40/41] As outlined below, this chain length dependence appears to reflect the relative probabilities that the termini of variously sized fragments of our flexible DNA model collide.

Cyclic or looped structures of DNA arise when the termini of a molecule or of a molecular fragment are confined to a particular three-dimensional arrangement in space. As illustrated in Figure 8, for a DNA helix comprising n virtual bonds, the end-to-end separation of the chain describes a vector **r** in the reference frame (x_{vb}, y_{vb}, z_{vb}) along the first bond (P_0 to P_1). Ring closure requires that **r** = 0. This vector,

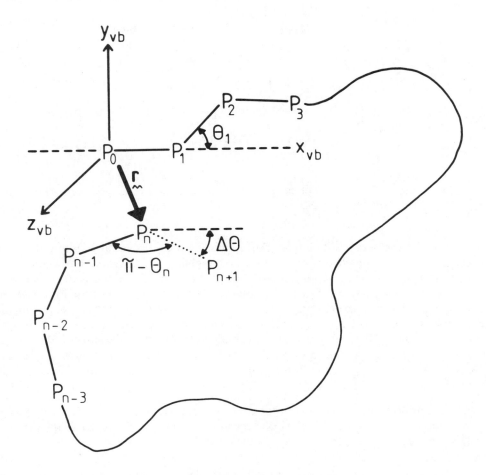

Figure 8. Polynucleotide segment of n virtual bonds connecting atoms 0 through n in a conformation approaching the requirements for cyclization or loop formation. Reference frame is defined along the first virtual bond. Valence angle supplements are denoted by θ. $\Delta\theta$ is the angle between the hypothetical n + 1 bond (dotted) and the first bond.

however, is nonzero in cases of loop formation. In the studies outlined below to estimate the probability of foldback loops, the appropriate **r** vectors describe a cylindrical shell of diameter 16 ± δd Å where d² = 2(x² + y²) and thickness 0 ± δz Å in the (x,y,z) *helical* coordinate system of the initial virtual bond. This separation insures that the terminal bases of the loop may associate via hydrogen bonding. Previous estimates of polynucleotide loop closure[42/43] based upon Jacobson-Stockmeyer theory[44] assume **r** to include all points on a spherical shell of ∼16Å diameter centered at P_0. This approximation ignores the geometrical constraints of base pairing at the ends of a loop.

The formation of DNA rings and loops of size n further requires that the terminal bonds of the chain are oriented in an arrangement that maintains the valence bond angles at atoms P_0 and P_n and also the torsion angles about bonds $P_0...P_1$ and

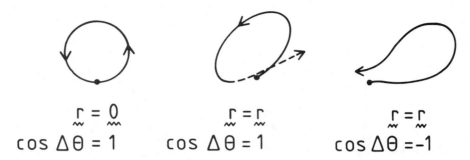

$\underset{\sim}{r} = \underset{\sim}{0}$ $\underset{\sim}{r} = \underset{\sim}{r}$ $\underset{\sim}{r} = \underset{\sim}{r}$

$\cos \Delta \theta = 1$ $\cos \Delta \theta = 1$ $\cos \Delta \theta = -1$

Figure 9. Schematic illustrating the distance (**r**) an orientational ($\cos \Delta \theta$) constraints associated with DNA ring closure, *anti* loop formation, and *syn* loop formation.

$P_{n-1}...P_n$ within acceptable limits. Fulfillment of the valence angle requirements in cyclization and loop closure constrains the imaginary $n + 1$ virtual bond connecting P_n to P_{n+1} in Figure 8 to approximately parallel angular orientations $\Delta \theta$ with respect to the initial virtual bond ($P_0...P_1$) of the chain. As illustrated in Figure 9, the parameter $\tau = \cos \Delta \theta$ must fall in the range $1-\delta\tau$ to 1 in both DNA closed circles and in so-called *anti*[45] or superhelical loops but must vary between -1 and $-1 + \delta\tau$ in *syn*[45] or hairpin loops. The assurance of acceptable torsion angles about the P_n juncture further requires that the intersecting planes containing virtual bonds n-1, n, and $n+1$ describe an angle ω_n within the range of allowed local molecular motions of our model.

Upon elaboration of Jacobson-Stockmayer theory to include conditions of angle compliance, the probability $Q(\mathbf{r})$ of occurrence of cyclic ($\mathbf{r} = 0$) or looped ($\mathbf{r} = \mathbf{r}$) DNA segments of length n is given by the product[46]

$$Q(\mathbf{r}) = [W_0(\mathbf{r})\delta r] [T_r(\tau)\delta\tau] [\underset{\eta}{\Sigma}\Omega_{r\tau}(\omega_n)\delta\omega_n]. \tag{1}$$

In this expression $W_0(\mathbf{r})$ is the spatial probability density function and δr is the admissable departure of r from the value required for cyclization or loop formation; $T_r(\tau)d\tau$ is the probability that τ assumes a specific value (± 1) when $\mathbf{r} = \mathbf{r}$ and $\Omega_{r\tau}(\omega_n)\delta\omega_n$ is the probability of occurrence of torsion angle ω_n for bond n when $\mathbf{r} = \mathbf{r}$ and $\tau = \tau$. Included in the summation of Equation 1 are the various admissible conformations of ω_n indexed by η. As described elsewhere the $T_r(\tau)$ term may be expanded in averaged Legendre polynomials $<P_k>_{r=r}$ with argument $\tau_r = 0$.[46] The averaged polynomials used in the expansion here are evaluated in terms of the scalar chain moments $<\tau^m r^{2p}>$ with $m = 0, 1, 2$ and $p = 0, 1, 2$. Torsional correlations are assumed to be random and $\underset{\eta}{\Sigma}\Omega_{r\tau}(\omega_n) \delta\omega_n$ is taken to be unity.

In contrast to the analysis presented here, cyclization or loop closure of DNA based upon Jacobson-Stockmayer theory ignores all angular correlations and assumes that $W_0(\mathbf{r})\delta r$ is a one-dimensional Gaussian distribution centered at the chain origin.[44] As a first correction in this paper, we assume $W_0(\mathbf{r})\delta\mathbf{r}$ at each chain length to be a three-dimensional Gaussian function centered about the persistence vector **a**. This ellipsoidal distribution is the basis function of the three-dimensional Hermite expansion functions presented in Figures 6 and 7. As a second correction we evaluate $T_r(\tau)\delta\tau$ in a Legendre polynomial expansion.[46]

Data describing the probability of DNA ring closure $Q(0)$ with chain length n are presented at various levels of approximation in Figure 10. As expected, the probability of loop formation based upon Jacobson-Stockmayer theory (curve 1), decreases linearly with chain length. As chain length increases in this scheme the spherical volumes associated with a one-dimensional Gaussian increase in size. Since all points within the spheres are equally likely and since the total probability of all conformations of a given sized chain is unity, the probability of occurrences at $r = 0$ decreases with n. In marked contrast the $Q(0)$ of cyclization based upon a three-dimensional Gaussian centered at a (curve 3) exhibits a maximum at 2^8 repeating units. As noted in the contour surfaces above for chains less than 2^8 residues, a locally "stiff"DNA backbone cannot easily fold back toward the chain origin. At chain lengths greater than 2^8 units where the molecule is more flexible and $r = 0$ is an attainable state, the spatial density distributions increase in volume. Following the argument given above for Gaussian spheres the probability of conformations where $r = 0$ also decreases with n beyond 2^8 in the ellipsoidal scheme. Curves 1 and 3 coalesce at 2^{10} units where the magnitude of a is now small compared to the volume enclosed by the distribution function.

As evident from curve 2 in Figure 10, introduction of the valence angle correlation term $T_0(1)\delta\tau$ into the one-dimensional Gaussian expression of curve 1 depresses Q over all n up to n = 2^{12}. A local maximum in Q now occurs at 2^7 residues but the most probable cyclized chain length remains at 2^0. The angular correction factors for all $n \leq 2^{12}$ similarly decrease the probability of loop closure in curve 3 to curve 4. Below chain length 2^8 $T_0(1)$ approaches 0 in the "stiff"DNA chain and $\log_{10}Q(0)$ lies far below the range of data plotted in Figure 10. According to curve 4 cyclization occurs with greatest and approximately equal probability in DNA rings of length 2^8 and 2^9.

In the absence of angular correlations the probability of condensing a locally "stiff" DNA backbone into loops of end diameter ~ 16Å parallels the behavior of curve 3 in Figure 10. Since the constraints upon r in loop closure are less stringent than those in cyclization, the computed loop probabilities are several orders of magnitude larger than those reported in Figure 10. The angular correlation factors $T_r(1)$ describing *anti* loop formation are less than unity and depress the probability of loop formation in a manner analogous to that described above for chain cyclization. In contrast, the angular correlation factors $T_r(-1)$ exceed unity at some chain lengths and hence enhance the formation of *syn* or hairpin loops. The ratio $T_r(-1)/T_r(1)$ exhibits a maximum at n = 2^7 and does not attain values of unity until chain lengths 2^{10}.

Summary

The above applications of polymer chain statistics to double helical DNA clearly demonstrate how the severely limited internal rotations of a "rigid" B-DNA backbone are magnified into the enormous flexibility of a long polynucleotide

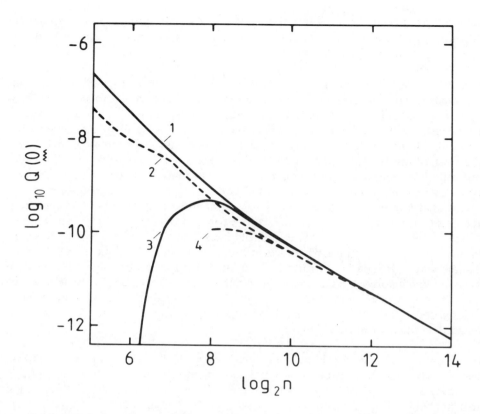

Figure 10. Cyclization probability Q(0) plotted against the number of virtual bonds in a flexible B-DNA helix. Curves 1-4 correspond to different approximations for Q: curve 1, spherical Gaussian with no angular correlations; curve 2, ellipsoidal Gaussian centered at **a** with no angular correlations; curve 3, spherical Gaussian with angular correlations; curve 4, ellipsoidal Gaussian with angular correlations.

chain. Minor rotational motions within each dimer unit of the chain account satisfactorily for the radii of gyration of DNA spanning a broad range of molecular weights. In addition, the three dimensional spatial distribution functions generated by this set of limited internal rotations follow various macroscopic descriptions previously ascribed to DNA of different chain lengths. If long enough, the "rigid" DNA can condense without the occurrence of severe bends or kinks into compact conformations.[32/47] Indeed, the probability of both cyclization and loop closure based upon this model of limited flexibility parallels the occurrence of such structures in experimental systems.

The detailed motions ascribed to the phosphodiester linkages in the above model, however, are only approximate. The conformational states chosen are realistic in the sense that base stacking is maintained and also that the bases may interact to some extent with the aqueous environment. As a first approximation, all states of the model are equally favored. In a more exact model, flexibility in all the rotatable bonds of the polynucleotide backbone as well as slight variations in valence bond angles and bond lengths must be considered. Furthermore, chain averages should be based upon the relative energies of the various configurations of the helical backbone.

The above computations of polynucleotide average properties are further approximate in the sense that they apply to an ideal unperturbed DNA double helix free of long range excluded volume and electrostatic effects. The average dimensions of very long B-DNA helices subject to these long-range repulsive forces are expected to exceed those of an unperturbed chain.[48] The probabilities of cyclization and loop formation in very long chains will also be lower than the data reported above. Excluded volume effects, however, are negligible in short, more "stiff" DNA helices and are not expected to alter the computed preference for rings and loops in the range of 2^8 - 2^9 residues.

Acknowledgement

The author is grateful to Professor Joachim Seelig at the Biozentrum of the University of Basel, Switzerland for his hospitality during the course of this work and to the J. S. Guggenheim Memorial Foundation for a fellowship. The work was also supported by a grant from the U.S.P.H.S. (GM-20861). Computer time was supplied by the University of Basel Computer Center.

References and Notes

1. Bloomfield, V. A., Crothers, D. M., and Tinoco, I., Jr., *Physical Chemistry of Nucleic Acids*, Harper and Row, New York, Chapter 5 (1974).
2. Schellman, J. A., *Biopolymers 13*, 217 (1974).
3. Kratky, O. and Porod, G., *Rec. Trav. Chim. Pays-Bas 68*, 1106 (1949).
4. Shimada, J. and Yamakawa, H., *J. Chem. Phys. 67*, 344 (1977).
5. Yamakawa, H., Shimada, J., and Fujii, M., *J. Chem. Phys. 68*, 2140 (1978).
6. Erfurth, S. C. and Peticolas, W. L., *Biopolymers 14*, 247 (1975).
7. Johnson, W. C., Jr. and Tinoco, I., Jr., *Biopolymers 7*, 727 (1969).
8. Wells, R. D., Larson, J. E., Grant, R. C., Shortle, B. E., and Cantor, C. R., *J. Mol. Biol. 54*, 465 (1970).
9. Patel, D. J. and Canuel, L., *Proc. Natl. Acad. Sci. USA 73*, 674 (1976).
10. Kearns, D. R., *Ann. Rev. Biophys. Bioeng. 6*, 477 (1977).
11. Griffith, J. D., *Science 201*, 525 (1978).
12. Vollenweider, H. J., James, A., and Szybalski, W., *Proc. Natl. Acad. Sci. USA 75*, 710 (1978).
13. Pilet, J., and Brahms, J. *Biopolymers 12*, 387 (1973).
14. Pilet, J., Blicharski, J., and Brahms, J., *Biochem. 14*, 1869 (1975).
15. Hogan, M., Dattagupta, N., and Crothers, D. M., *Proc. Natl. Acad. Sci. USA 75*, 195 (1978).
16. Bram, S. and Beeman, W. W., *J. Mol. Biol. 55*, 311 (1971).
17. Bram, S., *J. Mol. Biol., 58*, 277 (1971).
18. Sundaralingam, M., in *Structure and Conformation of Nucleic Acids and Protein-Nucleic Acid Interactions*, Sundaralingam, M. and Rao, S.T., Ed., University Park Press, Baltimore, pp. 487-524 (1975).
19. Arnott, S., Chandrasekaran, R., and Selsing, E., in *Structure and Conformation of Nucleic Acids and Protein-Nucleic Acid Interactions*, Sundaralingam, M. and Rao, S. T., Eds., University Park Press, Baltimore, pp. 577-596 (1975).
20. Printz, M. P. and von Hippel, P. H., *Proc. Natl. Acad. Sci. USA 53*, 363 (1965).
21. McConnell, B. and von Hippel, P. H., *J. Mol. Biol. 50*, 297 and 317 (1970).
22. Teitelbaum, H. and Englander, S. W., *J. Mol. Biol. 92*, 55 and 79 (1975).

23. Frank-Kamenetskii, M. D. and Lazurkin, Yu, S., *Ann. Rev. Biophys. Bioeng. 3*, 127 (1974).

24. McGhee, J. D. and von Hippel, P. H., *Biochem. 14*, 1297 and 1568 (1975).

25. In contrast to this interpretation, previous molecular theories of breathing assume the phenomenon to involve conformations significantly perturbed from the classical B-DNA structure. See, for example, Sobell, H. M., Reddy, B. S., Bhandary, K. K., Jain, S. C., Sakore, T. D., and Seshadri, T. P., *Cold Spring Harbor Symp. Quant. Biol. 42*, 87 (1978) and Yathindra, N., in *Proceedings of the International Symposium on Biomolecular Structure, Function, and Evolution*, Srinivasan, R., Ed., Pergamon, London (1979).

26. Kim, S.-H., Berman, H. N., Seeman, N. C., and Newton, M. D., *Acta Crystallogr. B29*, 703 (1973).

27. Lee, C.-H., Ezra, F. S., Kondo, N. S., Sarma, R. H., and Danyluk, S. S., *Biochem. 15*, 3627 (1976) and *16*, 1977 (1977).

28. Cozzone, P. J. and Jardetzky, O., *Biochem. 15*, 4853 and 4860 (1976).

29. Yathindra, N. and Sundaralingam, M., *Proc. Natl. Acad. Sci. USA 71*, 3325 (1974).

30. Olson, W. K., *Biopolymers 14*, 1775 (1975).

31. Olson, W. K., *Macromolecules 8*, 272 (1975).

32. Olson, W. K., *Biopolymers 18*, 1213 (1979).

33. Olson, W. K., *Biopolymers 17*, 1015 (1978).

34. Davis, R. C., and Tinoco, I., Jr., *Biopolymers 6*, 223 (1968).

35. Jordan, R. C., Brant, D. A., and Cesaro, A., *Biopolymers 17*, 2617 (1978).

36. Yevich, R. and Olson, W. K., *Biopolymers 18*, 113 (1979).

37. Yoon, D. Y. and Flory, P. J., *J. Chem. Phys. 61*, 5358(1974).

38. Hardman, N. and Jack, P. L., *Nuc. Acids Res. 5*, 2405 (1978).

39. Deumling, B., *Nuc. Acids Res. 5*, 3589 (1978).

40. Lee, C. S. and Thomas, C. A., Jr., *J. Mol. Biol. 77*, 25 (1973).

41. Pyeritz, R. E. and Thomas, C. A., Jr., *J. Mol. Biol. 77*, 57 (1973).

42. DeLisi, C. and Crothers, D. M., *Biopolymers 10*, 1809 (1971).

43. Rubin, H. and Kallenbach, N. R., *J. Chem. Phys. 62*, 2766 (1975).

44. Jacobson, H. and Stockmayer, W. H., *J. Chem. Phys. 18*, 1600 (1950).

45. Johnson, D. and Morgan, A. R., *Proc. Natl. Acad. Sci., USA 75*, 1637 (1978).

46. Flory, P. J., Suter, U. W., and Mutter, M., *J. Am. Chem. Soc. 98*, 5733 (1976).

47. Additional calculations demonstrate that the random occurrence of a sharp kink in the flexible DNA model is not consistent with the solution properties of B-DNA; $<s^2>$ is reduced substantially, Gaussian behavior is attained at much shorter chain lengths (2^9 - 2^{10}), and small (2^5 - 2^6) loops and rings are highly favored when kinks occur with a probability of 0.01.

48. Flory, P. J., *Principles of Polymer Chemistry*, Cornell University Press, Ithaca, N.Y., Chapter 12 (1953).

Conformational Flexibility of the Polynucleotide Chain

Alexander Rich, Gary J. Quigley, and Andrew H.-J. Wang
Department of Biology
Massachusetts Institute of Technology
Cambridge, Massachusetts 02139

One of the widely appreciated features of protein structures is the fact that the polypeptide chain can fold into different conformations. This includes tightly folded conformations such as found in the α helix or more extended conformations as is found in the β sheet. It is less widely appreciated that the polynucleotide chain also has conformational flexibility which results in significant modifications of the chain extension. Many of these differences are associated with different puckers of the five-membered furanose ring which is a component of both the DNA and RNA chains. Here we illustrate changes in the sugar ring conformation in a variety of structures, taking examples in oligonucleotides as well as in polynucleotides. The property of polynucleotide sugar rings to alter their pucker should be looked upon as a degree of conformational freedom. It is important to have this kind of flexibility in order to form complex structures beyond the simple double helix.

The Double Helix and Ring Pucker

Double helices can be formed with both DNA and RNA molecules. These double helices have similar features in that the bases are in the center of the molecule with the familiar Watson-Crick pairing, either adenine-thymine pairs in DNA or adenine-uracil in RNA. The sugar phosphate chains are on the outside of the double helix. Although the two types of double helices are similar, they differ in one fundamental feature; the pucker of the five-membered furanose ring is different in ribose and deoxyribose.[1] The differences in ring pucker are illustrated in Figure 1. The furanose ring is viewed edge on in a plane which is defined by the ring oxygen O1′ and carbons C1′ and C4′; C1′ has attached to it the ring nitrogen (N) of the purine or pyrimidine; C4′ has attached to it the additional carbon atom C5′ as indicated in the diagram. Both of these constituents, the base (N) and C5′, are located on the upper side of the ring in Figure 1. It is clear that the principal difference in the conformation in the sugar ring is whether the carbon atom C2′ or C3′ is above the plane of the ring. Atoms which are above the plane of the ring in Figure 1 are considered to be in the *endo* conformation.[2] Thus the conformation C2′ *endo* has C2′ on the upper side of the ring and this correspondingly forces C3′ to be on the lower side of the ring. In the other conformation C3′ is *endo* and C2′ is on the lower side of the ring. The major conformation of a deoxyribose chain is one in which the furanose ring has a C2′

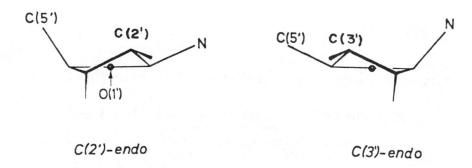

Figure 1. Diagram illustrating two different conformations of the ribofuranose ring. The plane of the ring is defined by three atoms: C1′ to which is attached the glycosidic bond indicated by N, O1′ and C4′ which is attached to the atom C5′. In this diagram, we are looking edge on at the plane defind by these three atoms. The remaining two atoms in the ring, C2′ and C3′ are located either above or below the plane of the ring. Atoms located above the plane of the ring are in the *endo* position. On the left C2′ is in the *endo* position and on the right C3′ is in the *endo* position. Although two examples of ring pucker are shown in this diagram, there are actually a number of intermediate states in which the displacement of C2′ or C3′ is not as great as that illustrated here. The detailed nomenclature for furanose ring pucker is complex[2]; here we elect to use only the simplified *endo* conformations.

endo conformation. In contrast, the normal polyribose chain conformation is C3′ *endo*. The principal reason for the difference is associated with the additional space occupied by the oxygen atom attached to C2′in ribose which is absent in the deoxy series where only a hydrogen atom is attached to C2′. It should be emphasized that even though these are the normal conformations of these sugars in their respective double helices, the energy barriers involved in changing sugar conformation are not very great.[3]

Because of the C2′ *endo* conformation of deoxyribose, the form of the DNA double helix is such that the base pairs are found on the helix axis in the familiar B form of the DNA. The bases occupy the axis of the molecule so that the familiar double helical DNA has a solid, rod-like appearance with bases in the center and two helical grooves running down the molecule. These are the major and minor grooves which occupy the space between the deoxyribose-phosphate chains (see Color Plates 2 and 6).

In contrast, the C3′ *endo* conformation of ribose makes significant modifications in the RNA double helix. The base pairs are no longer perpendicular to the helix axis, but are tilted at about 15-18°. Furthermore, they are set back away from the helix axis. In fact, there is a clear space in the center of the molecule of approximately 3 Angstroms in diameter. The RNA double helix thus looks more like a molecular ribbon wrapped around an imaginary cylinder 3 Angstroms in diameter in which there is now a very deep major groove and a comparatively shallow minor groove on the outside of the helix.

These forms of the familiar double helices are not invariant. Reducing the water content of the medium by adding alcohol readily converts the familiar B form of DNA into the A form in which deoxyribose adopts the C3′ *endo* conformation which is normally found in RNA double helices. The double helix then changes its conformation and looks more like double helical RNA (see Color Plates 3 and 7). It is noteworthy that in the conversion from the B to the A form of DNA, there is over an Angstrom difference in the phosphate-phosphate distance along each polynucleotide chain.

The reason for the change in phosphorus-phosphorus distance can be seen schematically in Figure 2 which shows the conformation of two forms of an adenosine nucleotide in a ribose polynucleotide chain. In the upper figure, the ribose is in the normal C3′ endo conformation and the phosphorus-phosphorus distance is near 5.9 Angstroms. In the lower part of Figure 2, the ribose is in a C2′ endo con-

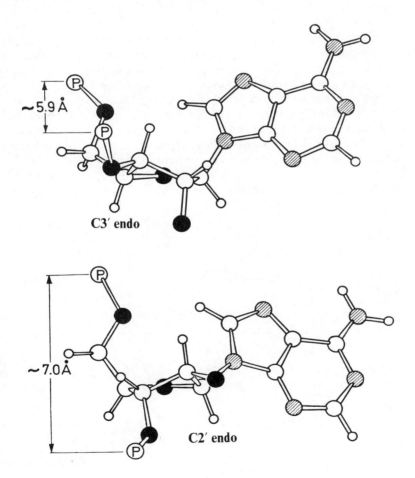

Figure 2. Illustration of two different conformations of adenylic acid. In the upper diagram with the C3′ endo conformation, the two phosphate groups are both above the plane of the ribose ring and are approximately 5.9 Angstroms apart. In the lower diagram, the C2′ endo conformation, the phosphate attached to O3′ is located below the plane of the ribose ring and the phosphates are now approximately 7 Angstroms apart. These two conformations are associated with considerable differences in the extension of the sugar phosphate chain.

formation (normally found in DNA) and the phosphorus-phosphorus distance is near 7 Angstroms. The change in distance is largely associated with a change in the pucker of the ring, although there are also small changes in the other dihedral angles in the backbone. The polynucleotide chain thus has a degree of conformational freedom which allows it to change the degree of extension in a significant way. This flexibility is used in a variety of ways in polynucleotide structures.

Polynucleotide Chain May Turn Corners With Changes in Ring Pucker

There are many examples in which the conformation of the sugar ring is changed. In some cases, the changes in pucker are associated with a change in the direction of a polynucleotide chain. An example of this is seen in Figure 3 which shows the crystal structure of adenyly-(3′,5′)-uridine (ApU) which formed a crystalline complex with 9-amino acridine.[4] In this structure, there are stacked columns of planar molecules in which the 9-amino acridine molecule alternates with an adenine-uracil base pair hydrogen bonding through the ring nitrogen of N7 of adenine. Uracil is at one end of the chain and the backbone has an extended conformation in which the molecule actually turns a corner so that its adenine residue is then hydrogen bonded to still another uracil residue (Figure 3). In the course of making this turn, one of the sugar residues adopts the unusual C2′ endo conformation as shown in the diagram. As seen at the right of the diagram, the uridine ribose O1′ is found at the bottom while it is found at the top of the adenosine ribose. This illustrates the extent to which the chain has radically changed direction in this structure. A somewhat similar conformation is seen in the structure of adenylyl-(3′,5′)-adenosine (ApA) which

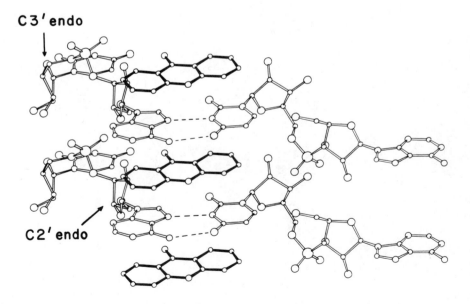

Figure 3. The crystal structure of the complex of ApU and 9 amino acridine. The polynucleotide chain of ApU is in an extended conformation and the chain turns a sharp corner. This can be seen by the different orientation of the two ribose residues in the oligonucleotide. The change in direction is associated with a change in ring pucker as shown.

crystallizes with proflavin in a similar manner.[5] However, it is not obligatory for the ribose in RNA to change pucker in coiled regions of the molecule. The examples cited above illustrate the fact that changes in pucker may be seen in some nucleotide structure determinations.

Tables I and II cite the structure of oligonucleotides which have crystallized by themselves (Table I) or together with intercalators (Table II). They are listed with description of the sugar conformation in various parts of the backbone. Although we have cited two examples in which oligonucleotides change sugar pucker where there is a change in the direction of the polynucleotide chain, there are several

Table I
Oligonucleotides

		5′ end	3′ end	Reference
DNA fragments	pdTpdT	C2′ endo	C2′ endo	15
	d-(pApTpApT)	C3′ endo (A)	C2′ endo (T)	16
RNA fragments	ApU	C3′ endo	C3′ endo	17
	GpC	C3′ endo	C3′ endo	18
	GpC	C3′ endo	C3′ endo	19
	ApA⁺	C3′ endo	C3′ endo	6
	A⁺pA⁺	C3′ endo	C3′ endo	6
	UpA	C3′ endo	C3′ endo	20, 21

Table II
Intercalator Complexes

	Nucleotide	Intercalator	Sugar Pucker 5′ end	3′ end	Reference
DNA fragments	dCpG	TPH⁺	C3′ endo	C2′ endo	7
RNA fragments	CpG	Acridine orange⁺	C3′ endo	C2′ endo	8
	rIodo UpA	Ethidium⁺	C3′ endo	C2′ endo	9
	rIodo CpG	Ethidium⁺	C3′ endo	C2′ endo	10
	rIodo CpG	9-Amino acridine⁺	C3′ endo	C2′ endo	11
	rIodo CpG	Acridine orange⁺	C3′ endo	C2′ endo	12
	rIodo CpG	Ellipticine⁺	C3′ endo	C2′ endo	12
	CpG	Proflavine⁺	C3′ endo	C3′ endo	13
	rIodo CpG	Proflavine⁺	C3′endo	C3′ endo	12
Non-helical	ApU	9-Amino acridine⁺	C2′ endo	C3′ endo	4
	ApA	Proflavine⁺	C2′ endo	C3′ endo	5

oligonucleotides in which there has been no change in the pucker even though the chain is in a fairly extended conformation. An example is seen in the structure of the trinucleotide ApApA.[6]

The Double Helix Changes Conformation Upon Intercalation

One of the conformational changes frequently encountered in double helical nucleic acids is that associated with the binding of a planar intercalator molecule which lodges between the base pairs. It does this without substantial disruption of the helix. Intercalation has two important effects: it introduces a gap between adjacent base pairs which are then separated by 6.8 Angstroms instead of the normal 3.4 Angstroms. This is due to the planarity of the intercalator which usually has unsaturated rings with π electrons which are 3.4Å thick. In addition, there is an unwinding of the double helix.

A series of structures have been solved involving intercalators, mostly in the ribonucleotide series but with one in the deoxynucleotide series (Table II). The structure of a double helical fragment of DNA together with an intercalator is shown in Figure 4. Here we see the structure of deoxy CpG which accommodates a terpyridine-platinum intercalator in the midst of a double helical fragment.[7] As seen in the diagram, the 5' end of the double helical segment adopts the unusual C3' *endo* conformation while the 3' end maintains the C2' *endo* conformation which is normal for double helical DNA.

Figure 5 shows the structure of the ribose dinucleoside phosphate CpG together with the intercalator acridine orange.[8] It can be seen that the 3' end of the double helical RNA fragment has adopted the unusual C2' *endo* conformation while the 5' end

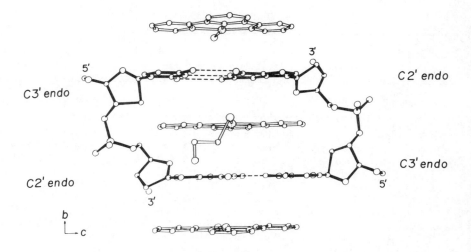

Figure 4. The structure of the deoxy CpG-platinum terpyridine intercalator complex. It can be seen that the DNA double helix has a planar intercalator lodged between the base pairs. This is associated with an unusual ring pucker on the 5' side of the oligonucleotide segment. This conformation allows an extension of the polynucleotide chain.

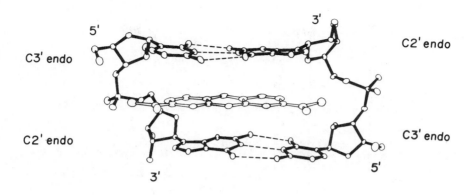

Figure 5. The structure of the CpG-acridine orange intercalator complex. It can be seen that there is an unusual conformation at the 3′ end of the oligonucleotide chain surrounding the intercalator. This mixed pucker conformation is associated with an extension of the polynucleotide chain.

maintains the normal C3′ *endo* conformation. A large number of intercalator structures have been solved in the ribonucleotide series as listed in Table II. Sobell and his colleagues were the first to point out that intercalation is associated with a modification of the pucker of the ribonucleotide chain on the 3′ side of the intercalator.[9-12] Although intercalation is generally associated with conformations similar to those seen in Figure 5, a number of alternative conformations are listed at the bottom of Table II. These are usually associated with the intercalators proflavin[5-12-13] and 9-amino acridine[4] both of which have the property of forming hydrogen bonds between the intercalator and the phosphate of the dinucleoside phosphate. In both cases, other modes of pucker are found. For example, in the complex of proflavin with the ribose CpG fragments, both residues have the normal C3′ *endo* conformation.[13]

For "simple" intercalators in which there is no hydrogen bonding to the phosphate residue, it is possible to make an interesting generalization about the way double helical DNA and RNA accommodate intercalator addition. This is illustrated schematically in Figure 6. At the left the DNA double helix is shown diagrammatically with the normal C2′ *endo* conformation in all residues, except for those on the 5′ side of the intercalator where the unusual C3′ *endo* conformation is adopted in a manner analogous to that which is illustrated in Figure 4. On the right the diagram shows the way in which double helical RNA accommodates an intercalator. All of the residues are in the normal C3′ *endo* conformation except for those residues on the 3′ side of the intercalator which adopt the unusual C2′ *endo* conformation. However, it can be seen that in the region immediately surrounding the intercalator (enclosed in the dashed line) the conformation of both the RNA and DNA chain are similar. Thus both molecules elongate to accommodate an intercalator by adopting a similar conformation. These conformational changes, as described by Sobell[9-12] explain the nearest neighbor exclusion. If a DNA or RNA double helix is saturated with an intercalator, the most that can be accommodated is one intercalator for every two base pairs. The reason for this is probably associated with the necessity for mixed pucker on either side of the intercalator.

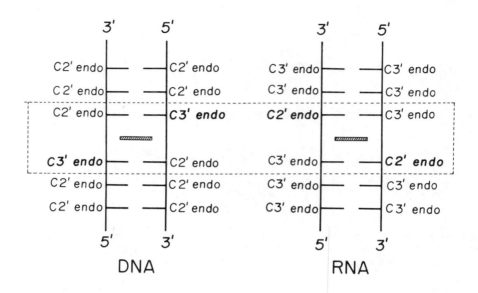

Figure 6. A schematic diagram illustrating the manner in which either DNA or RNA double helices change the pucker of the sugar residues in the region immediately surrounding a simple intercalator. Although the changes are different in the two types of double helices, the conformation in the region surrounding the intercalator enclosed by the dashed rectangle is the same in both cases.

Conformational Flexibility in Nucleic Acid Macromolecules: Yeast tRNA[Phe]

Examples of changes in the type of sugar pucker can be seen in the three-dimensional structure of yeast phenylalanine transfer RNA (tRNA[Phe]). Although most of the seventy-six nucleotides in this transfer RNA molecule adopt the C3′ endo conformation which is normal for a ribonucleotide chain, several residues are found to adopt the less common C2′ endo conformation. This is frequently associated with an interesting type of structural accommodation which is similar to those interactions described above in the oligonucleotide structures. Figure 7 is a diagram (see also Color Plate 1) which shows secondary and tertiary hydrogen bonding found in the yeast tRNA[Phe]. This schematic diagram is useful in interpreting Figures 8 through 11 which illustrate conformational changes in various parts of the molecule.[14]

Changes in Polynucleotide Direction

The principal sites using C2′ endo conformation in yeast tRNA[Phe] are listed in Table III. Nine examples are cited. These occur in two principal situations. In five cases (residues 7, 17, 18, 48 and 60) the C2′ endo conformation is adopted when the polynucleotide chain undergoes a distinct change in direction. For example, nucleotides 1 through 7 are involved in the double helical acceptor stem but residues 8 and 9 are extended and provide a linker region which attaches one end of the acceptor stem with the beginning of the D stem. In this region, there are several conformational changes, one of which is associated with the C2′ endo conformation of ribose 7 where the polynucleotide chain changes direction. Other examples are

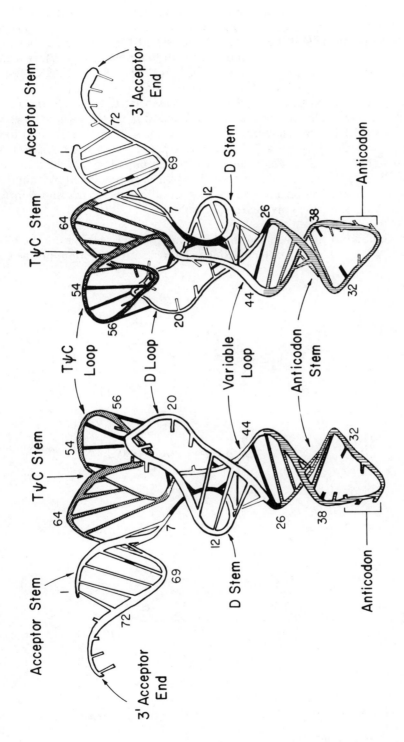

Figure 7. A schematic diagram showing two side views of yeast tRNA[Phe]. The ribose-phosphate backbone is depicted as a coiled tube, and the numbers refer to nucleotide residues in sequence. Shading is different in different parts of the molecules, with residues 8 and 9 in black. Tertiary hydrogen-bonding interactions between bases are shown as solid black rungs, which indicate either one, two or three hydrogen bonds. Those bases that are not involved in hydrogen bonding to other bases are shown as shortened rods attached to the coiled backbone.

found in regions where the backbone loops out away from the center where the bases are stacked. The dihydrouracil residues 16 and 17 are extended out away from the remainder of the molecule and their bases are not stacked with the other bases in tRNA. The backbone has a "bulge" or looping-out at that point (Figure 7). In order to accommodate this extended conformation, residues 17 and 18 adopt the C2′ *endo* conformation. In addition, ribose 47 in the varible loop is extended and its base thrusts away from the center of the molecule. This is associated with a C2′ *endo* conformation in ribose 48. Still another example is found in the T loop where bases 59 and 60 are excluded from the stacking of the other bases in the T stem and loop (Figure 7). This exclusion involves a C2′*endo* conformation in ribose 60. All of these conformational changes are thus associated with an abrupt change in the direction of the polynucleotide chain. This directional change is accommodated by adoption of a C2′ *endo* conformation and there is a substantial change in the phosphorus-phosphorus distance along the polynucleotide chain.

It should be pointed out, however, that there are many other examples in the loop regions of tRNA[Phe] in which the polynucleotide chain changes direction but the C2′

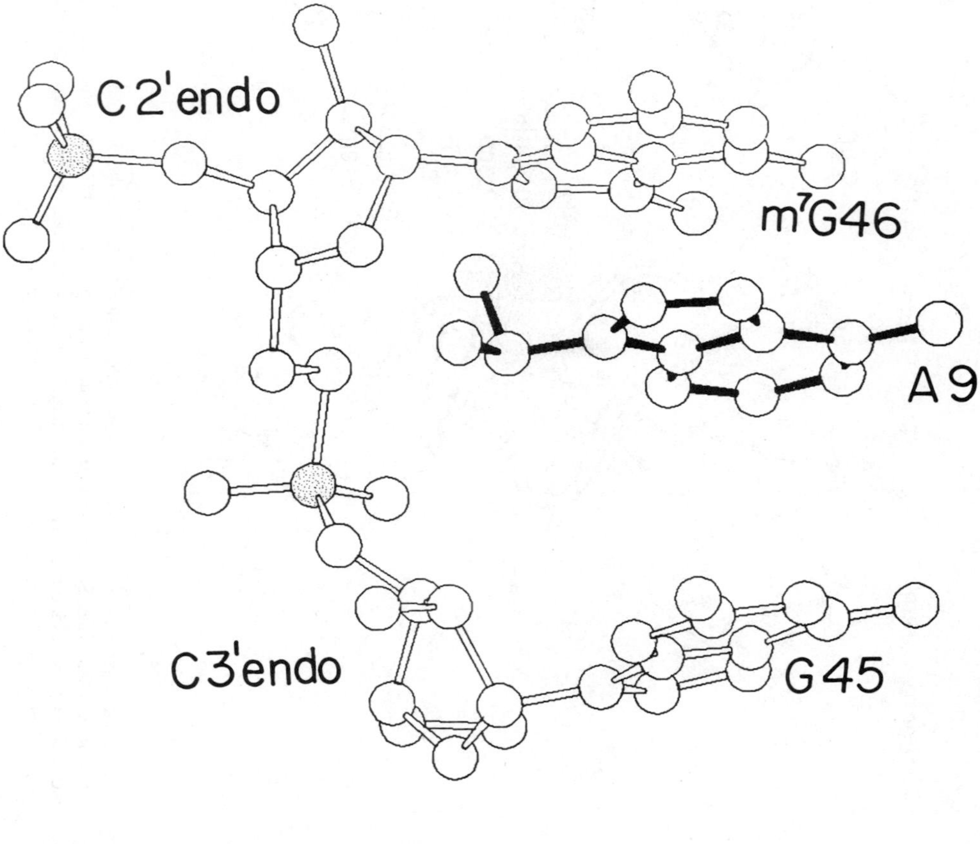

Figure 8. Intercalation in yeast tRNA[Phe]. Adenine of residue 9 intercalates between the bases of 45 and 46. This intercalation is accommodated by a change in the ribose pucker at the 3′ end of the segment. The ribose of m'G46 is in the unusual C2′ *endo* conformation.

endo conformation is not used. Thus, the change in pucker is a structural accommodation which may be adopted in the molecule, especially where the chain is extended, but it is not generally used when the polynucleotide chain undergoes a change in direction.

Intercalation in Yeast tRNA^Phe

There are two parts of the molecule in which extensive intercalation is found involving pairs of nucleotides. These are in the central region with intercalation involving nucleotides 8 and 9 as well as in the corner of the molecule where the T and D loops interact. Figure 8 shows the conformation adopted by the sugars of residues G45 and m⁷G46 which has the adenine ring of A9 intercalated between them. Residues 45 and 46 are not involved in a double helix; nonetheless, the conformation of m⁷G46 is C2′ *endo* in a manner analogous to the conformational changes which are seen for double helical ribonucleotide fragments surrounding intercalators. The C2′ *endo* conformation is adopted at the 3′ end of the intercalator region in Figure 8. Another example is shown in Figure 9 where residues U8 and A9 are found in the extended

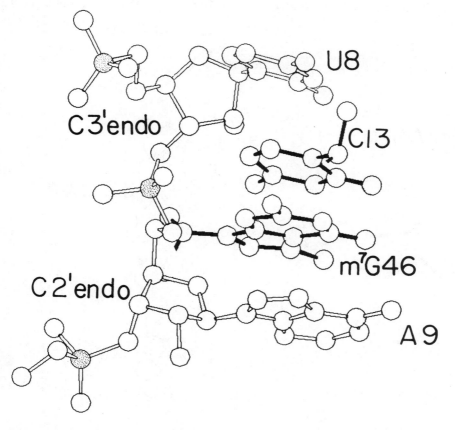

Figure 9. The nucleotides U8 and A9 have intercalated between them the bases C13 and m⁷G46. These bases are both hydrogen bonded to G23 in the tertiary structure of yeast tRNA^Phe. This intercalation is accommodated by a change in the conformation of ribose of A9.

segment which connects the acceptor stem with the D stem. It can be seen that A9 at the 3' end of the segment has adopted the C2' *endo* conformation even though it is not in a complementary double helical structure.

Figure 9 shows that intercalated between U8 and A9 are the bases m⁷G46 and C13 both of which are hydrogen bonded to G22. The base pair C13•G22 is part of the D stem. In Figure 9, it can be seen that the nucleotides U8 and A9 accommodate the additional distance associated with intercalation by adopting the C2' *endo* conformation in residue A9.

By comparing Figures 8 and 9, it can be seen that they are both portions of the same structure in which there are two interacting segments of the polynucleotide chain, both of which intercalate around each other. The polynucleotide chains are interleaved between each other so that each intercalates into the opposite member of the pair.

A similar pair of interleaved structures are found near the corner of the molecule where the bases of nucleotides 18 and 19 interact with the bases of 57 and 58. In Figure 10, it can be seen that m¹A58 and G57 are spread apart with residue G18 in-

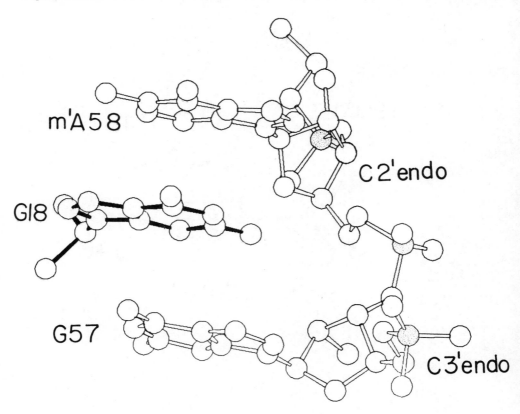

Figure 10. Diagram showing the intercalation of guanine 18 between G57 and m¹A57. The ribose of m¹A58 is in the C2' *endo* conformation to accommodate the intercalation.

tercalated between them. This intercalation is associated with C2′ *endo* conformation found in m¹A58. In this case, it also adopts the C2′ *endo* conformation at the 3′ end of the oligonucleotide segment which surrounds the intercalating guanosine of G18. This is quite similar to the double helical RNA intercalator models which are described in Table II, as the bases G57 and m¹A58 are hydrogen bonded to other residues, although not in Watson-Crick pairs.

An interesting variant is seen in Figure 11 which shows the intercalation of G57 between the bases of G18 and G19. This intercalating interaction is one of the important stacking interactions which stabilize the corner of the tRNA molecule and helps maintain the interaction of the T loop and D loop. However, in this example, the less common C2′ *endo* conformation is found in both riboses of G18 and G19. By analogy with the simple intercalator structures, one would expect a C2′ *endo* conformation to be found in G19 but not in G18. Here it is likely that the unusual C2′*endo* conformation of G18 is not associated so much with the intercalation of G57 but is probably related to the fact that residues 16 and 17 are excluded from the molecule as described above so that their bases do not stack with the others. Another

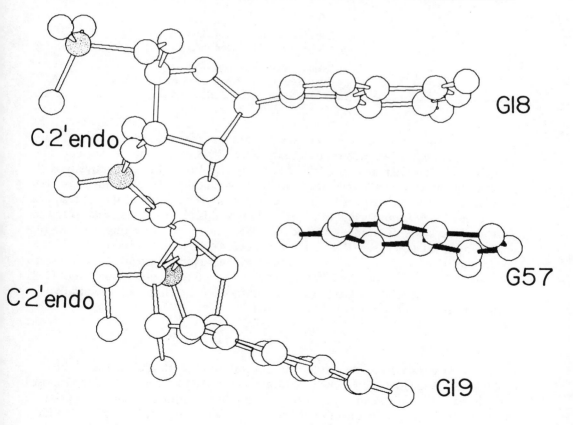

Figure 11. The intercalation of G57 between G18 and G19 is illustrated. Unlike the previous examples, the riboses of G18 and G19 are both in the unusual C2′ *endo* conformation. The C2′ conformation of G18 is probably associated with the unusual conformation of residues 16 and 17 immediately adjoining G18.

Table III
C2′ *Endo* Conformations in Yeast tRNA[Phe]

Residue number	Description
Intercalation:	
9	8-9 intercalation by C13
19	18-19 intercalation by G57
46	45-46 intercalation by A9
58	57-58 intercalation by G18
Change of Direction:	
7	Extended segment at bend in chain, juncture between acceptor stem and 8-9 connection
17	Looping out of backbone at residues 16 and 17
18	Looping out of backbone
48	Looping out of variable loop backbone at U47; juncture between variable loop and T stem
60	Extended segment at bend in chain, bases 59 and 60 excluded from T stem and loop stacking

interesting feature associated with this is the fact that the bases of G18 and G19 are not only separated from each other by a distance of 6.8 Angstroms, which would be necessary for the intercalation of G57, but they are also translated relative to each other so the bases are no longer strictly on top of each other. This translation of G18 relative to G19 in the plane of the bases, can only be accommodated if both residues are in the C2′ *endo* conformation. It is likely that this feature is also related to the fact that the bases of residues 16 and 17 are excluded from stacking with the remainder of the bases in the tRNA structure. Some tRNA molecules contain only one nucleotide in this region which is excluded from the stacking, instead of the two seen in yeast tRNA[Phe]. It is possible that in these cases, with only one residue, that G18 may have the normal C3′ *endo* conformation since the translation of G18 relative to G19 in the plane of bases may not be required.

Discussion

In the above, we have discussed the conformational flexibility found in the polynucleotide chains. This flexibility is inherent in the fact that there is a furanose ring in the chain which can adopt two different conformations which are associated with a varied extensibility of the chain. Structural studies on simple oligonucleotides reveal that conformational changes occur in regular double helical structures associated with intercalation as well as in other structures where polynucleotide chain undergo an abrupt change in direction. In the three-dimensional structure of yeast tRNA[Phe], examples are shown of both types of conformational changes.

Changes in pucker due to intercalation in general follow trends which are seen in intercalation in simple double helical RNA structures even though the tRNA examples do not involve simple RNA double helices. This implies that the adoption of a C2′ endo conformation together with an associated extension of the polynucleotide chain is not solely limited to double helical intercalation, but may in fact be of a more general nature as shown by the examples cited above.

Nucleic acid molecules have considerable conformational flexibility. We have only a hint of this flexibility in studying the structure of the double helix itself. However, when one begins to study the interaction of double helical structures with other molecules, especially intercalators, we see that there is a method for accommodating them involving changes in extensibility. In complex globular polynucleotides such as the transfer RNA molecule, a variety of changes in pucker are observed. The molecule adopts unusual conformations associated either with chain extension, changes in direction of the polynucleotide chains or with the accommodation of other bases which intercalate into the chain even though the chains are not involved in a double helical array. As the structure of more complex polynucleotide structures are solved, it is our expectation that this will be found to be a completely general feature. This element of conformational flexibility associated with changes in ring pucker is likely to be a constant feature of polynucleotide chains when they interact with other molecules including proteins, as well as when they interact with other polynucleotide chains.

Acknowledgements

This research was supported by grants from The National Institutes of Health, The National Science Foundation, The National Aeronautics and Space Administration and the American Cancer Society. We thank J. Simpson for help in preparing the manuscript. A. H. J. W. is supported in part by the M. I. T. Center for Cancer Research (Grant No. CA-14051).

References and Footnotes

1. Arnott, S. and Hukins, D, W. L., *Biochem. Biophys. Res. Comm. 47,* 1504-1509 (1972).
2. Sundaralingam, M. in *Structure and Conformation of Nucleic Acids and Protein-Nucleic Acid Interactions,* Ed by M. Sundaralingam and S. T. Rao, University Park Press, Baltimore, Maryland, pp. 487-524 (1974).
3. Yathindra, N. and Sundaralingam, M. in *Structure and Conformation of Nucleic Acids and Protein-Nucleic Acid Interactions,* Ed by M. Sundaralingam and S. T. Rao, University Park Place, Baltimore, Maryland, pp. 649-676 (1974).
4. Seeman, N. C., Day, R. O., and Rich, A., *Nature 253,* 324-326 (1975).
5. Neidle, S., Tayler, G., Sanderson, M., Shieh, H.-S. and Berman, H. M., *Nucleic Acids Res., 5,* 4417-4422 (1978).
6. Suck, D., Manor, P. C., and Saenger, W., *Acta. Crysta. B32,* 1727-1737 (1976).
7. Wang, A. H.-J., Nathans, J., van der Marel, G., van Boom, J. H. and Rich, A., *Nature 276,* 471-474 (1978).
8. Wang, A. H.-J., Quigley, G. J. and Rich, A. *Nucleic Acids Res.* (in press).
9. Tsai, C. C., Jain, S. C. and Sobell, H. M., *J. Mol. Biol. 114,* 301-315 (1977).
10. Jain, S. C., Tsai, C. C. and Sobell, H. M., *J. Mol. Biol. 114,* 317-331 (1977).
11. Sakore, T. D., Jain, S. C., Tsai, C. C., and Sobell, H. M., *Proc. Natl. Acad. Sci. USA, 74,* 188-192 (1977).

12. Sobell, H. M., in *The International Symposium on Biomoleculr Structure, Conformation, Function and Evolution.* (Ed. R. Srinivasan), Pergamon Press, Oxford, New York (in press).
13. Neidle, S., *et al.*, *Nature 269*, 304-307 (1977).
14. Quigley, G. J. and Rich, A., *Science 194*, 796-806 (1976).
15. Camerman, N., Fawcett, J. K. and Camerman, A., *Science 182*, 1142-1143 (1973).
16. Viswamitra, M. W., Kennard, O., Jones, P. G., Sheldrick, G. M., Salisbury, S., Falvello, L. and Shakked, Z., *Nature 273*, 687-688 (1978).
17. Seeman, N. C., Rosenberg, J. M., Suddath, F. L., Kim, J. J. P., and Rich, A., *J. Mol. Biol. 104*, 109-144 (1976).
18. Rosenberg, J. M., Seeman, N. C., Day, R. O., and Rich, A., *J. Mol. Biol. 104*, 145-167 (1976).
19. Hingerty, B., Subramanian, A., Stellman, S. D., Sato, T., Broyde, S. B., and Langridge, R., *Acta Crysta. B32*, 2998-3013 (1976).
20. Sussman, J. L., Seeman, N. C., Kim, S. H. and Berman, H. M., *J. Mol. Biol. 66*, 403-421 (1972).
21. Rubin, J., Brennan, T. and Sundaralingam, M., *Biochem. 11*, 3112-3129 (1972).

Accessible Surface Areas of Nucleic Acids and Their Relation to Folding, Conformational Transition, and Protein Recognition

Charles J. Alden and Sung-Hou Kim

Department of Biochemistry
Duke University Medical Center
Durham, North Carolina 27710

and

Department of Chemistry and Laboratory of Chemical Biodynamics
University of California
Berkeley, California 94720

Introduction

Most biologically functional macromolecules have an intrinsic ability to fold into well defined three dimensional structures in physiological environments. Once they are folded, each molecule presents to its solvent environment a characteristic surface structure, when averaged over a long time period. Although the solvent-accessible surface structure is in dynamic fluctuating states, as demonstrated by the exchange of tritium or deuterium between solvent water and internal protons of a protein,[1] the fluctuation is centered around a defined average structure, with a well defined average "solvent-accessible surface structure."

A three dimensional structure determined by X-ray crystallographic method is a time-averaged (usually for a period of several weeks or longer of data collection) and space-averaged (over million molecules in a crystal) structure. Thus the crystal structure can be considered as a "static" structure representing the average of the population of dynamically fluctuating structures. Based on this "static" structure, one can determine the surface structure that represents the population-averaged surface of the molecule which is accessible to solvent or other interacting ligands.

For proteins, which incorporate a wide spectrum of polar, aliphatic and aromatic side groups and which exhibit a variety of secondary structures, the question of solvent effects as a driving force in chain folding has long been discussed.[2/3] The concept of accessible surface area has been introduced to estimate the relative changes in solvent-accessible surface upon folding polymers from random coils to compact forms and also to assess the surface exposures of different classes of atoms within folded polypeptides.[4-8] For nucleic acids, by contrast, the concept of relative surface exposure has not been explored extensively. Thus we have examined the solvent-

331

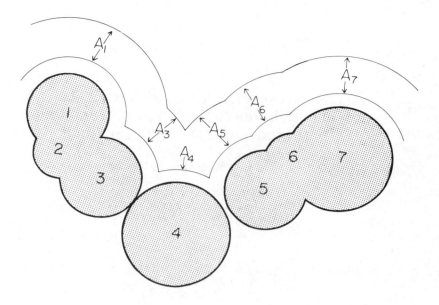

Figure 1. Schematic diagram of accessible surface areas. Around the examined molecule (shaded) is constructed an envelope defining the loci of points for the center of a spherical probe which may just touch the molecule. This construction is sliced by a set of parallel planes, and the perimeter (A) of the surface envelope is calculated for each atom. In the figure two separate envelopes, corresponding to different probe sizes, are shown. Note that as the probe radius increases, the accessibilities to it for atoms within a concavity decrease.

accessible surfaces of a variety of polynucleotides, (a) to evaluate the extent of surface burial on folding, (b) to compare relative exposures between hydrophobic and hydrophilic atoms, (c) to identify the most accessible atoms for intermolecular interactions, and (d) to relate helical conformational transitions to environmental changes. The details of the topics described here have been published.[9]

Calculation of "Static" Accessible Surface Area

The methodology employed in this study is very similar to that described by Lee and Richards[4] in their calculation of polypeptide surface areas. We imagine a spherical "probe" or solvent molecule of radius r_w free just to touch but not penetrate the van der Waals surface of the examined molecule. The closed surface defined by all possible loci for the *center* of the probe is defined as the accessible surface of the molecule, and those portions of the surface for which the probe touches only one atom comprise the accessible surface of that atom (Figure 1). For actual calculation, a molecule is sliced into sections with thickness h, then the accessible surface area for each slice is calculated. The calculation is an approximate integration, for which the *precision* will increase as h decreases. In general, the calculation is repeated several times, with slicings at different orientations, both to test the reproducibility of the calculations and to achieve a credible average value for each surface. The *accuracy* of these calculations depends only upon the values chosen for the van der Waals and probe radii (r_v and r_w), and of course upon the accuracy of the input coordinates. For details, see Lee and Richards[4] and Alden and Kim.[9]

Throughout, we use the terms "accessible surface area," and "exposure," synonymously, but to emphasize different aspects. Note that these surface areas involve contacts with only *one* probe molecule; interaction of the probe with other (bound) solvent molecules is not considered in these calculations

To determine best values for van der Waals radii, we used the intermolecular contact and volume survey of Bondi[10] for carbon, phosphorus, aromatic nitrogen and ester oxygen radii. Values for presumed spherical carbonyl oxygen, amino and methyl groups were derived from the volume decrements for each group, cited in the same survey, by the use of a computer program which calculates volumes in an approximate integral fashion, analogous to the surface accessibility program. The van der Waals radii used are listed in Table I. In some cases, we used modified radii to implicitly include hydrogen atoms for computational simplicity (Table I).

Maximal Accessible Areas of Polynucleotide Components

To gauge the change in surface exposure upon folding a molecule, it was necessary first to determine the maximum *possible* surface exposure of the components of that molecule, corresponding to the unfolded state. For polypeptides, this exposure has usually been determined by calculating the surface exposure of the middle residue of model tripeptides with all conformation angles *trans*.[4] For polynucleotides, with several variable torsion angles per residue, the all-*trans* conformation may not cor-

Table I
Van der Waals Radii Used In This Study

Symbol	Group Type	Radius (Å) Including hydrogen atoms explicitly	Including hydrogen atoms implicitly
ALC	Aliphatic carbon	1.70	2.00
ARC	Aromatic carbon	1.77	1.77
SOX	Sugar oxygen (ring and ester)	1.40	1.40
BOX	Carbonyl oxygen	1.64	1.64
POX	Phosphate oxygen	1.64	1.64
ARN	Aromatic ring nitrogen	1.55	1.55
AMN	Amino nitrogen	1.72	1.75*
PHO	Phosphorus	1.80	1.80
ARH	Aromatic ring hydrogen	1.01	--
ALH	Aliphatic hydrogen	1.17	--
AMH	Amino hydrogen	1.17	--
WAT	Water	1.40	1.40

*1.86 Å is suggested as more appropriate.

respond to the one (of ~50) most exposed conformation; in addition, it is known that two adjacent phosphoester bond torsion angles are reluctant simultaneously to adopt *trans* values.[11] Figure 2 illustrates the accessible surface exposure of the phosphate oxygens for the model case of dimethyl phosphate as a function of the conformation angles about the two ester bonds. Included also are indicators for the corresponding conformation angles about the ester bonds observed in the structure of transfer RNA.[12] We note that the all-*trans* conformation, avoided by polynucleotide phosphates, corresponds to a minimal exposure for the charged phosphate oxygens. In addition, the repulsive lone-pair interactions of the ester oxygens are believed to discourage adoption of this conformation in polymers.[13/14] Hence a more elaborate construction was invoked to determine maximal exposures

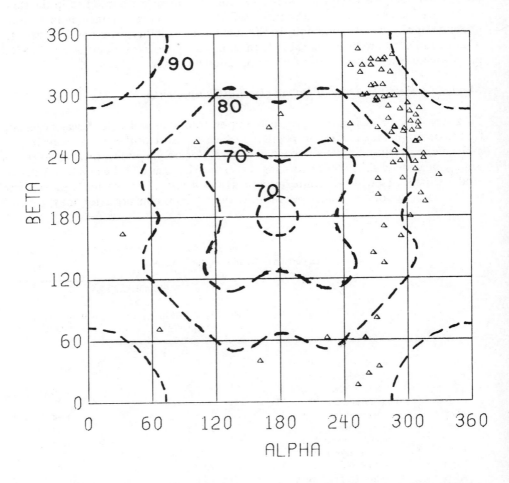

Figure 2. Accessible surface exposure of phosphate oxygens in dimethyl phosphate. The exposure of the phosphate oxygens was calculated for different conformations about the two ester bonds. Dashed lines represent contours of equal exposure; triangles represent the corresponding conformations in the backbone of the observed crystal structure of yeast phenylalanine transfer RNA.[12] The reluctance of nucleotide backbones to adopt an all-*trans* conformation may in part be due to the burial of charged phosphate oxygens that such conformations entail.

for the individual atoms (see Alden and Kim[9]). Maximal exposure for the deoxyribose atoms were calculated for both the C3'-*endo* and C3'-*exo* puckered forms. The total maximum exposures for the carbon and oxygen atoms were seen to be very similar, although the exposure of individual atoms were different for the two forms. Maximal exposures of the nucleoside bases were calculated for the bases assuming the normal *anti* conformation observed in polynucleotides. Table II lists the maximal surface exposures for the DNA components (calculated with explicit hydrogens) and for the RNA components (with implicit hydrogens). While it is quite possible that not all of the atoms in a nucleotide may simultaneously achieve their maximal exposure, this list gives a fair estimate of the relative degree of exposure for the various nucleotide atom types for the unfolded state. Thus we note that the average *maximal* accessible surface area for an unfolded DNA chain is approximately 490 Å^2 per nucleotide, with the bases accounting for nearly half (44 percent) of this total.

Polarity of Exposed Atoms

One difficulty in describing the relative exposures of polar and nonpolar groups is the assignment for the aromatic nitrogen atoms, which are quite polar and frequently form hydrogen bonds within the plane of the base, yet which contribute to the π orbitals of the ring and act as aromatic molecules when approached normal to the plane of the base. Arbitrarily dividing the contribution from such atoms equally among the polar and nonpolar classifications, we note that in the unfolded state polar atoms comprise slightly over half (54 percent) of the exposed surface of DNA, with approximately equal contributions from the bases and the backbone. For the backbone there is an almost equal exposure of hydrocarbon and oxygen atoms, while for the bases 42 percent of the atoms may be classed as non-polar. For the individual nucleoside bases, however, there is a pronounced range of maximal polar group exposure: 69 percent for G, 61 percent for C, 52 percent for T and 45 percent for A.

Surface of DNA Double Helices

Accessible surface areas for double- helical DNAs were calculated by summing the individual atomic exposures of the middle base-paired residues in various double-

Table II
Maximum Accessible Surface Areas in Å^2 for
The Components of Double Helical DNA and RNA
for ($r_w = 1.4\ \text{Å}$)

DNA backbone with C3'-endo sugars	275.01
DNA backbone with C3'-exo sugars	274.28
RNA	288.93
Adenine	213.66
Cytosine	195.84
Guanine	236.76
Thymine	212.67
Uracil	186.60

stranded, complementary trinucleotides. Helices examined were A- and B-DNA,[15] C-DNA[16] and D-DNA.[17] For each of these helices hydrogen atoms were included in the input coordinates, and areas were calculated with a slicing separation of 0.1 Å. In all cases except D-DNA, several different base sequences were examined, to test for sequence and composition dependence of surface accessibility. Values for (A,T)- and (G,C)- rich polymers were taken as the averages of the four distinct trinucleotide sequences consisting of one type of base pair. The results of these averagings are summarized in Table III.

Overall, the folding of a long DNA molecule into a double helix results in the burial of approximately two-thirds of its maximally exposed surface, a value quite similar to the resulst for globular proteins.[4/7] A comparison of the relative surface exposures of B-DNA with the unfolded molecules is shown in Figure 3. For most atom

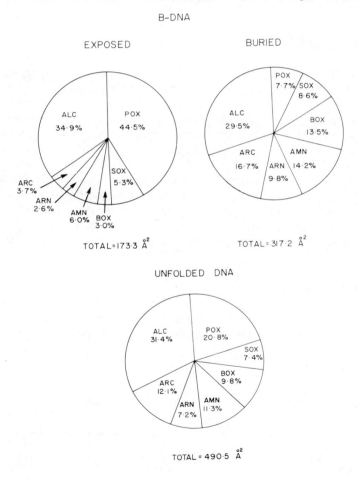

Figure 3. Group type exposures in unfolded and helical B-DNA. Solvent accessible surface areas were calculated using a 1.4Å probe radius for maximally unfolded DNA and B-DNA of average base sequence. The accessible surface buried in folding the extended chain into a double helix, is also shown. In folding of the polymer, the bases become mostly buried while the phosphate oxygens remain nearly fully exposed.

Table III
Group Type Exposures in $\AA^2$
For Two Residues of DNA Double Helices

(a) A,T Rich Polymers

	A-DNA	B-DNA	C-DNA	D-DNA
ALC	7.50	11.87	17.39	26.31
ARC	4.77	5.41	4.50	3.96
SOX	23.12	21.34	20.24	22.41
POX	143.92	152.52	152.06	128.12
BOX	17.48	11.90	10.21	8.44
ARN	11.38	8.99	9.12	7.20
AMN	1.40	1.58	1.91	2.34
PHO	0.96	0.36	0.14	0.30
ARH	5.02	2.35	1.40	0.00
ALH	126.47	122.70	125.64	129.17
AMH	7.98	11.28	11.23	11.67
TOTAL	350.00	350.30	353.84	339.92
POS	20.76	21.85	22.26	21.21
NEUT	143.76	142.33	148.93	159.44
NEG	185.48	186.12	182.65	159.27

(b) G,C Rich Polymers

ALC	6.46	6.36	7.12
ARC	6.32	12.33	11.28
SOX	22.00	14.96	14.10
POX	150.34	155.84	157.06
BOX	12.36	9.18	9.18
ARN	9.28	8.88	9.04
AMN	6.35	3.25	2.68
PHO	0.96	0.36	0.14
ARH	3.16	5.70	4.88
ALH	106.48	100.78	106.98
AMH	30.80	25.33	23.81
TOTAL	354.51	342.97	346.27

types, the reduction of accessible surface upon folding is close to their proportion of the total maximal surface area, with base atoms being buried relatively more than sugar atoms. The sole and striking contrast to this pattern occurs for the phosphate oxygens, whose exposure is reduced only slightly upon transition from the extended to the helical form. Thus the accessible surface in DNA double helices is more polar than for the random coils, with the phosphate oxygens alone accounting for nearly 45 percent of the total accessible surface area. The contribution from the bases now accounts for only about a fifth of the total surface in the double-stranded helices, while the sugars account for over a third of the surface.

Considering the observed effect of solvent upon DNA helix conformation, and the pronounced difference in gross shapes of the different forms, the similarities of their total exposures were somewhat surprising. Hence we calculated the surface accessibilities for A- and B-DNA over a wide range of probe radii (1.0 to 5.0Å), to determine if larger probes, such as hydrated metal ions or structured water aggregates, could discriminate between the different DNA shapes. The results of this test are shown in Figures 4 and 5. We note that as r_w increases, the *total* surface ac-

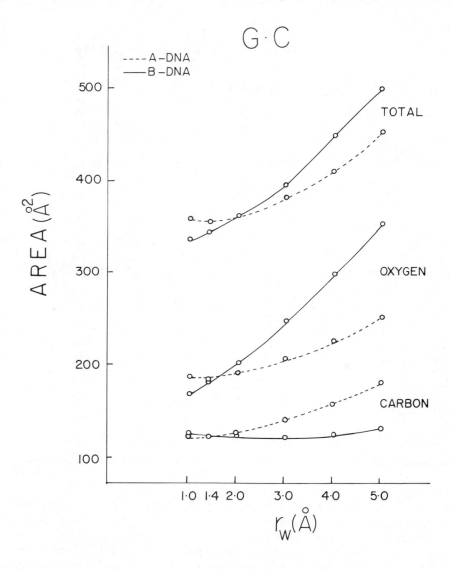

Figure 4. Accessible surface areas for two base-paired residues of poly(dG)•poly(dC) in the A- and B-helical forms, calculated varying the probe radius r_w. For the polymer, the B-DNA form entails relatively greater total exposure, and particularly a greater exposure of phosphate oxygens, for large probe radii. This result is consistent with the experimentally observed result that high water activity encourages adoption of the B-DNA conformation.

cessibility of B-DNA increases more rapidly than does that of A-DNA. Moreover, this difference is attributable almost entirely to a greater exposure of the phosphate oxygens in B-DNA, as the aliphatic carbon exposure becomes relatively greater for A-DNA. Curiously, DNA accessibility curves intersect just at the r_w corresponding to the radius of a single water molecule. On the other hand, when one examines the *base* surface accessibility for larger values of r_w (corresponding to bulky side groups of proteins, structured water, or hydrated metal ions) the accessibility of the major groove of A-DNA decreases abruptly, as does the minor groove exposure of B-DNA (see Figures 6 and 7).

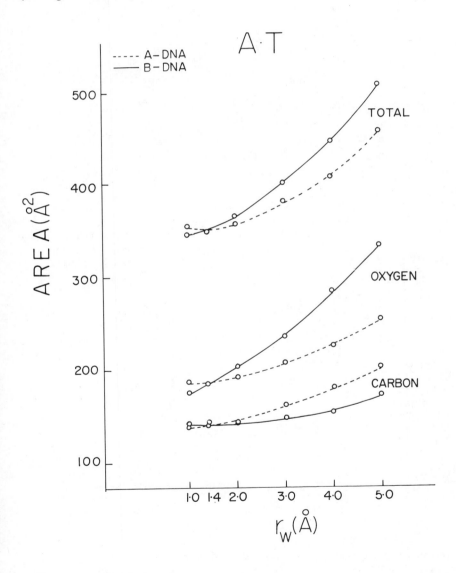

Figure 5. Accessible surface areas for two base-paired residues of poly(dA)•poly(dT) in the A- and B-helical forms, calculated varying the probe radius r_w. Rest of the details same as in Figure 4 legend.

For probe radii greater than 3 Å, only guanine amino groups in the minor (shallow) groove exhibit significant exposure in A-DNA, while thymine methyl and cytosine amino groups in the major (wide) groove dominate the base exposures of B-DNA. These groups may be expected to be the most significant for recognition by DNA-binding proteins.

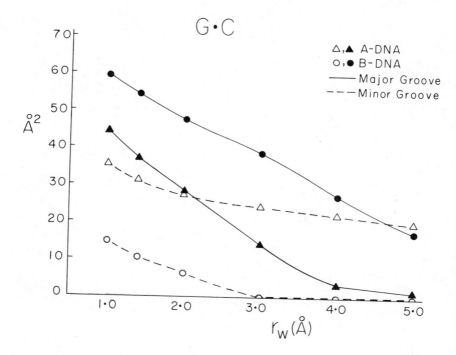

Figure 6. Accessible surface areas for DNA base atoms of the major and minor grooves of A- and B-DNA for G•C base pairs calculated varying the probe radius r_w. As r_w increases to large values, corresponding to extended water complexes or larger amino acid side chains, the only groups with significant exposures are the amino groups of C and G.

For (A,T)-rich polymers, adoption of the A, B, or C helical forms results in virtually identical total group-type exposures to a 1.4Å radius probe molecule, corresponding to a single water molecule. The eight-fold alternating- sequence D-DNA has a slightly smaller total exposure, and a significantly smaller (over 30 Å²) exposure of oxygens, reflecting the virtual collapse of the minor groove in that helix. For (G,C)-rich double helices there is a slightly greater variation (~12 Å²) in total exposures depending upon conformation. Between different conformations there are pro-

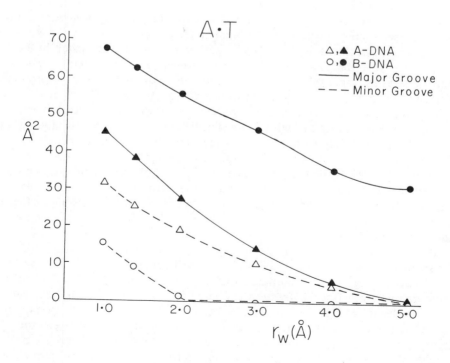

Figure 7. Accessible surface areas for DNA base atoms of the major and minor grooves of A- and B-DNA for A•T base pairs, calculated varying the probe radius r_w. As r_w increases to larger values (see legend for Figure 6), the only group with significant exposures is the methyl group of T.

nounced changes in the exposures of some atoms. In A-DNA, C1′ is quite exposed and C3′ is not, while the converse is true for B-DNA, reflecting the difference in sugar puckering modes. The exposures of all other backbone atoms are quite similar for the two helices. There is, however, a marked difference in exposure of the base atoms between the different polymers.

In B-DNA, the exposure of base atoms at the major groove is about 25 $Å^2$ (50 percent) greater than in A-DNA, while base atoms in the B-DNA minor groove have only about one-third the exposure of those in A-DNA. In particular, thymine methyl groups in the major groove are twice as exposed as in B-DNA as in A-DNA, while guanine amino groups in the minor groove are far more exposed in A-DNA. Within a given conformation the exposure of a particular base also is influenced slightly by the *sequence* of adjacent base pairs, with the most pronounced difference being a greater exposure of pyrimidines and smaller exposure of purines in alternating sequences compared to homopolymer tracts. However, the sequence-dependent variations in exposure are much less noticeable than are the variations with helix conformation. For isolated water molecules the major groove of A-DNA is only slightly more exposed than the minor groove, whereas in B-DNA, the major groove is five to six tims more exposed than the minor groove.

Surfaces for RNA Double Helix and Single Helix

Surface accessibilities for RNA structures were calculated using nonhydrogen atoms and a coarser slicing width (0.5 $Å$). Maximal surface exposures for the RNA components were calculated as described in earlier section and are listed in Table II. While slight differences result from the treatment of the aliphatic carbons and amino groups as single spheres, omitting separate hydrogens, the most pronounced difference arises from the additional ribose hydroxyl group. This is evident in the slightly enhanced sugar oxygen exposure, and decreased aliphatic carbon exposure, for unfolded RNA compared with DNA. The folding of RNA into the double helical A-RNA[18] structure results in a burial of approximately 63 percent of the maximal surface. As in the case of helical DNAs, the greatest reduction in exposure comes for the aromatic bases, while the relative contribution of the phosphate oxygens to the exposed surface is again over double that in the unfolded state. For double-helical A-RNA, polar groups comprise nearly 70 percent of the total accessible surface with sugar oxygens contributing about 10 percent more and aliphatic carbons correspondingly less, than for DNA (Figure 8).

The surface exposure of a single strand of polyribocytidylic acid (rC) in the 11-fold A-RNA conformation was compared with the observed 6-fold poly (rC) conformation obtained from fiber diffraction studies.[19] We note that the total exposure per residue is about 20 $Å^2$ (10 percent) less in the latter case, with the reduction being due almost entirely to a greater burial of carbon atoms; nitrogen and oxygen atoms have about equal exposures in both cases. This suggests that 6-fold poly (rC) is probably more stable than an A-RNA type conformation for a single strand. Potential energy calculations of dinucleoside monophosphates, extended to helical polymers, without consideration of solvent effects, also indicate a preference for approximately six-fold helices,[20] so there is reason to believe that the fiber structure is representative of

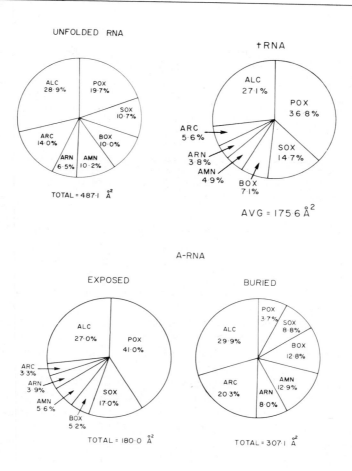

Figure 8. Group type exposures in extended and folded RNA. Solvent accessible surface areas were calculated using a 1.4Å probe radius for maximally exposed RNA (with random base sequence), the crystal structure of yeast phenylalanine transfer RNA,[12] double helical A-RNA (with random base sequence) and the buried surface in A-RNA. Despite the irregular folding and considerable chemical modification of the tRNA, its accessible surface has a distribution very similar to that of A-RNA.

a preferred helical form for single stranded polycytidylic acid and is not merely an artifact of lattice interactions in the crystallite.

Transfer RNA

The accessible surface for yeast phenylalanine transfer RNA in the orthorhombic crystal form was calculated using the coordinates of Sussman, *et al.*[12] Only non-hydrogen atoms (with their appropriate radii) were included, and a slicing separation of 0.5 Å was employed. To facilitate computation, the molecule was divided into three overlapping segments; the region of overlap was sufficiently broad that every atom was accompanied by its full set of neighbors in at least one set of calculations. Individual atomic accessible surface areas were averaged for three separate slicing orientations and summed for each residue and atom type and for the whole

Table IV
Backbone Exposures in tRNA Stems and Loops:
Average (and Standard Deviations) per Nucleotide in $\AA^2$

	tRNA helical*		tRNA non-helical		tRNA	A-RNA
Ribose C	41.6	(7.5)	32.2	(18.2)	38.2	48.7
Phosphate	72.1	(9.4)	48.6	(17.9)	63.5	75.8
Sugar 0	29.8	(4.6)	17.6	(12.3)	25.2	25.2

*A residue is classified as helical when it adopts a conformation characteristic of a standard double helix *and* when its base is stacked with both neighboring residues. Here residues 2-6, 11-14, 23-31, 35-44, 50-53, and 62-75 are designated as helical, all others (except 1 and 76) as nonhelical.

molecule, as listed in Table IV. The average exposure per residue, 175 $\AA^2$ is slightly less than the average for an interior residue of normal double helical A-RNA, despite the presence of many substituted groups on the surface and the unstacking of four bases as obseved in the crystal structure of this tRNA (for a review, see Kim[21]); apparently the extensive tertiary interactions more than compensate for any base stacking in the chain folding. Figure 8 shows the relative exposures of various group types in this tRNA; it may be noted that the overall distribution is quite similar to that of an A-RNA: bases are exposed slightly more and phosphate oxygens somewhat less than for the regular polymer. The near identity of alphatic carbon exposures is rather remarkable: it should be noted that the contribution of aliphatic carbons in the modified bases and sugars accounts for nearly 20 percent of the ALC portion, and 5 percent of the total accessible area of this tRNA, just offsetting the greater burial of the ribose carbons in the nonhelical residues compared to A-RNA. Inspection of the list of individual atomic exposures for this tRNA revealed that, without exception, every "modified" aliphatic group is exposed substantially, suggesting that the modification of bases significantly increases (by about 20 percent) the areas of groove surfaces for various proteins to recognize.

The exposure calculations for the phosphate oxygens and ribose carbons of the individual residues of tRNA reveal that the backbone exposures within helical regions are quite uniform and close to those in helical A-RNA, while significant variations are found only at residues forming sharp bends at the ends of stems and in loops. Furthermore, the backbone atoms in loops are found to be *less* exposed than those in the stems.

Two-Thirds of Nucleic Acid Surfaces are Buried on Folding

We have calculated the water-accessible surface areas of double-stranded DNAs, RNA and for one transfer RNA, and compared them to the maximal surface exposures for unfolded polynucleotides. In general, folding of the polymer reduces the surface exposure to about one-third of its maximal possible value, which is similar to the surface reduction calculated for folding of globular proteins.[4,7] Upon folding, there is a noted tendency for nonpolar groups (particularly the aromatic bases) to be buried, as expected, and for the charged phosphate oxygens to maintain near maximal exposure. As is also the case for proteins, those polar atoms which are buried form hydrogen bonds with other polar groups. Thus there is a general tendency for polynucleotides to obey the dictum of "polar out, nonpolar in" when folding from a random chain to a compact structure.

Driving Force in Nucleic Acid Folding is Quite Different from That of Proteins

For an average DNA residue, the surface area that becomes buried on folding maximally extended coils into the double-helical B-DNA form is $317\mathring{A}^2$, of which $126\mathring{A}^2$, or about half, is due to nonpolar atoms. Similarly for transfer RNA, neglecting changes in exposure caused by addition of modified groups, the surface area buried upon folding an extended coil into the native structure is about $23,700\mathring{A}^2$, of which about $12,200\mathring{A}^2$ corresponds to burial of hydrocarbons. Per nucleotide residue, these values correspond to 311 and $160\mathring{A}^2$, respectively. The hydrophobic effect of nonpolar surface burial is widely regarded as one of the dominant factors in protein folding. Based on a calibration of amino-acid solubilities *versus* side chain accessibilities, this effect has been estimated to contribute about $-24\text{cal/mole-}\mathring{A}^2$ to the net free energy of protein folding.[7/25] For nucleic acids, the burial of water-accessible area per unit mass ($0.95\mathring{A}^2$/dalton for DNA, $0.97\mathring{A}^2$/dalton for tRNA) is quite similar to that calculated for proteins (0.85 to $1.12\mathring{A}^2$/dalton).[7] In addition, the proportion of polar and nonpolar surface, both exposed and buried, are not greatly different for proteins[8] and for nucleic acids. However, despite these similarities, the thermodynamic properties of the two systems differ markedly: upon chain unfolding, nucleic acids exhibit positive values for entropy and enthalpy,[22-24] whereas those quantities are both negative for protein denaturation. If one assumes a similar energy/area ratio for hydrophobicity in nucleic acids as in proteins, it follows that other forces more than counteract its effect upon the energetic terms. Alternatively, it may be possible that the large dipole moments of the aromatic bases eliminate the hydrophobic effect for their surfaces. In either case, the dominant driving force(s) in nucleic acid folding must differ from that of proteins.

Comparison of the measured free energies, enthalpies and entropis of 19 oligonucleotides[22] with their buried or exposed surface areas yielded only a very approximate correlation. This is a direct consequence of the somewhat surprising result that, within a given helical conformation, the exposed (or buried) water- accessible area for a given base pair shows only minor variation (less than $8\mathring{A}^2$) with respect to the sequence of the neighboring base pairs, although the thermodynamic parameters for certain sequence isomers may vary by as much as a factor of two. For these oligonucleotides, the free energy/buried area ratio was approximately -1.2 ± 0.6 cal/mole-$\mathring{A}^2$, which is the same order of magnitude as the corresponding ratio for transfer RNA.[24] Both the small magnitude and the large scatter of these values also suggest that factors other than solvent-base interactions are the dominant determinants of the sequence-specific stability of these nucleotides.

Correlation of A → B Helical Transitions to DNA Hydration

A wide variety of experimental evidence has related DNA helical conformation transitions to altered solvent conditions. X-ray diffraction[26] and Raman spectroscopy[27] of DNA fibers, as well as infrared dichroism of DNA films[28/29] and dilute solutions[30] all clearly indicate marked conformational changes at defined levels of relative humidity or water activity. Similarly, gravimetric,[31] infrared,[32] calorimetric,[33] and density gradient centrifugation studies[34/35] clearly demonstrate distinct hydration of DNA. These studies combined indicate that the level of hydration required to maintain DNA in a double-helical structure, and for which all the primary DNA hydration sites are occupied, is about 9 to 10 waters per nucleotide. To convert DNA to the B helical form requires extra water, for a total of 13 to 18 per nucleotide.[30] These added waters have properties more like bulk water than the primary hydration shell[32] and may be considered as a secondary shell of hydration, not interacting directly with the DNA but forming water-water bridges.[36]

The surface accessibility calculations described above provide support for this mechanism for DNA conformation transition. The virtual identity of the total surface accessibilities for A- and B-DNA at $r_w = 1.4\text{Å}$, corresponding to a single water molecule, indicates that the primary hydration layers for both forms are quan-

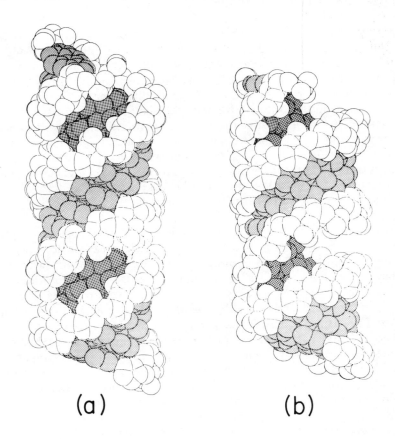

(a)   (b)

Figure 9. Illustrations of A-DNA double helix. View (a) is normal to the helix axis while view (b) represents the helix tilted by 27.0° to reveal maximum groove width. Dark shading indicates bases in the major groove and lighter shading indicates bases in the minor groove. In A-DNA the phosphate oxygens of the two strands are directed toward each other across the deep, narrow major groove while the minor groove is nearly flat. This contrasts with B-DNA (see Figure 10).

titatively similar. However, for larger values of r_w, the increased accessible surface for B-DNA relative to A-DNA, and particularly that portion associated with the phosphate oxygens (Figures 4 and 5), indicates that B-DNA is much more amenable to formation of extended solvent complexes about its charged phosphate groups. Sterically, this interpretation appears justifiable: in B-DNA both phosphate oxygens point radially outward from the helix, whereas in A-DNA one phosphate oxygen is directed inward toward the major groove (Figures 9 and 10).

This difference in hydration may be illustrated another way. For both A- and B-DNA, the volumes enclosed by the van der Waals surface of the molecule are virtually identical: about 450 Å^3 per two (paired) nucleotides. Constructing a cylinder

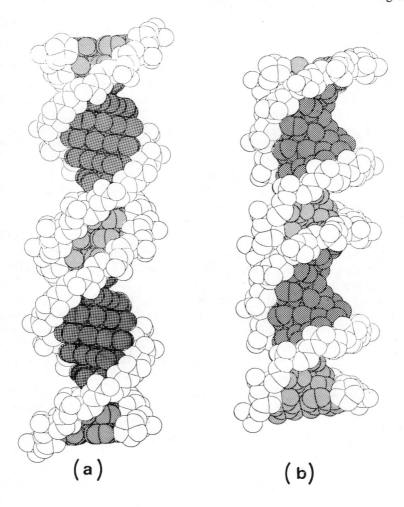

(a)   **(b)**

Figure 10. Illustrations of B-DNA double helix. View (a) is normal to the helix axis while view (b) represents the helix tilted by 31.5° to reveal the maximum groove width. The shading is the same as in Figure 9. Note that, unlike A-DNA (Figure 9) the phosphate oxygens are directed outward, delineating two grooves of approximately equal depth but unequal width.

about each helix, with radius r equal to the average of the radial phosphate oxygen coordinates plus one water diameter, the volume per double-stranded residue is V = $\pi r^2 h$, where h is the axial rise per residue of the particular helix. Subtracting 450 $\text{\AA}^3$ from this volume yields the groove volume accessible to water; dividing this volume by the mean volume of one water molecule in liquid (30 $\text{\AA}^3$) yields the number of waters per residue which may be contained in the grooves. Using r = 11.6 or 12.3$\text{\AA}$, and h = 2.56 or 3.38$\text{\AA}$, for A- and B-DNA respectively, the number of waters per nucleotide are 10.5 and 19.3 for the two polymers. Again, the calculation is crude but the result is quite compatible with the experimentally observed values of 9-10 and 18 waters per A- and B-DNA residue, respectively.[30] Thus, conversion from the B to the A form upon reduction of water activity (for a DNA capable of existing in either form) may be seen as a response to removal of excess water from the grooves, converting a structure with two concave grooves into a shorter structure with one flatter groove and one narrower groove (Figures 9 and 10).

It is now fairly well established, at least in hydrated films and fibers, that DNA rich in A•T pairs prefer the B conformations, while sequences rich in G•C pairs have a wider range of stability in the A and B forms.[26/29/37] Density gradient centrifugation studies[35] have linked the lower density of A,T-rich polymers to a higher level of hydration; this result is supported by a recent set of theoretical calculations which also indicate that an A•T pair can bind one or two more water molecules than a G•C pair.[37] In light of the multiplicity of coordinated geometries available to hydrated ions complexing with phosphate groups and the base-lined grooves, a unique steric description of a DNA-solvent superstructure appears unlikely. However, the influence of specific base atoms upon the secondary hydration layer (determined by examining atom exposures for $r_w > 1.4\text{\AA}$) can provide clues about the willingness of the environment to allow or enforce a particular DNA conformation. For A•T base pairs, the greatest exposure is exhibited by thymine methyl groups, and here only in the B conformation (over 80 percent of the total base exposure for $r_w \geq 4.0$ $\text{\AA}$). Since methyl groups are hydrophobic, their exposure in the major groove would tend to encourage water-water aggregation in the major groove and thus pressure the major groove to remain wide. For G•C pairs the most exposed groups are the amino groups of guanine on the minor groove (over 80 percent of the total base exposure in the A form, no exposure in the B form, for $r_w = 4.0\text{\AA}$) or cytosine on the major groove (over 70 percent of the total base exposure in B-DNA at $r_w \geq 4.0\text{\AA}$). Both groups are hydrophilic, therefore, extended water-DNA base interactions would result in either the A- or B-helical forms. Thus for (G,C)-rich DNA, the availability or lack of excess water will dictate the DNA conformation: under high water activity the major groove will fill and the phosphates will form more extensive hydration complexes   under low water activity, the major groove will collapse and the phosphate oxygens will become more buried.

Major Groove of B-DNA and Minor (Shallow) Groove of A-DNA Are Primary Recognition Surfaces for Proteins

As can be seen from Figures 6 and 7 as the radius of the probe becomes 3 $\text{\AA}$ or greater, the *primary* base exposures occur in the major groove of B-DNA and the minor groove of A-DNA. Specifically, the methyl group of thymine and amino group of cytosine provide the most of the accessible area on the major groove in B-DNA, and the amino group of guanine on the minor groove of A-DNA. Since most of the side chains of amino acids are considerably larger than single water molecules the above observation can be considered as suggesting that the *primary* recognition surface of B-DNA is the major groove and that of A-DNA is the minor groove. This is consistent with the observation that protein-DNA contacts occur predominantly in the major groove of DNA in the E. coli RNA polymerase - *lac* promoter complex[39] and in the *lac*-repressor-*lac* operator complex[40] if one interprets protection *as well as* enhancement of methylation as due to close contact of grooves with proteins.

Acknowledgements

This work was supported by grants from the National Institute of Health (CA-15802 and K04-CA-00352) and the National Science Foundation (PCM76-04248). C.J.A. is an N.I.H. research fellow.

References and Footnotes

1. Hvidt, A. and Nielsen, S. O. *Advan. Protein Chem. 21* 287-386 (1966).
2. Kauzmann, W. *Adv. Protein Chem. 14* 1-63 (1959).
3. Tanford, C. *J. Am. Chem. Soc. 84*, 4240-4247 (1962).
4. Lee, B. and Richards, F. M. *J. Mol. Biol. 55*, 379-400 (1971).
5. Shrake, A. and Rupley, J. A. *J. Mol. Biol. 79*, 351-371 (1973).
6. Finney, J. L. *J. Mol. Biol. 96*, 721-732 (1975).
7. Chothia, C. *Nature 254*, 304-308 (1975).
8. Richards, F. M. *Ann. Rev. Biophys. Bioeng. 6*, 151-176 (1977).
9. Alden, C. J. and Kim, S.-H. *J. Mol. Biol.*(in press).
10. Bondi, A. *J. Phys. Chem. 68*, 441-451 (1964).
11. Kim. S.-H., Berman, H. M., Seeman, N. C., and Newton, M. D. *Acta Cryst. B29* 703-710 (1973).
12. Sussman, J. L., Holbrook, S. R., Warrant, R. W., Church, G. M., and Kim, S.-H., *J. Mol. Biol. 123*, 607-630 (1978).
13. Sundaralingam, M. *Biopolymers 7*, 821-860 (1969).
14. Newton, M. D. *J. Amer. Chem. Soc. 95*, 256-258 (1973).
15. Arnott, S. and Hukins, D. W. L. *Biochem. Biophys. Res. Comm. 47*, 1504-1509 (1972).
16. Arnott, S. and Selsing, E. *J. Mol. Biol.98*, 265-269 (1975).
17. Arnott, S., Chandrasekaran, R., Hukins, D. W. L., Smith, P. J. C., and Watts, L. *J. Mol. Biol. 88*, 523-533 (1974).
18. Arnott, S., Hukins, D. W. L. and Dover, S. D., *Biochem. Biophys. Res. Comm. 48*, 1392-1399 (1972).
19. Arnott, S., Chandrasekaran and Leslie, A. G. W. *J. Mol. Biol. 106*, 735-748 (1976).
20. Hingerty, B. and Broyde, S. *Nucl. Acids Res. 5*, 127-137 (1978).
21. Kim. S.-H. "Advances in Enzymology" (ed. A. Meister) Vol. 46, pp. 279-315 (1978).
22. Borer, P. N., Dengler, B., Tinoco, Jr., I., and Uhlenbeck, O. C. *J. Mol. Biol. 86*, 843-853 (1974).
23. Edelhoch, H. and Osborne, Jr., J. C. *Adv. Protein Chem. 30*, 183-276 (1976).
24. Privalov, P. L. and Flimonov, V. V. *J. Mol. Biol. 122*, 447-464 (1978).
25. Chothia, C. *Nature 248*, 338-339 (1974).
26. Arnott, S., in "Organization and Expression of Chromosomes" (eds. V. G. Allfrey, E. K. F. Bautz, B. J. McCarthy, R. T. Schimke and S. Tissieres) Dahlem Konferenzen: Berlin (1976).
27. Erfurth, S. C., Bond, P. J. and Peticolas, W. I. *Biopolymers 14*, 1245-1257 (1975).
28. Pilet, J. and Brahms, J. *Biopolymers 12*, 387-403 (1973).
29. Brahms, J., Pilet, J., Tran, T. P. L. and Hill, L. R. *Proc. Nat. Acad. Sci. USA 70*, 3352-3355 (1973).
30. Wolf, B. and Hanlon, S. *Biochemistry 14*, 1661-1670 (1975).

31. Falk, M., Hartman, Jr., K. A., and Lord, R. C. *J. Amer. Chem. Soc. 84*, 3843-3846 (1962).
32. Falk, M., Hartman, Jr., K. A. and Lord, R. C. *J. Amer. Chem. Soc. 85*, 387-391 (1963).
33. Privalov, P. L. and Mrevlishvili, G. M. *Biofizika 12*, 22-29 (1967).
34. Hearst, J. E. *Biopolymers 3*, 57-68 (1965).
35. Tunis, M. J. B. and Hearst, J. E. *Biopolymers6*, 1345-1353 (1968).
36. Lewin, S., *J. Theoret Biol. 17*, 181-212 (1967).
37. Wells, R. D., Burd, J. F., Chan, H. W., Dodgson, J. B., Jensen, K. F., Nes, I. F., and Wartell, R. M. "CRC Critical Reviews in Biochem." 305-340 (1977).
38. Goldblum, A., Perahia, D. and Pullman, A. *FEBS Letters 91*, 213-215 (1978).
39. Johnsrud, L. *Proc. Nat. Acad. Sci. USA75*, 5314-5318 (1978).
40. Gilbert, W., Maxam, A. and Mirzabekov, A., in "Control of Ribosome Synthesis" (eds. Kjeldgaard and Maaloe) pp. 139-148 (1976).

The Coil Form of Poly(rU):
A Model Composed of Minimum Energy Conformers
That Matches Experimental Properties

S. Broyde
Biology Department
New York University
New York, New York 10003

and

B. Hingerty
Biology Division
Oak Ridge National Laboratory
Oak Ridge, Tennessee 37830

Introduction

In the study of the coil form of ribopolynucleotides, poly (rU) has attracted special attention. This is due to the fact that its properties are insensitive to temperature in the range 15°-45°C.[1] In this respect it differs from other sequences, whose characteristics come to resemble those of poly (rU) once the temperature is raised. The same is true of UpU versus the other ribodinucleoside monophosphates.[2] The optical[2-4] and NMR[5-8] evidence indicates that UpU and poly (rU) are largely unstacked above 15°C.

The detailed conformational characteristics of poly (rU) in solution are thus of interest because of their uniqueness at ordinary temperatures. A series of uracils, and possibly even one subunit, in single stranded regions of RNAs is likely to possess distinctive features of shape, which could be salient to their functions. For example, the region specifying termination of transcription at the end of the tryptophan operon of *Escherichia coli* has a 3' - terminal m-RNA transcript whose sequence is C-A-U-U-U-U$_{OH}$.[9] A series of as many as 6 - 8 uridine residues is, in fact, a common feature of the termination sites at the 3' end of m-RNA molecules.[10] These are not necessarily all encoded in the DNA.

A key work in the study of poly (rU) is that of Inners and Felsenfeld,[1] who measured the limiting characteristic ratio C_∞, the mean square unperturbed end to end distance divided by the product of the number of bonds in the chain and the mean square bond length. A value of 17.6 was obtained, which indicated that the coil was relatively extended and restricted in conformation.

A number of workers have calculated coil models that agree with this characteristic ratio. Inners and Felsenfeld[1] calculated a polynucleotide model that matched this value, employing conformers from Sundaralingam's 1969 survey of the crystallographic nucleotide and polynucleotide literature.[11] Another such calculation, based on the same data, was made by Delisi and Crothers,[12] who took into account the interdependence of the δ, ϵ and β,γ angle pairs. (See Fig. 1 for structure, numbering scheme and conformational angle designations for UpU.) Olson and Flory have made classical potential energy calculations for the polynucleotide backbone.[13] Using the statistical weights calculated for each minimum in the δ,ϵ and β,γ energy surfaces they reproduced the experimental characteristic ratio. In a recent work, Yevich and Olson have made extensive additional calculations which further characterize coils.[14] Tewari, Nanda and Govil calculate coil models with proper characteristic ratios,[15] employing the conformations obtained from quantum mechanical calculations.[16] However, the $\beta,\gamma = g\text{-},g\text{-}$ conformations are given added weights of about 2 kcal/mole. These calculated coil models do not agree with each other in conformational detail. Moreover, they were calculated for the polynucleotide backbone without considering the influence of the bases on conformation. Porschke has made the important observation, employing temperature jump techniques, that coils of different sequences have different conformational states.[18] It must be mentioned, however, that coils of poly(rU) and poly(rA) have similar characteristic ratios.[1]

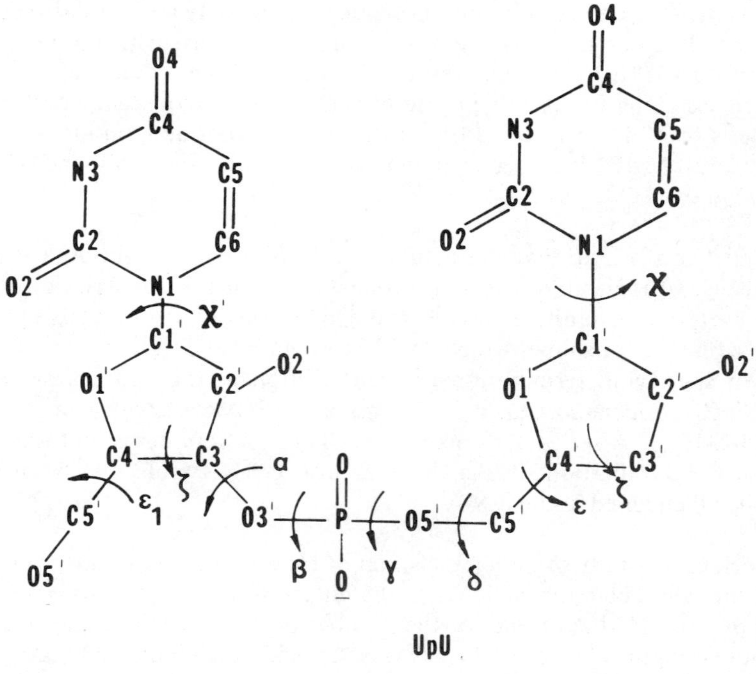

Figure 1. Structure, numbering scheme and conformational angle designations for UpU. The dihedral angles A - B - C - D are defined as follows: χ',χ: O1' - C1' - N1 - C6 α: P - O3' - C3' - C4' β: O5' - P - O3' - C3' γ: C5' - O5' - P - O3' δ: C4' - C5' - O5' - P ϵ,ϵ_1: C3' - C4' - C5' - O5'. The angle A - B - C - D is measured by a clockwise rotation of D with respect to A, looking down the B - C bond. A eclipsing D is 0°. Sugar pucker is described by the pseudorotation parameter P[25].

In the present study, our aim was to obtain a conformational model specific to the poly (rU) coil. This model was designed to match the experimental characteristic ratio[1] as well as conform to other criteria. The model is composed of minimum energy conformers of UpU. These were obtained by classical potential energy calculations in which all torsion angles and the ribose pucker were variable parameters. Its conformational characteristics match the NMR findings of Sarma and co-workers on UpU[7] and poly (rU).[8] In addition, its calculated persistence length is of the same order of magnitude as an experimental value observed for a perturbed denatured RNA chain,[17] while helical duplexes have persistence lengths that are very much larger.[19]

Methods

The potential energy calculations were carried out as detailed previously[20] including Van der Waals, E_{nb}, electrostatic, E_{el}, torsional, E_{tor}, and ribose strain, E_{st}, contributions to the energy, E.

$$E = E_{nb} + E_{el} + E_{tor} + E_{st} \tag{1}$$

$$E_{nb} = \sum_{i<j}\sum (a_{ij}r_{ij}^{-6} + b_{ij}r_{ij}^{-12}) \tag{2}$$

$$E_{el} = \sum_{i<j}\sum 332\, q_i q_j r_{ij}^{-1} D^{-1} \tag{3}$$

$$E_{tor} = \sum_{k=1}^{8} V_{o,k}(1 + \cos 3\theta_k) \tag{4}$$

$$E_{st} = \sum_{l=1}^{5} K\tau_l(\tau_l - \tau_{o,l})^2 \tag{5}$$

r_{ij} is the distance in angstroms between atoms i and j, q_i is the partial charge assigned to atom i, $V_{o,k}$ is the barrier to internal rotation for the dihedral angle k and θ_k is the value of that angle, $K\tau_l$ is a force constant, τ_l is the strained ribose bond angle, and $\tau_{o,l}$ is the value that angle adopts at equilibrium. Also, k denotes the eight independent dihedral angles and l the five deoxyribose bond angles. Partial charges were taken from Renugopalakrishnan, *et al.*[21] The dielectric constant (D) was assigned a value of 4 and the torsional barriers as well as the parameters a_{ij} and b_{ij} were taken from Lakshminarayanan and Sasisekharan.[22/23]

The energy of ribose was calculated previously by Dr. T. Sato, some of whose results have been reported by Sasisekharan.[24] In his work, the energy was minimized as a function of the pseudorotation parameter, P, the puckering amplitude, θ_m (notation of Altona and Sundaralingam[25]), and the bond angles of O1' - C1' - C2' (α_1) and O1' - C4' - C3' (α_2). These completely define the deoxyribose coordinates. For these calculations, τ_0 was taken to be 113.5° for C-O-C, 110.0° for C-C-O and 109.5° for C-C-C. K_τ values employed were, in kcal/mol rad², 66.5 for C - O - C, 59.9 for C - C- 0, and 54.0 for C - C - C. These are 70 percent of the values obtained experimentally,[26] and were devaluated by Sato in order to obtain a better fit between observed and calculated conformations. Results of Sato's calculations for ribose are presented

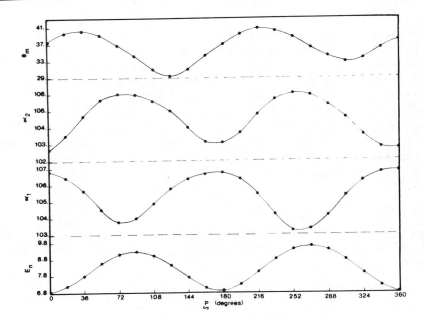

Figure 2. Energy, En, of ribose (kcal/mol); α_1, the bond angle O1' - C1' - C2' (deg); α_2, the bond angle O1' - C4' - C3' (deg); and θ_m the puckering amplitude (deg) as a function of P, the pseudorotation parameter. E_n was given previously in Stellman, *et al.*[49]

in Fig. 2. His results for deoxyribose have been given previously.[20] These energies and the other variables in Fig. 2 were incorporated in the energy calculations of the present work, and by linear interpolation permitted a continuous variation in ribose energy as a function of puckering. Bond lengths and angles were taken from Arnott, *et al.*[28] The eight backbone torsion angles and the ribose pucker were variables in a 9 parameter minimization of the energy, employing a modified version of the Powell algorithm.[29] Starting conformations included the minima obtained previously for UpU, calculated with fixed ribose pucker, as well as the low energy forms calculated for deoxydinucleoside monophosphates.[20/31]

The statistical weight w_i, of the *i*th minimum is given by:

$$w_i = A_i \exp(-\Delta E_i/RT)/Z \qquad (6)$$

where ΔE_i is the energy of the *i*th minimum, A_i is the relative area associated with it, and

$$Z = \sum_i A_i \exp(-\Delta E_i/RT) \qquad (7)$$

Olson has detailed procedures for calculating statistical weights from energy contour maps.[36] To evaluate the relative area, A_i associated with each minimum, the torsion angles other than β and γ were fixed at their values at the minimum. β and γ were separately varied in one degree intervals and the energy was calculated at each point until the 1 kcal/mole contour was located. The contours are approximately elliptical and have sheer walls beyond 1 kcal/mole. Indeed, the 1 kcal/mole and the 3 kcal/mole edges virtually coincide when the other torsion angles are kept fixed. The characteristic ratio, C_∞ was computed by the method of Flory,[32/33] expanded by Olson[34-37] for polynucleotides, with the matrix method of Eyring.[38]

The persistence vector calculation is done similarly, as described by Olson[39] and Yevich and Olson.[15]

Results and Discussion

UpU Minimum Energy Conformations

Table I presents low energy conformers of UpU. In these new calculations the ribose pucker as well as the backbone dihedral angles were variable parameters. Furthermore, the trans and g- domains of the C4' - C5' torsion ϵ were explored, in addition to the g + region that had been studied previously with ribose pucker fixed.[30] With C-3'-endo type pucker, the three lowest energy forms have ϵ = g+ and the 03'-P and 05'-P torsions β,γ are respectively t, g-, (skewed), g-,t and g-, g- (A form). These are at energies of less than 1 kcal/mole. The β,γ = g+, g+, ϵ = g+ and the Watson-Crick[37] (β,γ = g-,t, ϵ = t) conformations are at somewhat higher energies. With C-2'-endo pucker, there are three conformers with almost equally low energy. In two of them ϵ is trans, together with β,γ rotations of t,g+ and g+,g+. In the third form β,γ = t, g- and ϵ is g+. Other minimum energy conformations are the Watson-Crick and the B form, although they are of higher energy. Conformers with ϵ = g- were not found below 3 kcal/mole.

Statistical weights of these conformers are given in Table II. Also shown are the ranges and the relative areas of the 1 kcal/mole contours in the β,γ plane associated with each minimum. The β,γ = t, g- conformation has the highest statistical weight in each puckering domain.

ORTEP drawings of the three lowest energy conformations in each puckering region are shown in Figure 3. The interesting point is that these conformers are predominantly not stacked. The A form has bases highly overlapping and essentially coplanar. The β,γ = g-, t conformer 2 also has considerable stacking, but the alternate g-,t conformation 4 (not shown) is unstacked. This contrasts with, for example, the low energy conformations of dApdA, which are predominantly stacked.[20]

Experimental Observations on UpU and Poly (rU)

Our goal in the present work was to mesh these and other theoretical studies with experimental observations by others on UpU and poly rU, to obtain a model for the coil form of poly (rU). The NMR findings of Sarma and co-workers [7] reveal that the conformation of UpU changes little on elevating the temperature from 20°C to 89°C. These NMR studies indicate the following approximate conformational features for UpU in solution: ribose pucker 53 percent C-3'-endo, 47 percent C-2'-endo; ϵ = g +, 83 percent. The NMR properties of poly (rU) are virtually identical with those of UpU in this temperature range[8] (although a different form of the polymer exists below 15°C[39]). Consequently, the low energy forms of UpU are

Table I
Minimum Energy Conformations of UpU[a]

#	χ'	ε₁	α	β	γ	δ	ε	χ	P	ΔE	Description: β, γ; ε
						C-3'-endo Region					
1.	15	56	282	225	319	103	52	56	13	0.	t, g-; g+
2.	44	78	287	320	170	190	57	10	6	0.5	g-, t; g+
3.	3	58	210	316	278	180	51	17	8	0.9	g-, g-; g+; A
4.	45	62	204	326	144	203	50	4	12	1.9	g-, t; g+
5.	45	62	190	41	84	196	76	26	14	2.6	g+, g+; g+
6.	11	172	199	290	170	212	136	10	-5	3.6	g-, t; t
						C-2'-endo Region					
1.	31	55	289	197	48	224	141	11	180	0.0	t, g+; t
2.	21	176	279	91	84	251	176	54	185	0.04	g+, g+; t
3.	33	59	289	202	301	124	59	77	170	0.1	t, g-; g+
4.	15	178	211	271	172	192	150	78	202	2.1	g-, t; t
5.	27	59	202	255	317	170	62	81	179	3.1	g-, g-; g+; B

[a] ΔE is the energy difference in kcal/mol between the local minimum and the global minimum in each puckering domain. Dihedral angles and P are in degrees.

Table II
Calculated Statistical Weights, w, of
UpU Minimum Energy Conformations

Minimum #	β Range, (°)	γ Range, (°)	Relative Area	w
		C-3′-endo Region		
1.	219-228	316-325	.27	.67
2.	317-323	167-174	.14	.15
3.	309-322	275-282	.30	.16
4.	322-329	140-147	.16	.016
5.	37-46	80-86	.12	.0037
6.	281-298	168-172	.23	.0014
		C-2′-endo Region		
1.	191-208	43-51	.45	.27
2.	83-105	82-87	.37	.21
3.	181-211	297-307	1.0	.51
4.	262-280	168-177	.54	.0096
5.	244-261	311-324	.74	.0025

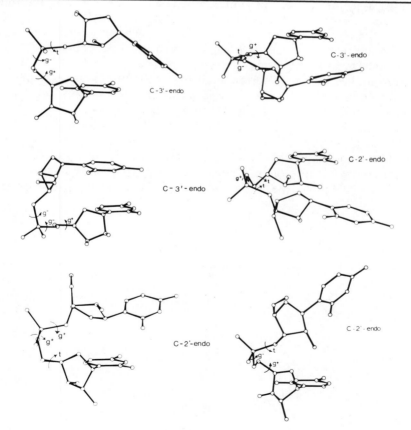

Figure 3. Lowest energy conformers of UpU.

similar to the conformations of poly (rU). The present poly (rU) conformational model was designed to match these data. In addition, it must match the measured characteristic ratio C_∞ of 17.6 obtained for poly (rU).[1] Another experimental quantity taken into account (although not obtained for poly (rU) coils) is the approximately 75 Å persistence length measured for a denatured perturbed RNA chain.[17] The theoretical study by Olson of the dependence of the unperturbed dimensions of polynucleotides on the orientation of the phosphodiester bonds[43] was also considered. She finds that the experimentally observed high values of C_∞ can be matched theoretically if conformers with β = trans make an important contribution to the coil form.

Characteristic Ratios

The first step in our model building effort was to compare the characteristic ratio of an energy weighted assembly of all conformers listed in Table I with the measured value. (Conformer 5 with C-3′-endo pucker was not included because it is sterically disfavored in polymers[44/45].) The calculated characteristic ratio in this case is 4.9. Quantum mechanical conformational calculations also yield low results.[14] Our low value is due to contributions by conformers with very low extensions. The extension is reflected in h, the rise per nucleotide residue along the helix axis (assuming the construction of a regular helix from each conformational building block). Table III gives h for the three lowest energy conformers in each puckering region. Negative values of h indicate a left handed helix. It is seen that the minima β,γ = g-, t and g+, g+ are very compact forms. Therefore, these are improbable in the extended coil, although they may be important under other conditions; for example, the turn in the anticodon loop of tRNAs is negotiated via the conformation β,γ = g-, t (C-3′-endo minimum 2) and involves a U base.

In the next step, we selected from among our low energy conformers combinations that would agree with the NMR results:[7] 53 percent C-3′-endo, 47 percent C-2′-endo, 83 percent ϵ = g+. Since the g- region of ϵ is not represented among conformers below 3 kcal/mole mole, the 17 percent non g+ conformers were assigned to the

Table III
Rise Per Residue of UpU Minimum Energy Conformation

Minimum #	$\beta, \gamma; \epsilon$	h
	C-3′-endo Region	
1.	t,g-; g+	3.78
2.	g-,t; g+	.80
3.	g-,g-; g+	2.64
	C-2′-endo Region	
1.	t,g+; t	-3.83
2.	g+,g+; t	-0.15
3.	t,g-; g+	5.51

trans domain. The likeliest candidate for this role on energetic grounds is conformer 1, with C-2'-endo pucker, which has β,γ = t, g+. The remaining 30 percent C-2'-endo contribution is assigned to conformer 3, β,γ = t, g-. It has the highest statistical weight for C-2'-endo pucker and is very extended (h = 5.51Å). For the C-3'-endo contribution, the β,γ = t, g- and the A forms are the energetically plausible choices. The characteristic ratio was then calculated for a series of polymers in which the energy weightings, ΔE_ψ, were changed so that the C-2'-endo conformers were present in the above proportions and the percent A form varied from 0 percent to 53 percent, in tandem with the C-3'-endo β,γ = t,g- conformation. Results are given in Figure 4, Curve I. We see that the experimental characteristic ratio is matched with about 1 percent A form and 52 percent β,γ = t, g-. The present coil model is thus composed of 82 percent conformers with β,γ = t, g-, 52 percent being C-3'-endo and 30 percent C-2'-endo. The remaining 18 percent has 17 percent, β,γ

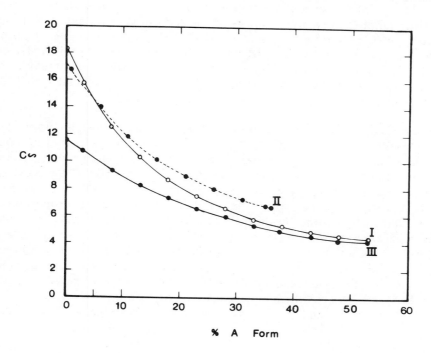

Figure 4. Characteristic ratio, C_∞, vs. percent A Form.

Curve I:
β,γ = t, g+, ϵ = t, C-2'-endo: 17 percent
β,γ = t, g-, C-2'-endo: 30 percent
balance of conformational blend:
β,γ = t, g-, C-3'-endo and A Form.

Curve II:
β,γ = g-, t, ϵ = t, C-3'-endo: 17 percent
β,γ = t, g-, C-2'-endo: 47 percent
balance of conformational blend:
β,γ = t, g-, C-3'-endo and A Form.

Curve III:
β,γ = g+, g+, ϵ = t, C-2'-endo: 17 percent
β,γ = t, g-, C-2'-endo: 30 percent
balance of conformational blend:
β,γ = t, g-, C-3'-endo and A Form.

= t, g+, ϵ = t and 1 percent A form. The pseudo-energies, ΔE_ψ, for the conformers in this model were calculated from equation 6 and are given in Table IV. The relative areas of Table II and the above weights were employed. Comparing ΔE_ψ with ΔE of Table I, we find that relative energies are adjusted by less than $\sim$1 kcal/mole.

Numerous other conformational combinations of those listed in Table I, all coinciding with the NMR data, were examined to determine if the correct characteristic ratio could be achieved. This involved a search employing other ϵ = trans conformers in combination with the two β,γ = t, g- forms and/or with the β,γ = g-, t forms. As expected, no polymers containing the low extension conformers of Table III could achieve the necessary high C_∞. Figure 4, Curve III shows, for example, how the characteristic ratio varies with percent A form if the ϵ = trans conformation with β,γ = t, g+ is replaced with the C-2'-endo β,γ =g+, g+ conformer 2. It is possible to calculate various two state poly (rU) coil models with the correct characteristic ratio. One example consists of the A form in combination with the C-2'-endo β,γ = t, g+ conformation, at either $\sim$25 percent or $\sim$92 percent A form. However, these do not match the NMR results. Our earlier two state coil model for poly (dA)[46] also does not quite match the recently obtained NMR data for dApdA.[47] We are presently calculating multi-state models for the poly dA coil that agree with the NMR findings.

Another satisfactory coil model for poly (rU) was obtained which employed the C-3'-endo Watson-Crick conformer 6, β,γ = g-, t, ϵ = t as 17 percent of the conformational blend, together with 36 percent β,γ = t, g-, C-3'-endo and 47 percent β,γ = t, g- C-2'-endo. It shares the predominant characteristics of the previous model in being about 83 percent β,γ = t, g-. Figure 4, Curve II, shows C_∞ vs. percent A form for a polymer in which the A form -t, g- (C-3'-endo) proportion is varied from 0 to 36% A form, the other conformers being held as stated above. The experimental characteristic ratio is achieved at 0 percent A form. This model is less likely in view of the lower statistical weight of the Watson-Crick conformation versus that of the β,γ = t, g+, ϵ = t form in its puckering domain. The 3.8 kcal/mole difference between ΔE_ψ (Table IV) and ΔE (Table I) of the Watson-Crick conformation also shows this. A combined model consisting of both the Watson-Crick conformation (8.5 percent) and the β,γ = t, g+, ϵ = t (8.5 percent) form also

Table IV

Conformational Properties of Poly rU Coil Models
that Match Experimental Observations

Model #	$\beta, \gamma; \epsilon$	Pucker	Percent	ΔE_ψ, kcal/mole
1	t, g-; g+	C-2'-endo	30	0.72
	t, g+; t	C-2'-endo	17	0.59
	t, g-; g+	C-3'-endo	52	-0.39
	g-,g-; g+ (A)	C-3'-endo	1	2.03
2	t, g-; g+	C-2'-endo	47	1.27
	g-, t; t	C-3'-endo	17	-0.17
	t, g-; g+	C-3'-endo	36	0.18

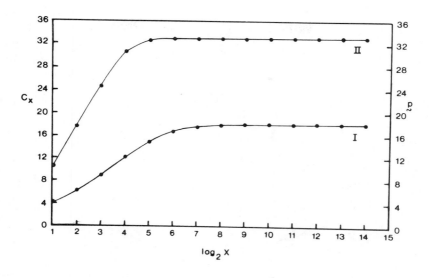

Figure 5. Characteristic ratio, C_x, (Curve I) and persistence vector magnitude $|P|$ (Curve II, as a function of degree polymerization x for a polymer composed of the following conformers (Model 1):

$\beta, \gamma = $ t, g-, C-2′-endo,   30 percent
$\beta, \gamma = $ t, g-, C-3′-endo,   52 percent
$\beta, \gamma = $ t, g+, $\epsilon = $ t, C-2′-endo,   17 percent
$\beta, \gamma = $ g-, g-, C-3′-endo,   1 percent

reproduces C_∞. Table IV summarizes the conformations in the present coil models.

The calculated characteristic ratio reaches 99 percent of its limiting value with x, the number of residues in the polymer $= 2^9$ (Figure 5, Curve I). By contrast, helices require $2^{12}-2^{13}$ residues.[39]

Persistence lengths

The persistence lengths of these poly (rU) coil models are consistent in order of magnitude with the 75 Å experimental determination of this quantity for denatured RNA.[17] The computed limiting persistence vector has a magnitude of 31.5 Å for the first model. For the second model, the persistence vector magnitude is 25.2 Å. The measured quantity is probably larger than the calculated value because the latter is for an unperturbed polymer. The persistence vector converges to 99 percent of its limiting value with 2^5 residues (Figure 5, Curve II) while helices require 2^{11} subunits.[39]

Models

Figure 6 shows ORTEP[41] drawings of segments of the two models, containing the three conformers present in each. The bases are unstacked in the first model (except for the 1 percent A form, which is not shown). There is stacking at the Watson-Crick

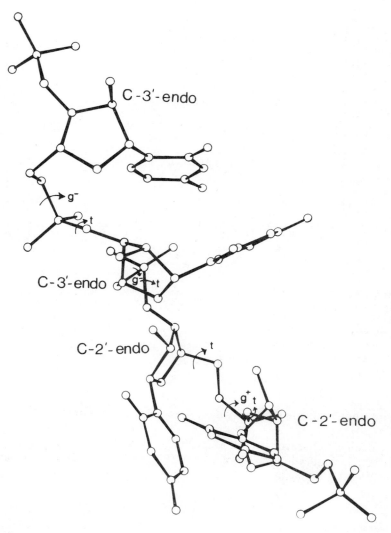

Figure 6a. Segment of poly rU coil model 1. Two subunits have β,γ = t,g-, ϵ = g+, one subunit has β,γ = t,g+, ϵ = t.

conformations of the second model, constituting 17 percent of the blend. The NMR data suggest that 8 ± 5 percent of the bases are stacked, so either model or a combination of both can roughly agree with these results; however, the first model is better. The location of P atoms in a poly (rU) coil containing approximately correct proportions of the three conformers in the first model is shown in Figure 7. Figure 8 is a space filling model of the tetramer shown in Figure 6a.

Conclusion

The present coil model of poly (rU) has the following features: (1) It is composed of low energy conformers of UpU. Specifically, the β,γ = t, g- conformations, which

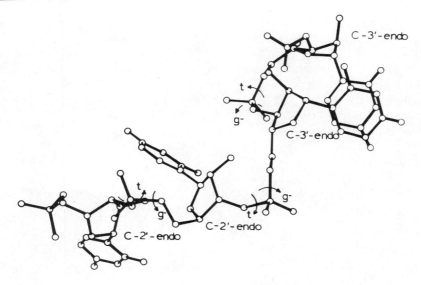

Figure 6b. Segment of poly rU coil model 2. Two subunits have $\beta,\gamma = $ t,g-, $\epsilon = $ g+, one subunit has $\beta,\gamma = $ g-,t, $\epsilon = $ t.

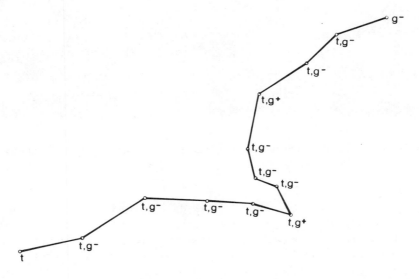

Figure 7. Phosphorus atoms in a segment of poly rU coil model 1.

have the highest statistical weights in each of their puckering domains constitute 82 percent of the building blocks. The remainder is 17 percent $\beta,\gamma = $ t, g+, $\epsilon = $ t and 1 percent A form. The t, g+ conformation is the second most probable conformer with C-2'-endo pucker. A second, less likely model has $\beta,\gamma = $ g-, t, $\epsilon = $ t (Watson-Crick form), C-3'-endo pucker as 17 percent of the blend. These results agree with theoretical studies by Olson[44] showing that coils must possess a large proportion of conformers with $\beta = $ trans. The structure of pdTpdT in the crystal,[48] which has β,γ

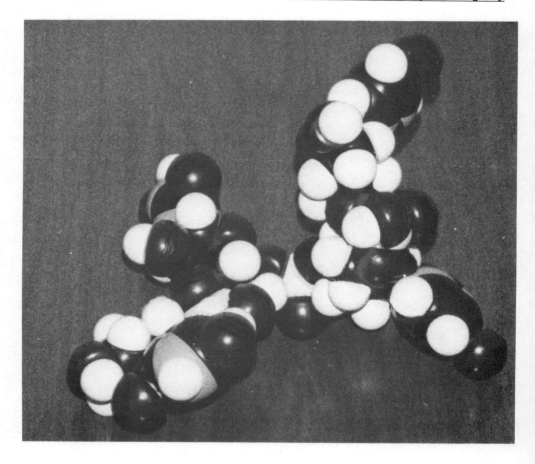

Figure 8. Corey - Pauling - Koltun space filling molecular model of conformers shown in Figure 6a.

= t, g- is also pertinent. (2) It matches NMR data on UpU[7] and poly (rU)[8] in solution. (3) The calculated characteristic ratio agrees with the value of 17.6 measured for poly (rU) coils.[1] (4) The calculated persistence lengths are of the same order of magnitude as was measured for denatured RNA and DNA.[17] Persistence lengths of helical, double stranded DNA, on the other hand, are about 600 Å.[19] (5) Calculated characteristic ratios and persistence vectors in this model converge to their limiting values at 2^9 and 2^5 residues, respectively, while helices require many more subunits for convergence.[39]

The unique conformational features of a series of uracils are different from the A type helices found in ordered RNA single strands, double strands and RNA-DNA hybrids. Since a series of uracils is typically found at the transcription termination site of m-RNAs, the distinctive shape may be recognized by RNA polymerase as the signal to release the RNA transcript from its DNA template.

Acknowledgement

We thank Wilma Olson for much helpful advice. Research sponsored jointly by the U.S. Public Health Service under NIH Grant 5ROIGM 24482-02 and the Office of Health and Environmental Research, U.S. Department of Energy, under contract W-7405-eng-2b with the Union Carbide Corporation.

References and Footnotes

1. Inners, L. and Felsenfeld, G. *J. Mol. Biol. 50*, 373 (1970).
2. Warshaw, M. and Tinoco, I. *J. Mol. Biol. 20*, 29 (1966).
3. Richards, E., Flessel, C. and Fresco, J. *Biopol. 1*, 431 (1963).
4. Simpkins, H. and Richards, E. *Biopol. 5*, 551 (1967).
5. Ts'o, P. O. P., Kondo, N., Schweizer, M. and Hollis, D. *Biochemistry 8*, 997 (1969).
6. Alderfer, J. and Ts'o, P. O. P., *Biochemistry 16*, 2410 (1977).
7. Lee, C., Ezra, F., Kondo, N., Sarma, R. H. and Danyluk, S. *Biochemistry 15*, 3627 (1976).
8. Evans, F. and Sarma, R. H. *Nature 263*, 567 (1976).
9. Wu, A. and Platt, T. *Proc. Natl. Acad. Sci. USA 75*, 5442 (1978).
10. Gilbert, W. in *RNA Polymerase*, Losick, R. and Chamberlin, M., eds., Cold Spring Harbor Laboratory (1976), p. 193.
11. Sundaralingam, M. *Biopol. 7*, 821 (1969).
12. Delisi, C. and Crothers, D. *Biopol. 10*, 1809 (1971).
13. Olson, W. K. and Flory, P. *Biopol. 11*, 25 (1972).
14. Tewari, R., Nanda, R. and Govil, G. *Biopol. 13*, 2015 (1974).
15. Yevich, R. and Olson, W. *Biopol. 18*, 113 (1979).
16. Pullman, B., Perahia, D. and Saran, A. *Biochim. Biophys. Acta 269*, 1 (1972).
17. Mingot, F., Jorcano, J., Acuna, M. and Davila, C. *Biochim. et Biophys. Acta 418*, 315 (1976).
18. Porschke, D., *Biochemistry 15*, 1495 (1976).
19. Godfrey, J. and Eisenberg, H. *Biophys. Chem. 5*, 301 (1976).
20. Broyde, S., Wartell, R., Stellman, S. and Hingerty, B. *Biopol. 17*, 1485 (1978).
21. Renugopalakrishnan, V., Lakshminarayanan, A., and Sasisekharan, V., *Biopol. 10*, 1159 (1971).
22. Lakshminarayanan, A. and Sasisekharan, V. *Biopol. 8*, 475 (1969).
23. Lakshminarayanan, A. and Sasisekharan, V. *Biopol. 8*, 489 (1969).
24. Sasisekharan, V. *Jerusalem Symp. Quant. Chem. Biochemistry 5*, 247 (1973).
25. Altona, C. and Sundaralingam, M. *J. Am. Chem. Soc. 94*, 8205 (1972).
26. Snyder, R. G., and Zerbi, G. *Spectrochim. Acta 23*, 391 (1967).
27. Stellman, S., Hingerty, B., Broyde, S., and Langridge, R. *Biopol. 14*, 2049 (1975).
28. Arnott, S., Dover, S., and Wonacott, A. *Acta Cryst. B. 25*, 2192 (1969).
29. Powell, M. *Computer J. 7*, 155 (1964).
30. Broyde, S., Wartell, R., Stellman, S., Hingerty, B. and Langridge, R. *Biopol. 14*, 1597 (1975).
31. Hingerty, B. and Broyde, S. *Nucleic Acids Res. 9*, 3429 (1978).
32. Flory, P. J., *The Statistical Mechanics of Chain Molecules*, Interscience Publishers, N. Y., (1969), pp. 22-25, 114-117, 281-286.
33. Flory, P. *Proc. Natl. Acad. Sci. USA 70*, 1819 (1973).
34. Olson, W. and Flory, P. *Biopol. 11*, 1 (1972).
35. Olson, W. and Flory, P. *Biopol. 11*, 57 (1972).
36. Olson, W. *Biopol. 14*, 1775 (1975).
37. Olson, W. *Macromolecules 8*, 272 (1975).
38. Eyring, H., *Phys. Rev. 39*, 746 (1932).
39. Olson, W. *Biopol.* In Press.
40. Crick, F. and Watson, J. *Proc. Roy. Soc., Ser. A. 223*, 80 (1954).
41. ORTEP: A Fortran Thermal Ellipsoid Plot Program for Crystal Structure Illustrations. Caroll K. Johnson, Oak Ridge National Laboratory, Oak Ridge, Tennessee.
42. Young, P. and Kallenbach, N. *J. Mol. Biol. 126*, 467 (1978).
43. Olson, W. *Biopol. 14*, 1797 (1975).

44. Olson, W. *Nucleic Acids Res. 2*, 2055 (1975).
45. Yathindra, N. and Sundaralingam, M. *Proc. Natl. Acad. Sci. USA 71*, 3325 (1974).
46. Hingerty, B. and Broyde, S. *Nucleic Acids Res. 5*, 3249 (1978).
47. Cheng, D. and Sarma, R. H. *J. Am. Chem. Soc. 99*, 7333 (1977).
48. Camerman, N., Fawcett, J. K., and Camerman, A. *J. Mol. Biol. 107*, 601 (1976).
49. Stellman, S., Hingerty, B., Broyde, S., and Langridge, R. *Biopol. 14*, 2049 (1975).

Modelling of Drug-Nucleic Acid Interactions Intercalation Geometry of Oligonucleotides

Helen M. Berman
The Institute for Cancer Research
The Fox Chase Cancer Center
7701 Burholme Avenue
Philadelphia, Pennsylvania 19111

and

Stephen Neidle
Department of Biophysics
University of London Kings College
London, WC2B 5RL
England

Introduction

The interaction of drugs with nucleic acids can involve a number of distinct processes,[1] ranging from covalent bonding to various types of non-bonded interaction. Probably the most extensively explored category is concerned with drug molecules possessing planar aromatic chromophores; the intercalating model of Lerman[2] provided the overall conceptual framework and major impetus for much of this work. The Lerman hypothesis states that these planar groups can be bound in a "sandwich" manner in between adjacent base pairs of double-stranded nucleic acids. The past few years have seen the beginnings of attempts to extend the Lerman model, so as to provide atomic-level structural information on intercalation. These attempts have brought into focus the major problems to be solved by such approaches:

•what is the geometry of intercalation, in terms of changes in nucleic acid conformation?
•is there indeed a singular geometry for intercalation?
•what effect does the structure of the planar chromophore have on the geometry?
•what are the structural differences between RNA and DNA intercalation?
•what is the structural basis of unwinding and of neighbor exclusion?
•what is the structural basis for sequence-preference in binding?

In the case of single-stranded nucleic acids, the geometry of drug binding is even less well defined than for a duplex situation, because of lack of the constraints imposed by Watson-Crick base pairing.

Several physico-chemical techniques have been used to attack these problems; this contribution discusses an X-ray crystallographic approach. The pioneering studies in this area of Sobell and his associates[3-5] employed planar drugs with ribonucleosides as model systems. (The crystal structure of a deoxydinucleoside complex[15] had recently been reported; a discussion of its conformational implications must await a detailed account of the structure.) Studies in our laboratories have used similar models; accordingly here we will concentrate on what these analyses have revealed about the ways in which drugs and mutagens interact with single- and double-stranded RNA. It is perhaps timely to define the limitations and potential pitfalls of this "model" nucleic acid approach, in terms of the structures evaluated to date.

The Crystallographic Data

There is now a substantial body of crystallographic data available for complexes between a variety of planar chromophores and self-complementary ribonucleosides, some of which are given in Table I. As a group these structures have shown themselves to be difficult to analyze crystallographically because:

(1) they contain many atoms in the asymmetric unit.
(2) they contain a high percentage of planar groups. Thus the "random" arrangement of atoms that is necessarily assumed for the application of current direct methods procedures is no longer valid.
(3) the planar chromophores often display large thermal motions, which limit the resolution of the data obtainable.
(4) the crystals contain large amounts of ordered and disordered water molecules.

One solution to some of these crystallographic problems is to chemically insert a heavy atom into the structure; however, in doing so one does introduce even greater limitations on the ultimate precision of the atomic coordinates. For this reason, in our laboratories we have chosen to study complexes without any heavy atoms. However, even when one does select this route and consequently obtains a result with high resolution, low R factor and relatively high precision the implications of these crystallographic results with respect to macromolecular structures are inherently limited by the fact that the crystals contain *dinucleosides* which may not be the minimal unit for the description of drug-intercalated *polynucleotides*. The reasoning behind this statement will become apparent as we examine in detail the conformational features of these complexes.

Table I

Structure	Contents of Asymmetric Unit	Observed Data	Resolution	R Factor	Reference
i⁵ UpA-Ethidium Br	156 atoms	2017	1.34Å	.20	3
i⁵ CpG-Ethidium Br	150	3180	1.14	.16	4
CpG-Proflavine	82	4115	.85	.11	6
ApU-9 amino acridine	75	2874	.85	.07	8
i⁵ CpG-9 amino acridine	248	2251	1.34	.19	9
ApA-Proflavine	94.5	3431	.85	.11	7

Models for Binding to RNA Duplexes

Conformational Features

The crystal structure of CpG-proflavine[6] is at first glance in a different conformational class from that of iodo CpG-ethidium Br[4] and iodo UpA-ethidium Br[3] (Figure 1). But is this really true? Examination of Table II reveals that indeed the ethidium structures differ from the proflavine one in that in the former, the pucker of ribose sugar rings are C3′ *endo* at the 5′ end and C2′ *endo* at the 3′ end while in the latter, both ribose sugars are C3′ *endo*. However, when one compares all the other conformation angles in the structures with one another they are remarkably similar. Furthermore, they resemble those in A-RNA[10] more than B-DNA a fact which is not too surprising considering that all these structures are ribodinucleosides. If one can have an acceptable intercalation geometry (with base pair separations of at least 6.8Å and with or without mixed sugar puckering) which conformational features are responsible for the stretching open of the base pairs? From the crystallographic results, shown in Table II, it appears that the torsion angles around C5′-O5′ (δ) and the 3′ glycosidic angle (χ) are substantially increased from the values found in A-RNA. In order to verify that these two angles alone can effect the necessary base

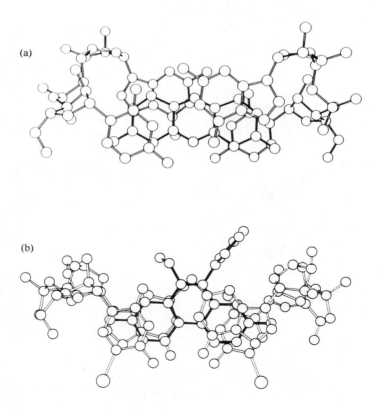

(a)

(b)

Figure 1. Views perpendicular to the plane of the base pairs of (a) CpG-proflavine and (b) iCpG-ethidium bromide.

(a)

pair separations for intercalation, we have simulated the process utilizing computer graphics.[11] Starting with the coordinates of A-RNA, the two torsion angles (δ and χ) were simultaneously increased to the values found in CpG-proflavine (Figure 2) and an acceptable intercalation geometry was produced. Using this same procedure, it was possible to generate several more reasonable structures with the δ angles in the range 225-235° and the χ angles ranging from 75° to 110°. We did the same calcula-

(b)

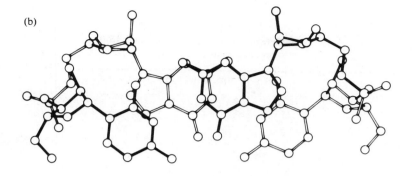

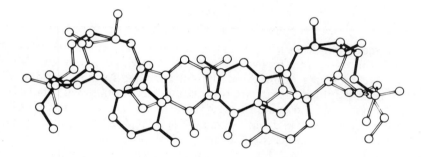

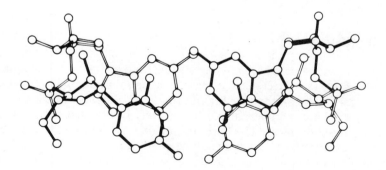

Figure 2. (a) The opening up of a model ribodinucleoside from an A-RNA conformation to that observed in CpG-proflavine. The views are parallel to the base pairs.
1) the dinucleotide in A RNA conformation
$\delta = 175°$, $\chi(3') = 14°$
2) $\delta = 200°$, $\chi(3') = 45°$
3) $\delta = 225°$, $\chi(3') = 80°$
(b) Same as (a) but viewed perpendicular to base pairs.

Table II
Conformational Features of Some
Ribodinucleotide-Dye Complexes

	Conformational Angles (in degrees)									Sugar Pucker
	χ	ζ	α	β	γ	δ	ϵ	ζ	χ	
CpG-proflavine	18	75	204	292	287	234	53	79	87	C3′ endo → C3′ endo
i⁵UpA ethidium bromide	26	98	207	286	291	236	52	133	99	C3′ endo →
	14	95	218	302	276	230	70	118	100	C2′ endo
i⁵CpG ethidium bromide	29	87	226	281	286	210	72	131	101	C3′ endo →
	24	84	225	291	291	224	55	134	109	C2′ endo
A RNA	13	83	213	281	300	175	50	83	13	
B DNA	85	157	159	261	321	209	31	157	85	

tion with a model dinucleoside that has mixed sugar pucker, C3′ *endo* (3′, 5′) C2′ *endo,* and again found that by increasing δ and χ, intercalation structures could be produced. We conclude from these simulations that for the ribodinucleosides studied crystallographically there are two essential conformational changes that occur when the intercalation complex forms: χ at the 3′ end and the C5′-05′ torsion angle increase by at least 50° from the values found in A-RNA. Additionally, the 3′ ribose sugar may adopt either a C2′ *endo* or C3′ *endo* conformation and other torsion angles may exhibit small but not significant deviations from those in A-RNA. This is not to say that other intercalation geometries are not possible but to date this genus with its two subsets of sugar geometries are all that have been observed at atomic resolution.

Base Pair Orientation and ''Unwinding''

If the changes in sugar pucker are not responsible for opening up the base pairs, then perhaps they cause the differences in base pair orientations that are apparent in Figure 1. In that figure we see that for the ethidium complexes, which have mixed sugar pucker, the base pairs are almost completely overlapped such that the base turn angle is essentially zero. (The base turn angle is the angle between the vectors connecting the C1′ atoms of each base pair when projected on the average base plane viewed from a point perpendicular to this plane.) On the other hand, in the proflavine complex where all the ribose sugars are in the C3′ *endo* conformation, this angle is approximately 34°. In order to determine whether or not the sugar pucker is related to the base turn angle we have used computer graphics to manipulate model intercalation geometries with either the same or mixed sugar puckers. The backbone conformation angles in the duplex were varied slightly and each strand was moved with respect to the other with constraints imposed so that they maintained their two-fold symmetry and acceptable Watson-Crick hydrogen bonding geometry. (As has been pointed out by Levitt[12] there is considerable flexibility in the base pair geometry.) We found that for the models with both mixed and the same ribose sugar puckers it was possible to produce models with base turn angles that varied from 0 to 34° (Table III). It would appear then that this parameter is *not* dependent on the conformation of the sugar but rather on some combination of small variations in the base pair geometry and/or the backbone conformations. The value it adopts is most likely dependent on the nature of the planar chromophore.

Table III
Structural Details of Some Model Intercalation Geometries

	Conformational Angles (in degrees)									
Model	$\chi(5')$	α	β	γ	δ	ϵ	$\chi(3')$	Sugar Pucker	Base Twise	Base Turn
1	13	213	281	300	225	50	80	C3' → C3'	10	15
2	13	213	281	300	225	50	80	C3' → C3'	4	27
3	13	220	280	300	230	70	95	C3' → C2'	16	11
4	13	210	300	280	235	70	95	C3' → C2'	15	25

Why is the base turn angle such an important parameter? In a polynucleotide with a single helix axis this angle is a measure of the nucleotide turn angle and can thus serve as a measure of the helical unwinding. Whether or not these dinucleoside models imply unwinding can only be determined by building models of larger oligonucleotide units. Therefore, in models of DNA-drug intercalation in which the helix axis is deliberately constrained[13/14], it is possible to relate the base turn angle directly to the unwinding angle. However, since dinucleosides are not double helices it may be only fortuitous when this turn angle is related to the helical unwinding.

Oligonucleotide Model Building

In order to determine how these dinucleosides relate to RNA intercalation we have built tri and tetranucleotides based on the dinucleoside crystal structure results. The ground rules we applied for this exercise are that (1) the conformation at the intercalation site must adhere to the basic geometry observed in the crystal structures of the complexes, i.e., all the torsion angles are held to the values found in A-RNA and only δ and χ were increased. (2) The two-fold relationship of the strands is maintained. (3) Watson-Crick base pairing is maintained within the limits observed in crystal structures and/or as suggested by Levitt.[12] Thus, twisting and tilting within each base pair is permissible.

In the first of two models we built, the objective was to fit the intercalation geometry as closely as possible into a "normal" RNA fragment. Operationally, this means that we are attempting to superimpose a normal RNA dinucleoside on the intercalated dinucleoside so that the resultant structure is:

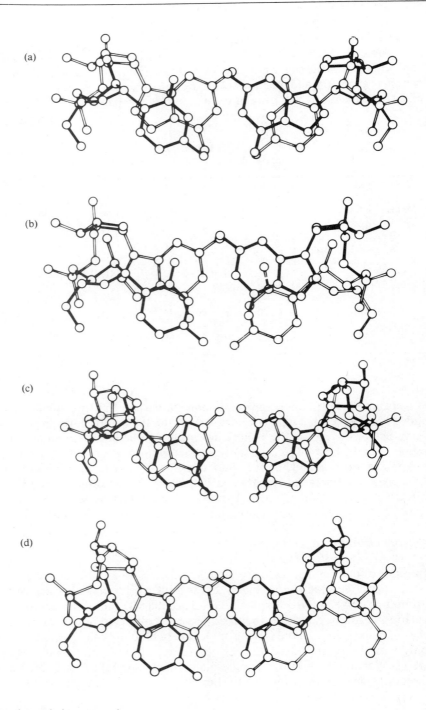

Figure 3. Four intercalation geometries
(a) C3′ endo C3′ endo small base turn angle
(b) C3′ endo C3′ endo larger base turn angle
(c) C3′ endo C2′ endo small base turn angle
(d) C3′ endo C2′ endo larger base turn angle

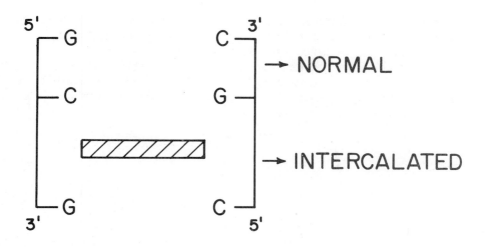

Examination of Table II shows that the χ (5′) angle of the cytosine in, for example, CpG-proflavine is 17° or very close to the A-RNA value, whereas the same angle is 85° for the guanosine. *The base-paired nucleosides are thus asymmetric* in intercalated structures. The consequences of this asymmetry on model building are shown in Figures 4a, b. Figure 4a shows that if we overlap the guanosine ribose rings it is impossible to build a duplex oligomer. Figure 4b illustrates an attempt to superimpose the cytosine groups. While the guanine bases approximately overlap, the ribose sugars do not. It is obvious from the figure that some drastic conformational alterations of the A-RNA structure are necessary at the site adjacent to guanosine. Thus we conclude that it is not possible to have a normal RNA conformation adjacent to the experimentally determined intercalated one. Less obvious is that if the intercalation geometry is to be held fixed, and Watson-Crick geometry maintained, then alterations to the site adjacent to the cytosine are also needed. This was done using both manual and computer model building and the results are shown in Figure 5 and tabulated in Table IV. Residue 1, the one adjacent to the 5′ cytosine, is in the trans, gauche⁻ conformation for β,γ with an unusually low δ value. Residue 3 is in the gauche⁻, trans conformation for β,γ with a trans value for ϵ. Both are unusual but energetically feasible conformers. The adjacent nucleosides are symmetric with respect to their χ angles which have values closer to the A-RNA ones. Full relaxation of the conformation to normal RNA values is possible at the next site. Other geometrical features of the model are that (1) the adjacent base pairs are about 4Å apart rather than the more usual 3.4Å, (2) there is a close contact (possibly a hydrogen bond) between the guanosine 2′ hydroxyl and the ribosyl oxygen (O1′) of the cytosine and (3) the phosphate oxygen at the intercalation site and the one at the guanosine adjacent site are 4Å apart. We note here that this latter phosphate geometry is ideal for Mg^{++} ion coordination. In summary, this model which naturally results in extended exclusion intercalation into RNA demonstrates that the "basic" intercalation geometry must also include the adjacent sites that have significant conformational changes.

A second model was built that would allow for neighbor exclusion binding; i.e., intercalation at every other site. The results are shown in Figure 6 and Table IV. Unlike the extended exclusion model, the adjacent site conformations are symmetrical; both are in the gauche⁻, trans conformation similar to residue 3 of model 1. In this model, the χ angles necessarily alternate and the phosphate oxygen atoms at residues 2 and 3 are 4Å apart. However, in this model this feature would also ap-

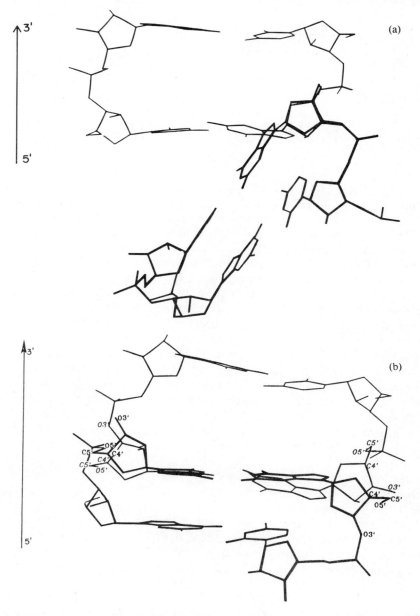

Figure 4. Attempts to fit a fragment of RNA to the intercalated site
(a) exact superposition of ribose of guanosine residues of CpG int. and GpC normal.
(b) best fit of cytosine residues of CpG int. and GpC normal.

Table IV

The Conformation Angles in Degrees
of the Intercalated Tetranucleotide

(a) The extended exclusion model (model 1)

	$\chi(5')$	α	β	γ	δ	ϵ	$\chi(3')$	Sugar Pucker	Base Twist
Residue 1	29	274	210	303	106	64	13	C3' endo	18°
Residue 2	13	213	281	300	235	50	85	C3' endo	4°
Residue 3	85	178	275	148	210	175	30	C3' endo	18°

(b) The neighbor exclusion model (model 2)

	$\chi(5')$	α	β	γ	δ	ϵ	$\chi(3')$	Sugar Pucker	Base Twist
Residue 1	85	171	286	143	196	176	13	C3' endo	31°
Residue 2	13	213	281	300	235	50	85	C3' endo	23°
Residue 3	85	172	288	143	196	176	13	C3' endo	31°

pear at every other site. So the unequal χ *angles* at the intercalation site give rise to a very distinctive (and presumably recognizable) phosphodiester geometry. Another feature of the model is that the base pairs are extremely twisted both at the intercalation site and at the adjacent site. Whether or not this type of structure would actually exist can only be answered experimentally.

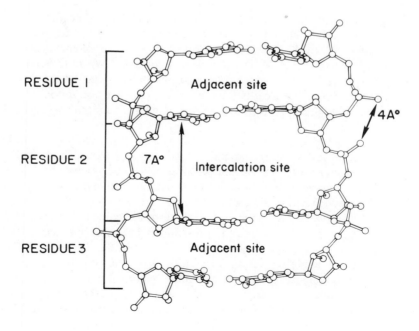

Figure 5. An extended exclusion model for GpCpGpC.

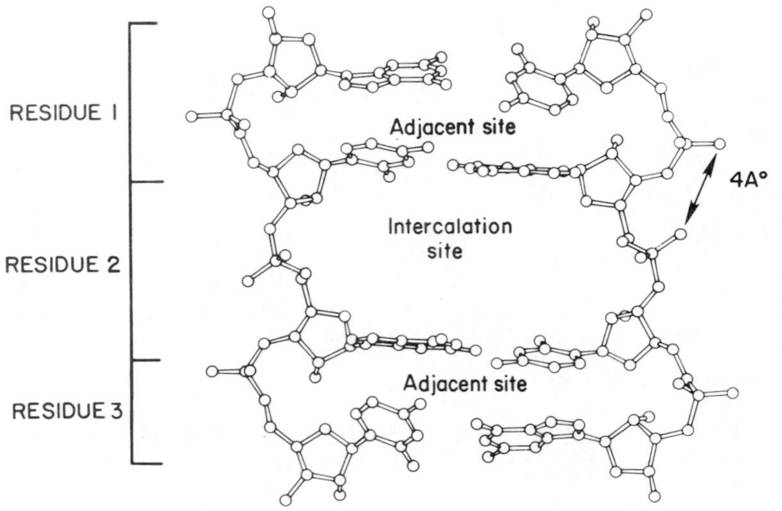

Figure 6. A neighbor exclusion model for GpCpGpC.

Our model building experiments with the dinucleoside geometries indicate that the base turn angle is dependent on both the conformation of the nucleoside backbone and the hydrogen bonding geometry of the base pair; it is probably not a good measure of unwinding. However, with these tetranucleoside models it is possible to observe helical unwinding as we show in Figure 7. The internucleoside turn angle over the four residues in a single strand of GpCpGpC in the A-RNA conformation is approximately 90° (30° between each residue). For the extended exclusion model this angle is considerably less as shown in Figure 7b. We note here that this turn angle is related only to the conformation of oligonucleotide. The obvious conclusion from examination of Figure 7 is that the extended exclusion model of GpCpGpC is unwound, despite the fact that the base turn angle at the *intercalation site* is large.

What then do we conclude from these crystallographic and model building studies? The first is that, despite many claims to the contrary, all the crystal structures of the ribodinucleosides studied to date by both Sobell and ourselves belong to the *same* conformational class in that there are two structural changes which allow intercalation to take place. The angles δ and χ (3′) are increased from the values found in A-RNA. Within this class the sugar at the 3′ terminal can be C3′ *endo*, C2′ *endo*, or anything in between. (Of course the presence of C2′ *endo* puckering makes the incorporation of these dinucleotides into B-DNA considerably easier.) The second is that the base turn angle is a very poor measure of helical unwinding, especially since none of these dinucleosides are "mini double helices." However, model building in which the intercalation geometry is held rigorously to that which is observed experimentally shows that these models do indeed imply quite substantial unwinding.

Recognition Properties of the Model Tetranucleotides

Some important properties emerge with respect to the potential interactions of these model tetranucleotides with drugs, small ions and proteins. It is clear from these

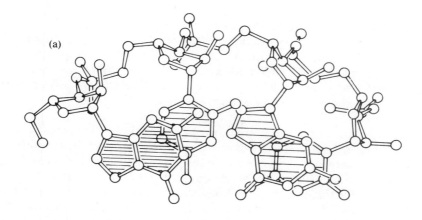

(a)

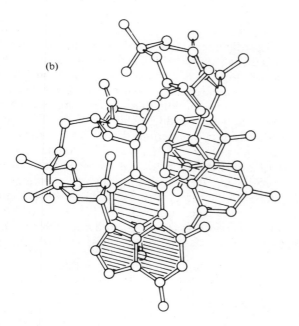

(b)

Figure 7. (a) A tetranucleotide in the A RNA conformation.
(b) The "extended exclusion" conformation for the GpCpGpC.

studies that because of the asymmetry of the χ angles at the intercalation site it is not possible to "polymerise" the geometry so that intercalation can occur at every site. *In other words, it is not necessary to invoke mixed sugar puckering in order to produce either intercalation or neighbor exclusion.* That is not to say that mixed sugar puckering does not occur in RNA or DNA. Indeed the small energy difference between the two conformations makes it not at all unlikely. Our model building studies also predict that while neighbor exclusion binding is certainly possible for RNA, it is less likely than a more extended exclusion binding.

In the course of these studies it also became apparent that the O2′ hydroxyl of the ribose sugar places severe restrictions on the conformational possibilities of the sites adjacent to the intercalation site. The conformational features of these neighboring sites in RNA may be sufficiently different from those in DNA as to explain why certain bifunctional drugs such as echinomycin that must necessarily span across the nucleotide backbone bind to DNA and not RNA.[16-18]

Another consequence of the alternation of the χ angles at the intercalation site is the effect it produces on the spacing of the phosphate groups between residues; the spacing between phosphate oxygen atoms in residues 1 and 2 is 6Å, whereas the spacing is 4Å between residues 2 and 3. This 4Å distance creates a geometry that is ideal for direct Mg^{2+} coordination. It is easy to see then how metal ions in competition for such sites could inhibit electrostatic binding of the drug at higher salt concentrations. Additionally, the potential for enzyme recognition of this site should not be overlooked.

Binding to Single Strands

Planar chromophores such as proflavine and ethidium bromide are also known to bind to single stranded RNA such as tRNA.[19-22] Little is known about the geometry of such binding. The first crystallographic determination of a drug-nucleoside complex[8] was of a structure that had the potential for but did not exhibit intercalative binding into a duplex. Instead, 9-amino acridine molecules were stacked between Hoogsteen base paired adenine and uracil residues. This structure therefore provided one kind of model of the type of drug interactions that might be operative in, for example, tRNA. More recently in our laboratories, we have examined the structure of a complex between a dinucleotide, ApA,[7] (which of course has no potential for Watson-Crick pairing) and proflavine. The adenine residues are paired as in acidic polyA[+23] and the proflavine molecules are stacked above and below these pairs in the sequence base pair, proflavine, proflavine, base pair. The striking feature of this particular structure, however, is its conformation. As is obvious from examination of Table V, the dinucleoside phosphate in this structure has little in common with

Table V
Torsion Angles for the Nucleotide Unit
in Various Structures

	$\chi(5')$	α	β	γ	δ	ϵ	$\chi(3')$
Prof-ApA	-119	272	290	293	175	168	71
Prof-CpG	17	201	290	289	231	52	85
RNA-11	13	213	281	300	175	50	13
UpA 1	12	206	81	82	203	55	37
UpA 2	19	224	164	271	192	54	44
ApA⁺	8	223	283	298	161	53	28
A⁺pA⁺	28	209	77	93	188	56	26
ApU-9 amino-acridine	76	222	100	86	202	63	72

the other dinucleoside phosphates studied to date and does not appear to conform with the rigid nucleotide concept.[24] Indeed, the only conformation angles that resemble the nucleotide geometry in RNA are β, γ, and δ. The explanation for this must lie in the association with the proflavine which means that our task of understanding the conformations of nucleic acids when they are in the environment of other molecules has become considerably more complicated.

Conclusions

It is apparent that, as yet, the crystallographic analyses of drug-dinucleoside complexes have not enabled the problems, posed at the beginning of this article to be unequivocally and definitively answered. Nonetheless, the structural results from a variety of these complexes do show a marked uniformity of conformational behavior, which leads one to suspect that similar features, may in part at least, be displayed at the polynucleotide level. This belief has prompted several model-building exercises. Such extrapolations from the dinucleoside systems must be performed with great care and their limitations clearly understood if they are to have any validity at all.

Acknowledgement

This research was supported in part by grants from NIH GM-21589, CA06927, RR05539, CA22780, the Cancer Research Campaign, a grant from NATO and an appropriation from the Commonwealth of Pennsylvania.

We thank Suse Broyde for her help in evaluating the oligonucleotide models and W. Stallings for discussions of various parts of this manuscript.

References and Footnotes

1. Peacocke, A. R., in "The Acridines" (ed. Acheson, R. M.) Wiley, New York (1973).
2. Lerman, L. S., *J. Mol. Biol. 3*, 18-30 (1961).
3. Tsai, C. C., Jain, S, C., and Sobell, H. M., *J. Mol. Biol. 114*, 301-315 (1977).
4. Jain, S. C., Tsai, C. C., and Sobell, H. M., *J. Mol. Biol. 114*, 317-331 (1977).
5. Sobell, H. M., Tsai, C. C., Jain, S. C. and Gilbert, S. G., *J. Mol. Biol. 114*, 333-365 (1977).
6. a. Neidle, S., Achari, A., Taylor, G. L., Berman, H. M., Carrell, H. L., Glusker, J. P., and Stallings, W. C., *Nature 269*, 304-307 (1977).
   b. Berman, H. M., Stallings, W., Carrell, H. L., Glusker, J. P., Neidle, S., Achari, A., and Taylor, G. *Biopolymers* (in press).
7. Neidle, S., Taylor, G., Sanderson, M., Shieh, H. S., and Berman, H. M., *Nucleic Acids Res. 5*, 4417-4422 (1978).
8. Seeman, N. C., Day, R. O., Rich, A., *Nature 253*, 324-326 (1975).
9. Sakore, T. D., Jain, S. C., Tsai, C. C. and Sobell, H. M., *Proc. Natl. Acad. Sci. USA 74*, 188-192 (1977).
10. Arnott, S., Smith, P. J. C., Chandrasekaran, R., in "Handbook of Biochemistry and Molecular Biology" (ed. Fasman, G. D.) Chemical Rubber Co., Cleveland, Ohio, 3rd Ed., Vol. 2, Sec. B, 411-422 (1976).
11. Berman, H. M., Neidle, S., and Stodola, R. K., *Proc. Natl. Acad. Sci. USA 75*, 828-832 (1978).
12. Levitt, *Proc. Natl. Acad. Sci. USA 75*, 640-644 (1978).
13. Alden, C. S. and Arnott, S., *Nucleic Acids Res. 2*, 1701-1717 (1975).
14. Alden, C. S. and Arnott, S., *Nucleic Acids Res., 4*, 3855-3861 (1977).

15. Wang, A. H. J., Nathans, J., van der Marel, G., van Boom, J. H., and Rich, A., *Nature 276*, 471-474 (1978).
16. Waring, M. J. and Wakelin, L. P. G. *Nature 252*, 653-657 (1974).
17. Wakelin, L. P. G. and Waring, M. J. *Biochem. J. 157*, 721-740 (1976).
18. Ughetto, G. and Waring, M. J. *Mol. Pharmacology 13*, 579-584 (1977).
19. Urbanke, C., Romer, R., and Mausa, G. *Eur. J. Biochem. 33*, 511-516 (1973).
20. Liebman, M., Rubin, J., Sundaralingam, M. *Proc. Natl. Acad. Sci. USA, 74*, 4821-4825 (1977).
21. Dourlent, M. and Helene, C. *Eur. J. Biochem. 23*, 86-95 (1971).
22. Finkelstein, T. and Weinstein, I. B. *J. Biol. Chem. 242*, 3763-3768 (1967).
23. Rich, A., Davies, D. R., Crick, F. H. C., and Watson, J. D. *J. Mol. Biol. 3*, 71-86 (1961).
24. Yathindra, N. and Sundaralingam, M. *Biopolymers 12*, 297-314 (1973).

DNA Structure and its Distortion by Drugs

D. M. Crothers, N. Dattagupta and M. Hogan
Department of Chemistry
Yale University
New Haven, Connecticut 06520

Several papers in this volume attest to a strong current interest in DNA structure and its distortion by drugs. In our contribution we summarize our use of the technique of transient electric dichroism to study DNA structure and how it is altered by drug binding.

Electric Dichroism of DNA

In an electric dichroism experiment one applies an electric field to a solution of macromolecules, and observes the change in absorbance of polarized light. If the molecules in the solution are electrically anisotropic, for example because of a permanent or induced dipole moment, the electric field causes them to orient. If, in addition, the molecules are optically anisotropic, orientation produces a change in the absorbance of polarized light.

Rod-like DNA molecules provide a simple illustration of this technique. Application of an electric field causes them to orient with their long axis parallel to the field. Since the 260 nm transition is π - π^*, it is polarized in the plane of the base, and therefore is approximately perpendicular to the helix axis. Hence orientation should cause a decrease in the absorbance of light polarized parallel to the electric field.

The absorbance change is expressed quantitatively through the reduced dichroism ϱ,

$$\varrho = \frac{A_\parallel - A_\perp}{A}$$

in which $A_\parallel$ and $A_\perp$ are respectively the absorbance for light polarized parallel and perpendicular to the electric field, and A is the absorbance without a field. The reduced dichroism of a regular helical molecule depends on two quantities: The fractional orientation Φ, and the angle α between the transition moment and the orientation axis. Specifically[1,2]

$$\varrho = 3/2(3\cos^2\alpha - 1)\Phi \tag{1}$$

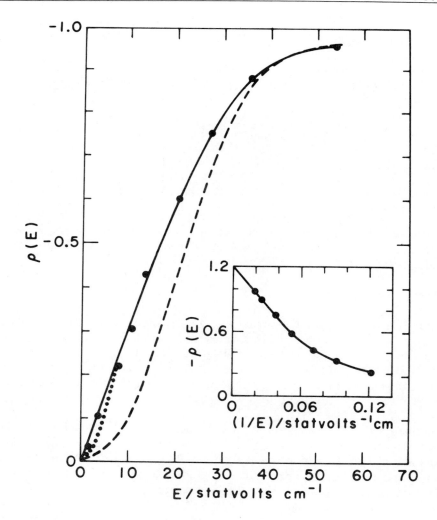

Figure 1. Dependence of the reduced dichroism ϱ on electric field strength E, compared with the calculated variation assuming a permanent (....) or induced (———) dipole moment. The solid line gives the calculated curve assuming the ion flow[3] model. The experimental points are for a 250 base pair fragment of calf thymus DNA[3] [Na$^+$] = 2.5 mM, T = 12°C.

The two parameters α and Φ in equation (1) are separated by extrapolating ϱ to infinite field, and hence perfect orientation, with $\Phi = 1$. An example is shown for DNA in Figure 1. Generally, it is found that ϱ is linear in 1/E at high values of the field, so that the extrapolated intercept at 1/E = 0 is unambiguous. Linear variation of ϱ with 1/E implies that the apparent molecular dipole moment is independent of the field. However, as we have argued[3] this does not necessarily imply that the molecule contains a permanent dipole moment. Indeed, given the two-fold symmetry of the two sugar-phosphate chain in DNA, a permanent dipole moment is implausible. It has been suggested that the field-induced polarization of DNA saturates at low fields, producing a constant dipole moment[4]. Alternatively, we have proposed that orientation results from asymmetry of the counterion atmosphere induced by the flow of ions through the solution.[3] This mechanism also predicts an orienting force which is linear in the field, as observed. Involvement of the ion atmosphere is verified by the dependence of the apparent dipole moment on the ionic strength of the solution.[3]

The amplitude of the extrapolated dichroism observed for DNA in Figure 1 is about -1.2, significantly different from the expected value of -1.5 if the base transition moments were strictly perpendicular to the orientation axis. It might be argued that this results from slight bending of the DNA molecules in solution. However, it can be shown that certain drug molecules bound at low levels to these same DNA molecules exhibit an extrapolated dichroism of -1.5, which would be impossible were there significant bending. Hence we conclude that the electric field not only orients the molecules, but also straightens them against the thermal bending forces.

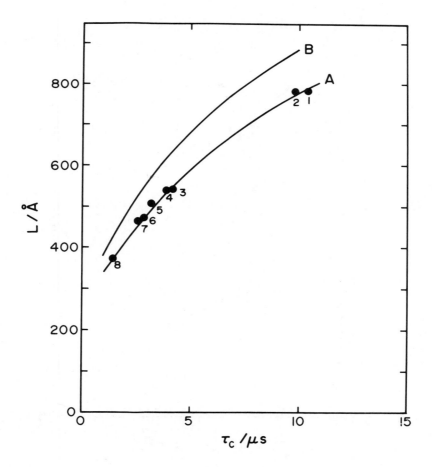

Figure 2. Variation of the orientational relaxation time τ_c with DNA length (L). Curve A gives the theoretical curve calculated using Broersma's relationship for a cylinder[8], and curve B gives the results if an ellipsoid of revolution is assumed.

The observed limiting dichroism of -1.1 to -1.2 implies that the base transition moments are tipped at an angle of about 17° from the plane perpendicular to the DNA helix axis. It is believed that these transition moments lie roughly along the short axis of a DNA base pair, and thus should be roughly coincident with the C_2 symmetry axis of the helix. However, since the C_2 axis must be perpendicular to the helix axis, we conclude that an individual base transition moment must be tipped about 17° from the C_2 symmetry axis, and therefore that the short axis of each base in a base pair is not coincident with the C_2 axis. This conclusion is not compatible with the classical B form structure of DNA[5], or with any structure in which both bases in a pair lie in a plane containing the C_2 axis.

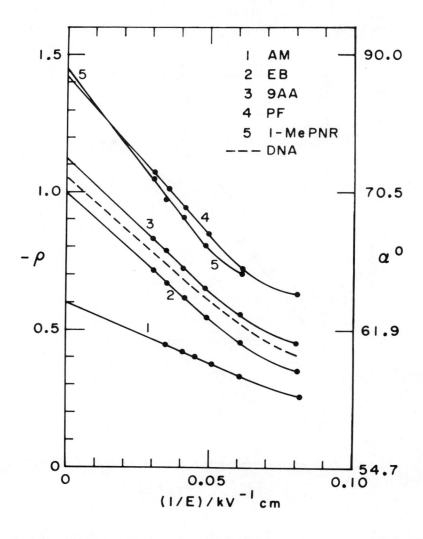

Figure 3. Representative field extrapolation for drug-DNA complexes in 2.5 mM Na⁺solution, 11°C[10] (1) AM = actinomycin, 440 nm (2) EB = ethidium, 520 nm; (3) 9AA = 9-amino acridine, 427 nm; (4) PF = proflavine, 430 nm (5) 1-MePNR = 1-methyl phenyl neutral red, 530 nm.

The only DNA model we know of which is consistent with our observations is that proposed by Levitt[6] on the basis of energy minimization calculations. He suggested that the base pair is not planar, but twisted. The result is a structure in which the bases in a pair have a propeller shape, with each transition moment axis tipped about 17° from the perpendicular plane, as required by our measurements.

The other main parameter measured in our electric dichroism experiments is the time required for orientation in the field. The relationship between the observed exponential time constant τ_c and the rotational diffusion constant D_r depends on whether the molecule orients in a single preferred direction (permanent dipole moment orientation) or whether the dipole moment can be induced in either direction along the helix axis. The equations are:[7]

$$\tau_c = 1/2D_r \qquad \text{(permanent moment) (2a)}$$
$$\tau_c = 1/6D_r \qquad \text{(induced moment) (2b)}$$

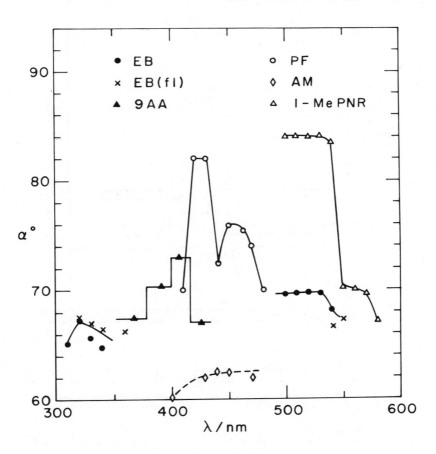

Figure 4. Wavelength (λ) dependence of the angle α measuring the orientation of the observed transition moment to the DNA helix axis[10]. Conditions as in Figure 6. x = ethidium dichroism detected by fluorescence.

A

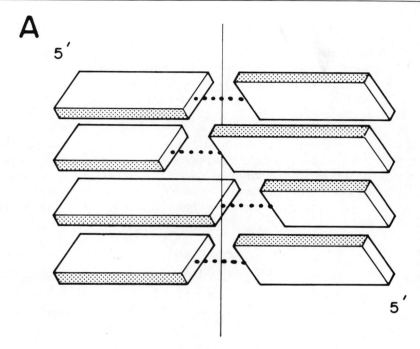

B

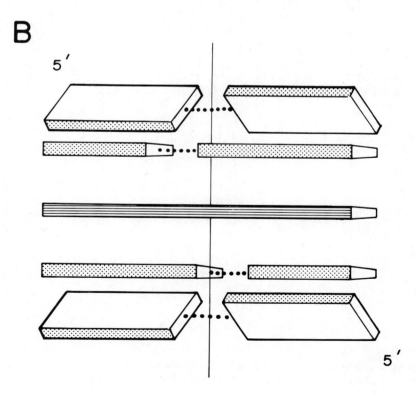

C

5$'$

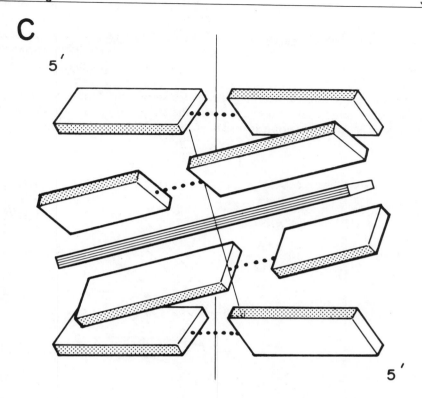

5$'$

Figure 5. A model for intercalation into propeller twisted DNA. (A) Schematic representation of the twisted DNA structure suggested by Levitt[6]. (B) Hypothetical binding intermediate which depicts the unfavorable contacts between flattened base pairs and adjacent propeller-twisted base pairs. (C) Model structure consistent with the dichroism data. Unfavorable contacts have been reduced by unstacking the 5$'$ pyrimidines. The complex is presented as being wedge-shaped as viewed along the short base pair axis to emphasize the variability seen for the twist of the complex[10].

We verified that Equation 2b describes the field-induced orientation for DNA by comparing the measured times with values calculated from the known molecular dimensions.[3] The excellent agreement between experiment and the hydrodynamic theory of Broersma[8] is shown in Figure 2. The theory predicts that τ_c varies with L^3, so one expects to be able to detect small variations in L from measurement of τ_c.

Dichroism of Intercalated Drugs

In recent work we have extended these transient electric dichroism experiments to drug-DNA complexes.[9-11] Figure 3 shows typical dichroism values for several different drugs.[10] As shown in Figure 4, the angular orientation α of the transition moment to the helix axis depends on wavelength. These results show clearly that the intercalated chromophores are not perpendicular to the helix axis.

Figure 5 shows a schematic model that illustrates the tilt (θ_1) and twist (θ_2) found for intercalated drugs. We found that θ_1 is roughly 20 $\pm$ 8°, and θ_2 is 10 $\pm$ 8°. We propose that the tilt θ_1 is an adaptation to the unfavorable contacts that develop in a model with the DNA base pairs twisted (Figure 5A) when an adjacent base pair is flattened by the drug (Figure 5B). Tilting the drug allows a gradual variation of the angle of stacking of purines through the intercalation site (Figure 5C).

Length changes in intercalated complexes also show considerable variability. Figure 6 shows the results found compared with the simple expectation value of 3.4Å per bound drug. 9-Amino acridine is notable for its small length increment.

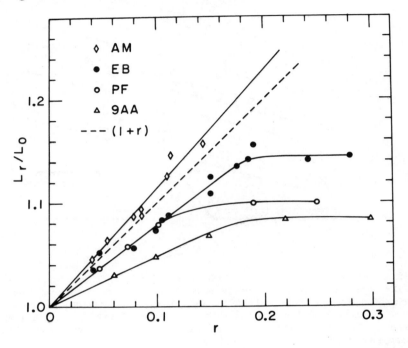

Figure 6. Dependence of the length of drug-DNA complexes on degree of binding r (drug molecules per base pair). The dotted line corresponds to the idealized 3.4Å length increase per drug.

Kinking of DNA by Irehdiamine

Irehdiamine is a steroidal diamine which induces considerable DNA hyperchroism upon binding. We investigated the complex using the dichroism method, and found results consistent with a substantial distortion of DNA structure.[9] Small amounts of the drug were found first to decrease the length of bacterial DNA molecules, followed by a length increase when the binding approached saturation. Furthermore, as shown in Figure 7, there was a substantial reduction in the DNA dichroism with increasing degrees of binding.

A DNA length decrease followed by an increase is consistent with a kinked structure for the irehdiamine complex as proposed by Sobell, et al..[12] Furthermore, the hyperchroism induced on binding is strongly suggestive of loss of base stacking, as are the proton NMR changes recently observed by Patel[13] for an analogous system. However, one feature of our results is not consistent with the detailed model propos-

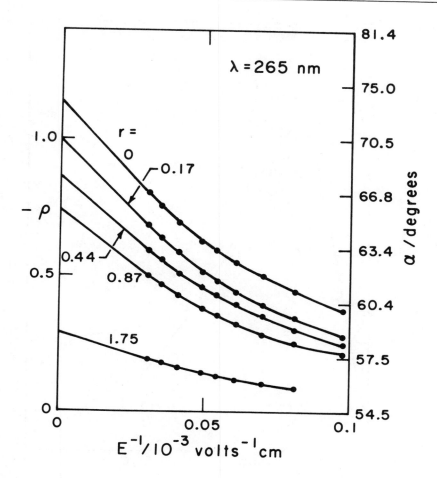

Figure 7. Field dependence of the reduced dichorism ϱ at different ratios r of added IDA per DNA base pair[9].

ed by Sobell, *et al.*: The superhelix in their β-kinked structure has a C_2 symmetry axis nearly coincident with the base pair short axis. (Actually, it lies between the two base pairs that form a section between kinks.) Hence in their model the base transition moments should be approximately perpendicular to the helix axis. Instead, we find that they are tipped by about 30° from the perpendicular plane.

Allosteric Transformation of Calf Thymus DNA by Distamycin

The final example which we present of distortion of DNA by a bound drug is an unusual case in which we observed a long range, cooperative alteration of DNA structure. We found this effect for binding of the drugs distamycin and netropsin to calf thymus DNA.[11] As shown in Figure 8, the Scatchard isotherm for drug binding (determined by a phase partition technique) is strongly cooperative (sloping upward at low r) for calf thymus DNA, but not for *E. coli* DNA.

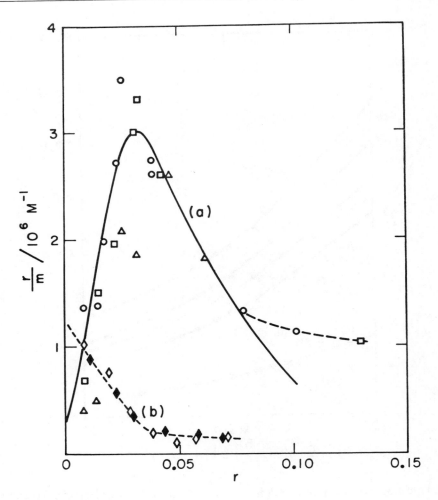

Figure 8. Scatchard plot of the binding equilibria of calf thymus (a) and *E. coli* (b) DNAs with distamycin at 11°C in 66 mM Na⁺ buffer, pH 6.5. Data were determined by a phase partition method[11]. The solid curve (a) was calculated from a statistical mechanical theory of cooperative allosteric transition of calf thymus DNA from its initial form to another structure that has higher affinity for distamycn.

The cooperative binding of distamycin to calf thymus DNA is accompanied by strong alteration of its structure, as summarized in Figure 9 and 10. Figure 9 shows that a striking length increase, and subsequent decrease, accompany increasing binding. Figure 10 shows that diatamycin binding, at a ratio of one drug to 30 base pairs, has a pronounced effect on the binding geometry of ethidium. The dichroism values observed for eithidium in the distamycin (r = 0.03)-DNA-ethidium (r = 0.1) complex extrapolate to -1.5. Hence the transition moments of virtually *all* the bound ethidium residues become perpendicular to the helix axis when there is only one distamycin for every 30 base pairs. This contrasts with a measured[10] tilt of 23° for the long axis of bound ethidium when no distamycin is present.

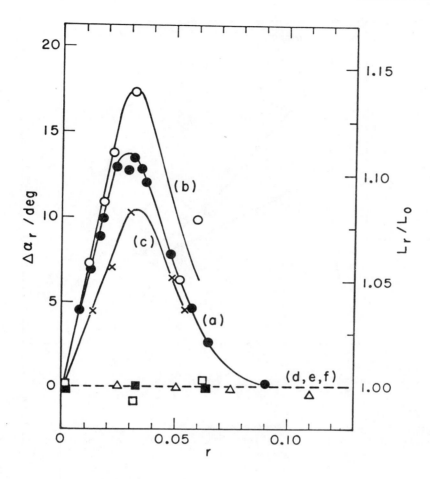

Figure 9. Apparent fractional length changes (Lr/Lo) (a, c, d, f) and ethidium orientation angle changes ($\Delta\alpha_r$) as a function of distamycin binding for calf thymus (a, b, c) and *E. coli* (d, e, f) DNA samples. T = 11°C, buffer 2.5 mM in Na$^+$, pH 7.0. (a) ●, length change for calf thymus DNA in absence of ethidium (b) O, ethidium orientation angle change with calf thymus DNA, $r_{ethidium}$ = 0.1 and variable $r_{distamycin}$ (c) x, length change for calf thymus DNA in the presence of ethidium, $r_{ethidium}$ = 0.1; (d) △, length change of *E. coli* DNA without ethidium (e) □, length change of *E coli* DNA with ethidium, $r_{ethidium}$ = 0.1; (f) ■, ethidium angle change of *E. coli* DNA, $r_{ethidium}$ = 0.1.

We have interpreted these results as implying a long range allosteric conversion of DNA from one structural form to another.[13] The results can be explained by a model in which form I, found in absence of drugs, is more favorable in free energy by about 20 cal per base pair. Form II has a distamycin affinity higher by about a factor 15 than form I, which accounts for the structural conversion induced by the drug. The cooperative nature of the transition is accounted for by an interfacial free energy of about 3-4 kcal/mol. As a consequence of this boundary free energy, less than 1 percent of the DNA is calculated to be in Form I in absence of the drug.

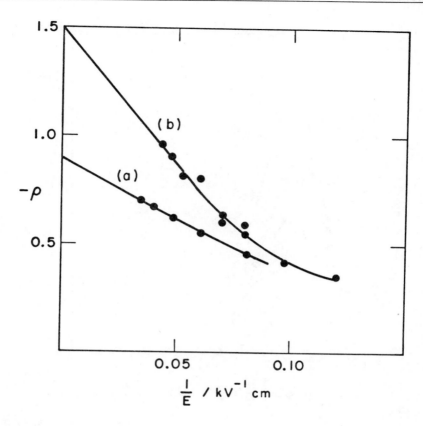

Figure 10. Extrapolation of the observed dichroism of ethidium ($r_{ethidium}$ = 0.1) to infinite field in the absence (a) and presence (b) of distamycin, $r_{distamycin}$ = 0.03. Calf thymus DNA samples, conditions as in Figure 9.

These observations of long range structural alteration of DNA by a bound drug support the possibility that DNA function can be influenced by proteins bound some distance away, as suggested by experiments on the mutual influence of two arabinose promoters.[14] However, since other DNAs did not exhibit the same structural change upon distamycin binding, we suspect that the details of base modification or sequence may be significant for the transmission of long range effects.

References and Footnotes

1. O'Konski, C. T., Yoshioka, K. and Orttung, W.H. *J. Phys. Chem. 63*, 1558-1565 (1959).
2. Allen, I. S. and Van Holde, K. E. *Biopolymers 10*, 865-881 (1971).
3. Hogan, M., Dattagupta, N. and Crothers, D. M. *Proc. Nat. Acad. Sci., USA 75*, 195-199 (1978).
4. Neumann, E. and Katchalsky, A. *Proc. Nat. Acad. Sci., USA 69*, 993-997 (1971).
5. Arnott, S. and Hukins, D. W. L. *Biochem. Biophys. Res. Commun. 47*, 1504-1509 (1972).
6. Levitt, M. *Proc. Nat. Acad. Sci., USA 75*, 640-644 (1978).
7. Tinoco, I., Jr. *J. Am. Chem. Soc. 77*, 4486-4489, (1955).
8. Broersma, S. *J. Chem. Phys. 32*, 1626-1631 (1960).

9. Dattagupta, N., Hogan, M. and Crothers, D. M. *Proc. Nat. Acad. Sci. USA 75*, 4286-4290 (1978).
10. Hogan, M., Dattagupta, N. and Crothers, D. M. *Biochemistry 18*, 280-288 (1979).
11. Hogan, M., Dattagupta, N. and Crothers, D. M. *Nature 278*, 521-524 (1979).
12. Sobell, H. M., Tsai, C., Gilbert, S. G., Jain, S. C. and Sakore, T. D. *Proc. Nat. Acad. Sci., USA 73*, 3068-3072 (1976).
13. Patel, D. and Canuel, L. L. *Proc. Nat. Acad. Sci., USA 76*, 24-28 (1979).
14. Hirsch, J. and Schleif, R. *Cell 11*, 545- 550 (1977).

Structure and Dynamics of Poly(dG-dC) in Solution Steroid Diamine•Nucleic Acid Complexes and Generation of an "Alternating B-DNA" Conformation in High Salt

Dinshaw J. Patel
Bell Laboratories
Murray Hill, New Jersey 07974

Introduction

The present understanding of deoxyribonucleic acid conformations at the poly-nucleotide duplex level is based primarily on fiber X-ray diffraction studies of natural and synthetic DNA's as a function of humidity and counterion.[1] It was of in-terest to extend such studies to investigate the structure and dynamics of nucleic acids in solution. The early studies using circular dichroism spectroscopy monitored DNA conformational transitions in films[2/3] and in solution[4/5] but lacked the probes necessary to differentiate between contributions due to hydrogen bonding, base stacking and the torsion angle changes in the sugar-phosphate backbone.

The observation of narrow resolvable proton resonances from the base and sugar rings and phosphorus resonances from the backbone has permitted the successful application of nuclear magnetic resonance (NMR) spectroscopy to elucidate the solution properties of self-complementary oligonucleotide duplexes.[6-11] By contrast, the resonances of DNA are broadened out beyond detection due to a combination of the large dispersion of chemical shifts and the slow tumbling times of rod-like polymers in solution. This contribution extends our recent efforts to investigate the structure and dynamics of synthetic DNA's of defined sequence by high resolution NMR spectroscopy.

Poly(dG-dC) in Low Salt Solution

Introduction

The thermodynamic and kinetic investigations of Baldwin and coworkers on poly(dA-dT) demonstrated that the polynucleotide backbone for synthetic DNA's with an alternating purine-pyrimidine sequence folds to give smaller branched duplexes which melt independently of each other.[12/13] It occurred to us that the spec-tra of poly(dA-dT) should be considerably simplified compared to natural DNA of the same molecular weight due to the structural equivalence of the base pairs and the flexibility induced by the rapid migration of the short branched duplex regions. This has resulted in the observation of well resolved proton resonances for poly(dA-dT) in the duplex state in aqueous solution.[14/15] The Watson-Crick hydrogen-bonded protons, the nonexchangeable protons distributed throughout the base pairs and the sugar rings, as well as the backbone phosphates, can be independently monitored through the duplex-to-strand conversion of the synthetic DNA to yield structural and kinetic information on this transition.[16] These studies have been extended to a comparison of the structural parameters for synthetic DNA and RNA duplexes with the same alternating inosine-cytidine sequence[17] and adenosine- uridine sequence[18] as well as the poly(dG-dT) duplex stabilized by wobble base pairs.[19]

We have undertaken a systematic NMR study of poly(dG-dC) in order to characterize the spectral parameters of this polynucleotide in solution. These parameters serve as controls for additional studies on the effect of metal ions and drugs on the helix-coil transition of this synthetic DNA in solution.

The salt dependence of the transition midpoint (T_m) of poly(dG-dC) follows the relationship[20]

$$T_m = 14 \log Na^+ (mM) + 81$$

so that in 10 mM NaCl, 1 mM Na phosphate, 1 mM Na EDTA buffer, the predicted transition midpoints are 96°C in the absence of added NaCl and 131°C in the presence of 4M NaCl. We have therefore monitored the proton and phosphorus resonances during the melting transition of poly(dG-dC) in 10 mM NaCl, ^2H_2O, up to a temperature of 110°C.

Hydrogen Bonds

We have monitored the guanosine H-1 exchangeable proton of the synthetic DNA in the spectral region downfield from 10 ppm. A typical spectrum of poly(dG-dC) in 10 mM NaCl, 1 mM phosphate, 1 mM EDTA, H_2O, pH 6.3 at 43.5°C is presented in

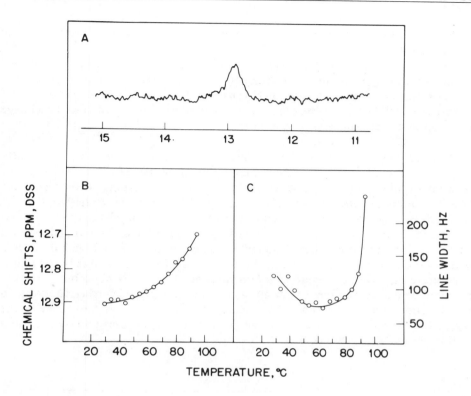

Figure 1. (A) The continuous wave 360 MHz proton NMR spectrum (11 to 15 ppm) of 37.5 mM poly(dG-dC) $S_{20,\omega}$ = 6.9, in 10 mM NaCl, 1 mM phosphate, 1 mM EDTA, H_2O pH = 6.3 at 43.5°C. The temperature dependence of the chemical shift (B) and line width (C) of the guanosine H-1 exchangeable proton in the synthetic DNA.

Fig. 1A. A single resonance is observed at 12.9 ppm and its chemical shift (Fig. 1B) and line width (Fig. 1C) have been monitored between 30° and 95°C. The resonance shifts upfield with increasing temperature and broadens out on raising the temperature from 80° to 95°C.

Duplex formation by poly(dG-dC) in the temperature range 30° to 90°C was unequivocally demonstrated by the observation of the guanosine H-1 exchangeable proton between 12.7 and 12.9 ppm in this temperature range (Fig. 1).[21] The chemical shift of 12.9 ppm for poly(dG-dC) at 30°C is upfield from the corresponding values of 13.15 to 13.20 ppm for the internal base pairs in the tetranucleotide duplexes dC-dG-dC-dG[11] and dG-dC-dG-dC[22] published previously. This discrepancy probably results from fraying of the terminal base pairs in oligonucleotide duplexes[23] so that the internal base pairs do not experience the entire upfield ring curret contribution[24] from the terminal base pairs.

The temperature dependence of the chemical shift of the guanosine H-1 proton in poly(dG-dC) exhibits a moderate slope between 30° and 60°C followed by a pronounced slope between 60° and 90°C. (Fig. 1B) The upfield shift with temperature in the region 30° to 90°C could result from an increase in the fraction of loops and branches with increasing temperature and/or changes in base pair overlaps in the premelting region.

The line widths of the guanosine H-1 proton of poly(dG-dC) are constant between 50° and 80°C. (Fig. 1C). They broaden below 50 °C presumably due to generation of a less flexible duplex (less branch formation) at the lower temperature and broaden above 90°C due to the onset of the melting transition[25] and rapid exchange of the exposed guanosine H-1 proton with solvent H_2O. (Fig. 1).

We have searched for additional exchangeable protons in poly(dG-dC) in the spectral region 5.5 to 8.5 ppm. An exchangeable resonance is detected at 6.61 ppm in the spectrum of the synthetic DNA at 78.9°C. (Fig. 2). Previous NMR studies of dG + dC containing tetranucleotide duplexes[11/22/26] permitted the observation of the cytidine NH_2 protons at low temperature with slow rotation about the C-N bond resulting in separate resonances for the hydrogen-bonded (8 to 9 ppm) and exposed (6.5 to 7.0 ppm) protons. These resonances coalesced into a broad peak on raising the temperature further. By contrast, the guanosine NH_2 protons were broad at low temperature due to intermediate rotation rates about the C-N bond but narrowed and could be detected at ∿6.7 ppm on raising the temperature further due to rapid rotation about the C-N bond. The exchangeable resonance at 6.61 ppm in the poly(dG-dC) spectrum is assigned to the guanosine NH_2 protons undergoing rapid rotation about the C-N bond. (Fig. 2).

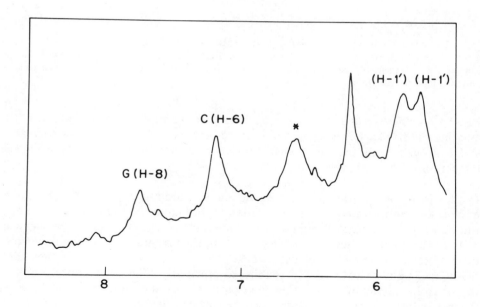

Figure 2. The continuous wave 360 MHz proton NMR spectrum (5.5 to 8.5 ppm) of 37.5 mM poly(dG-dC), $S_{20,\omega}$ = 6.9, in 10 mM NaCl, 1 mM phosphate, 1 mM EDTA, H_2O, pH 6.3 at 78.9°C. The guanosine NH_2 protons are designated by *.
The sharp resonance at ≈6.2 ppm is a spinning side band of H_2O.

Helix-Coil Transition

The 360 MHz Fourier transform NMR spectra (4.5 to 8 ppm) of poly(dG-dC) in 10 mM NaCl, 1 mM phosphate, 1 mM EDTA, ^2H_2O, pH 6.3 at 89°, 97° and 103°C are presented in Fig. 3. The base (guanosine H-8 and cytidine H-5 and H-6) and sugar (H-1′ and H-3′ linked to guanosine and cytidine rings) protons are well resolved in the duplex state (89°C) and shift downfield as average resonances with increasing temperature (Fig. 3). We are unable at this time to differentiate the sugar protons linked to the guanosine from those linked to the cytidine residues. The chemical shifts of the base and sugar protons of poly(dG-dC) in 10 mM NaCl are plotted in Fig. 4. The cytidine H-5 and H-6 protons and the two sugar H-1′ protons undergo upfield shifts of ≥0.3 ppm on duplex formation, while the guanosine H-8 and two sugar H-2′, 2″ protons shift upfield by 0.15 to 0.25 ppm. The remaining sugar protons shift by <0.1 ppm during the duplex to strand transition. (Fig. 4, Table I).

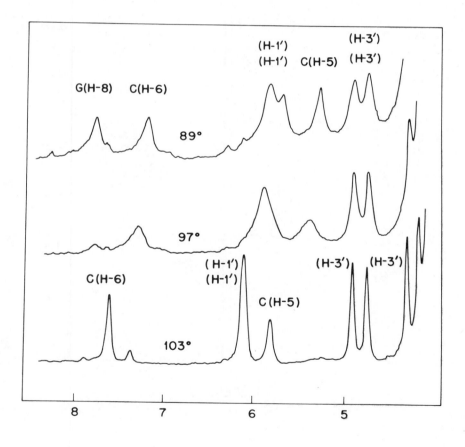

Figure 3. The temperature dependence of the 360 MHz Fourier transform proton NMR spectra (4.5 to 8 ppm) of 37.5 mM poly(dG-dC), $S_{20,\omega} = 6.9$, in 10 mM NaCl, 1 mM phosphate, 1 mM EDTA, ^2H_2O, pH 6.3 at 89°, 97° and 103°C.

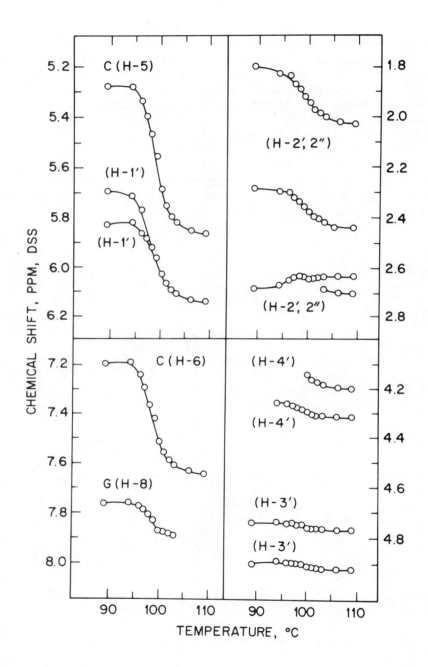

Figure 4. The temperature dependence (88° to 110°C) of the base and sugar proton chemical shifts of 37.5 mM poly(dG-dC), $S_{20,\omega}$ = 6.9, in 10 mM NaCl, 1 mM phosphate, 1 mM EDTA, ^2H_2O, pH 6.3.

Table I
Chemical Shifts (ppm) of poly(dG-dC) Nonexchangeable Base and Sugar Resonances
at 89° and 109°C and the Chemical Shift Difference ($\Delta\delta$)
in This Temperature Range

Protons	$\delta(89°C)$	$\delta(109°C)$	$\Delta\delta$
G(H-8)	7.765	~7.90	0.135
C(H-5)	5.280	5.870	0.590
C(H-6)	7.200	7.645	0.445
(H-1′)	5.696	6.144	0.448
(H-1′)	5.834	6.144	0.310
(H-3′)	4.742	4.770	0.028
(H-3′)	4.900	4.926	0.026

The melting transition of poly(dG-dC) in 10 mM NaCl exhibits a midpoint (T_m) $\cong$ 98.5°C and width $T_{3/4}$ - $T_{1/4}$ $\cong$ 2.8°C in 10 mM NaCl solution. (Fig. 4). By contrast, poly(dA-dT) which contains one less Watson-Crick hydrogen bond exhibits a T_m = 45°C in 10 mM NaCl solution.[27] The cytidine H-5 resonance of poly(dG-dC) which shifts by 0.59 ppm between 89° and 109°C, broadens in the vicinity of the midpoint of the melting transition. (Figs. 3 and 5). By contrast, the sugar H-3′ protons which undergo negligible shifts narrow gradually on proceeding from duplex to strands (Figs. 3 and 5). The excess width of the cytidine H-5 resonance (55.5 Hz, at T_m = 98°C, Fig. 5) in poly(dG-dC) can be related to the duplex dissociation rate constant (k_d) by the relationship

$$\text{Excess Width} = 4\pi f_d^2 f_s^2 (\Delta\delta)^2 (\tau_d + \tau_s)$$

where the fraction of duplex (f_d) and strand (f_s) = 0.5, the chemical shift difference ($\Delta\delta$) = 360 × 0.59 Hz and the lifetime of the duplex (τ_d = k_d^{-1}) and strand (τ_s) states are equal at the transition midpoint. We evaluate a duplex dissociation rate constant k_d = 1.3 × 10³ sec⁻¹ for poly(dG-dC) in 10 mM NaCl, 1 mM phosphate buffer at T_m $\approx$ 98°C.

Base Pair Overlaps

The chemical shifts of the cytidine H-5 and H-6 and the guanosine H-8 base protons in the duplex state of poly(dG-dC) are 0.1 to 0.15 ppm upfield from the corresponding values for the central base pairs of the dC-dG-dC-dG-dC-dG hexanucleotide duplex.[11] The chemical shifts in the duplex state reflect the ring current contributions from adjacent base pairs and are a measure of base pair overlap geometries.[24,29] The observed differences in the duplex state chemical shifts at the oligonucleotide and polynucleotide level for alternating (dG-dC) sequences suggests that the base pair overlaps in the center of short oligonucleotide duplexes with frayed ends differ somewhat from those observed in the longer duplex regions of polynucleotides.

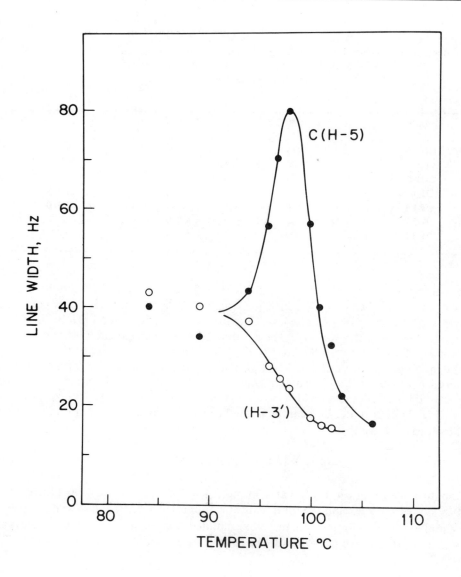

Figure 5. The temperature dependence (84° to 106°C) of the cytidine H-5 and downfield H-3′ line widths of 37.5 mM poly(dG-dC), $S_{20,\omega}$ = 6.9, in 10 mM NaCl, 1 mM phosphate, 1 mM EDTA, ^2H_2O, pH 6.3.

We have estimated the calculated upfield ring current shift on formation of the poly(dG-dC) duplex based on ring currents from nearest neighbor and next nearest neighbor base pairs[24/29] and the atomic diamagnetic anisotropy contributions[30] from adjacent base pairs for a B-DNA overlap geometry.[31] These are summarized in Table II for the guanosine H-1 and H-8 and the cytidine H-5 and H-6 protons. The total calculated shift (Table II) compared favorably with the experimental values (Table I) at the guanosine H-8 and cytidine H-5 protons but underestimates the observed value at the cytidine H-6 proton which is directed toward the backbone phosphate. Additional contributions from polarizability effects[32] will have to be included into the calculations when the contours become available.

Glycosidic Torsion Angles

The sugar proton chemical shifts are sensitive to changes in the glycosidic torsion angles of the attached base[33] and are less sensitive to ring current contributions[29] from adjacent base pairs. The experimental data suggest that the sugar H-1' protons monitor, in part, variations in the glycosidic torsion angles during the helix-coil transition (Fig. 4).

The upfield shifts on duplex formation decrease in the order C(H-5) > C(H-6) > purine(H-8) for both poly(dI-dC)[17] and poly(dG-dC) (Fig. 4). Also, the sugar H-1' protons undergo the largest upfield shift amongst the different sugar protons for both synthetic DNA's on duplex formation. These data suggest that poly(dI-dC) and poly(dG-dC) adopt similar structures in their duplex states in solution.

Table II

Calculated Upfield Shifts at Base Protons
of Poly(dG-dC) on Duplex Formations

Protons	Nearest Neighbor Ring Current Upfield Shift, ppm[a]	Nearest Neighbor Atomic Diamag. Anisot. Upfield Shift, ppm[b]	Next Nearest Neighbor Ring Current Upfield Shift, ppm[a]
G(H-1)	0.55	0.45	0.2
G(H-8)	0.0	0.15	0.05
C(H-6)	0.05	0.15	0.05
C(H-5)	0.50	0.10	0.05

[a]Calculation based on Arter and Schmidt (1976).
[b]Calculation based on Giessner-Prettre and Pullman (1976).

Backbone Phosphates

The backbone phosphates in the synthetic DNA have been monitored using ^{31}P NMR spectroscopy as a function of temperature. The 145.7 MHz Fourier transform NMR spectrum [2 to 6 ppm upfield from $(CH_3O)_3PO$]of poly(dG-dC) in 10 mM NaCl, 1 mM phosphate, 1 mM EDTA, ^2H_2O, pH 6.3 at 42.4°C is presented in Fig. 6A. A single resonance is observed at 4.27 ppm and its chemical shift is plotted between 30° and 110 °C in Fig. 6B. The resonance shifts gradually downfield between 30° and 85°C and then undergoes a much larger downfield shift between 90° and 110°C. (Fig. 6B).

Variations in the 03'-P and 05'-P polynucleotide backbone torsion angles can be probed by recording ^{31}P NMR spectra of nucleic acids. The two most striking examples are the 6 ppm spread of a few backbone phosphates in tRNA[34] and shifts of

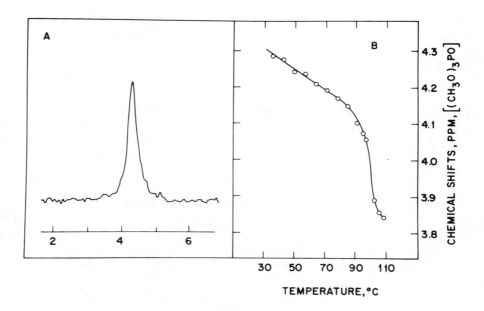

Figure 6. (A) The 145.7 MHz Fourier transform NMR spectrum (2 to 6 ppm) upfield from trimethylphosphate) of 37.5 mM poly(dG-dC), $S_{20,\omega}$ = 6.9, in 10 mM NaCl, 1 mM phosphate, 1 mM ED-TA, ^{2}H$_2$O, pH 6.3 at 42.4°C. (B) The temperature dependence of the chemical shift of the ^{31}P resonance between 30° and 110°C.

1.5 and 2.5 ppm to lower field of the internucleotide phosphates at the intercalation site of actinomycin D into nucleic acid duplex.[7][11]

The chemical shift of the ^{31}P resonance of poly(dG-dC) as a function of temperature can be divided into two distinct regions. (Fig. 6B). The resonance gradually shifts downfield in the duplex state between 30° and 80°C. Previous studies have demonstrated that the extent of branching of synthetic DNA's with alternating purine-pyrimidine sequences in the duplex state varies with temperature[13] and the ^{31}P chemical shifts may reflect this variation. The ^{31}P resonance undergoes a much larger downfield shift during the melting transition (90° to 110°C) (Fig. 6B). This probably reflects the formation of an increasing population of unstacked strands where the 03′-P angle changes from a *gauche* to a *trans* configuration.[35-37]

Poly(G-C)

We have also recorded the 145.7 MHz ^{31}P NMR spectra of the corresponding synthetic RNA, poly(G-C), in 10 mM cacodylate, 1 mM phosphate solution as a function of temperature. We observe two partially resolved resonances separated by ∼0.5 ppm in the spectrum of poly(G-C) at 37°C (Fig. 7A), in contrast to the single resonance observed in the corresponding spectrum of poly(dG-dC) in solution (Fig. 6A). The two ^{31}P resonances exhibit different temperature coefficients of their chemical shifts in the premelting temperature range (Fig. 7B). We suggest that small differences in the O-P torsion angles at the internucleotide phosphates linking CpG and GpC may account for the unexpected observation of two partially resolved resonances for poly(G-C) in solution.

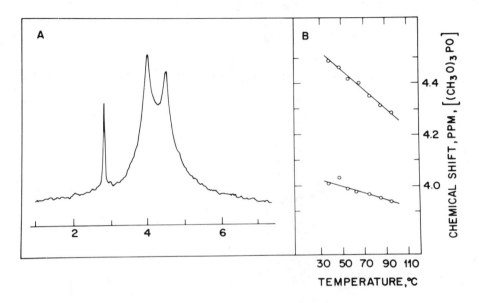

Figure 7. (A) The 145.7 MHz Fourier transform NMR spectra (1 to 7 ppm upfield from trimethylphosphate) of 39.7 mM poly(G-C), $S_{20,\omega}$ = 10.2, in 10 mM NaCl, 1 mM phosphate, 1 mM EDTA, ^2H_2O, pH 6.4 at 37°C. (B) The temperature dependence of the ^{31}P resonances between 30° and 110°C.

Poly(dG-dC) in High Salt Solution

Introduction

The most striking example of a conformational transition in synthetic DNA is the demonstration by Pohl and Jovin of a salt induced cooperative reversible structural change in the poly(dG-dC) duplex state in aqueous solution.[20] The circular dichroism spectrum of poly(dG-dC) at low salt is similar to DNA of high dG + dC content but essentially inverts on addition of 4M NaCl or 1.5 M MgCl$_2$.[20] The high salt circular dichroism spectrum was also observed for poly(dG-dC) in 55 percent ethanol - 45 percent water solution,[38] or on addition of the antibiotic mitomycin to the synthetic DNA in aqueous solution.[39] The trypanocidal drug ethidium bromide binds to the low salt form of poly(dG-dC) more tightly than the corresponding high salt form.[40] Crystals of the self-complementary tetranucleotide dC-dG-dC-dG grown from low and high salt exhibit different space groups with a reversible transition between the two forms in the crystalline state.[41] The structure of the high salt form of poly(dG-dC) in solution has not been characterized though Olson suggests that it may be a right-handed double helix with vertical base pair stacking[42] and Sobell et al. suggest that it may be β-kinked B-DNA.[43]

We compare below the NMR parameters for the low salt (10 mM NaCl) and high salt (4M NaCl) forms of poly(dG-dC) in solution. The research of Pohl and Jovin[20] has established that the transition midpoint occurs at a salt concentration of 2.5 M NaCl. Further, the activation energy of 22 Kcal/mole in both directions requires that the interconversion between low and high salt forms will be slow on the NMR time scale.[20]

Hydrogen Bonds

The Watson-Crick guanosine H-1 proton of poly(dG-dC) in low (10 mM NaCl) and

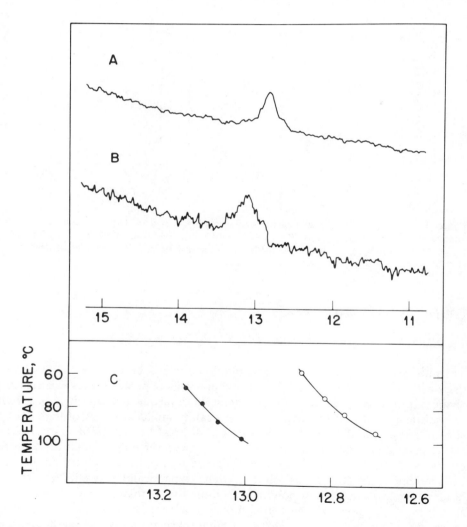

Figure 8. The continuous wave 360 MHz proton NMR spectra (11 to 15 ppm) of 37.5 mM poly(dG-dC), $S_{20,\omega} = 6.9$, in the absence (A) and presence of 4M NaCl (B) in 10 mM NaCl, 1 mM phosphate, 1 mM EDTA, H_2O, pH 6.3 at 68.5°C. The chemical shifts of the guanosine H-1 resonance of poly(dG-dC) in no salt (o) and 4 M salt (•) solution are plotted in (C).

high (4M NaCl) salt resonate at 12.84 ppm and 13.10 ppm, respectively at 68.5°C. (Figs. 8A and 8B). This establishes that the base pairs are intact for poly(dG-dC) in high salt solution. A comparison of the low salt spectrum (Fig. 8A) with the high salt spectrum (Fig. 8B) as a function of temperature (Fig. 8C) establishes the presence of an 0.25 ppm downfield shift for the synthetic DNA in high salt solution. The guanosine H-1 proton is located in the center of the base pair and its chemical shift is a sensitive indicator of base pair overlap geometries and also of the strength of the hydrogen bond. A downfield shift could be indicative of a stronger hydrogen bond and/or a change in base pair overlap geometries resulting in a decrease of the ring current contributions from adjacent base pairs.

Base Pair Overlaps

The chemical shifts of the nonexchangeable base protons of poly(dG-dC) in low and high salt are compared in Fig. 9. The guanosine H-8 proton is readily identified since it can be deuterated at high temperature in ^2H_2O solution. The cytidine H-6 proton at

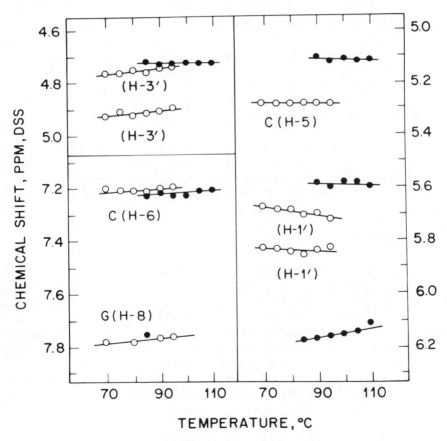

Figure 9. The temperature dependence (70° to 110°C) of the base and sugar proton chemical shifts of 37.5 mM poly(dG-dC), $S_{20,\omega}$ = 6.9, in 10 mM NaCl, 1 mM phosphate, 1 mM EDTA, ^2H_2O, pH 6.3 in the absence (o) and presence (•) of 4M NaCl solution.

7.2 ppm is separated by 1 ppm from the spectral region 4.6 to 6.0 ppm where the cytidine H-5, sugar H-1′ and sugar H-3′ protons are observed. The cytidine H-6 and guanosine H-8 protons exhibit the same chemical shift in low and high salt solution. By contrast, due to slow exchange between low and high salt forms of poly(dG-dC) we are uncertain as to which resonance between 4.6 and 6.2 ppm can be unambiguously assigned to the cytidine H-5 resonance in high salt solution. We shall assume that the resonance at 5.11 ppm corresponds to cytidine H-5 for poly(dG-dC) in high salt since it is the closest to the cytidine H-5 resonance (5.29 ppm) in low salt. It is unlikely that the cytidine H-5 resonance in the two forms would be very different when the cytidine H-6 resonance exhibits similar chemical shifts. This reasoning suggests somewhat greater overlap between the cytidine H-5 proton and adjacent base pairs on proceeding from low to high salt forms of the poly(dG-dC) duplex. We aim to establish this assignment by studying the NMR parameters for poly(dG-br^{5}dC) as a function of salt in the near future.

Glycosidic Torsion Angles.

The sugar H-1′ protons of poly(dG-dC) are observed at 5.70 and 5.84 ppm in low salt at 90°C and at 5.58 and 6.16 ppm in high salt at this temperature (Fig. 9). Thus, either one or both sugar H-1′ protons are undergoing large shifts on proceeding from the low salt duplex to the high salt duplex. Since the sugar H-1′ proton chemical shifts predominantly monitor the glycosidic torsion angles[33] the NMR results suggest that one or both sugar glycosidic torsion angles change on conversion of poly(dG-dC) from the low salt to the high salt state.

We also observe an interesting shift at the sugar H-3′ protons in that these protons are separated by 0.16 ppm in the low salt form but coalesce in the high salt form. (Fig. 9). The select shift of one of the sugar H-3′ protons correlates with the large shift of at least one of the sugar H-1′ protons of poly(dG-dC) between low and high salt forms.

Backbone Phosphates

The ^{31}P NMR spectrum of poly(dG-dC) in 4 M NaCl exhibits two well resolved resonances separated by 1.25 ppm (Fig. 10B). One of these resonances exhibits the same chemical shift as poly(dG-dC) in low salt [∼4.1 ppm from (CH$_3$O)$_3$PO, Fig. 10A) while the other is shifted downfield to 2.82 ppm relative to standard (CH$_3$O)$_3$PO]. The similar chemical shifts of the dCpdG and dGpdC internucleotide phosphates for poly(dG-dC) in low salt solution (Fig. 10A) suggests similar rotation angles about the O-P bonds for both types of phosphates The observation of two separate ^{31}P resonances for poly(dG-dC) in high salt (Fig. 10B) suggests that the O-P rotations angles at either dCpdG or dGpC internucleotide phosphates has changed resulting in a 1.25 ppm downfield shift.

We comment below on the unequal areas observed for the two ^{31}P resonances of poly(dG-dC) in 4 M NaCl solution (Fig. 10B). The branched structure of poly(dG-dC) contains duplex regions along with single-stranded loop regions and branch points. The internucleotide phosphates in single-stranded regions should fall at the

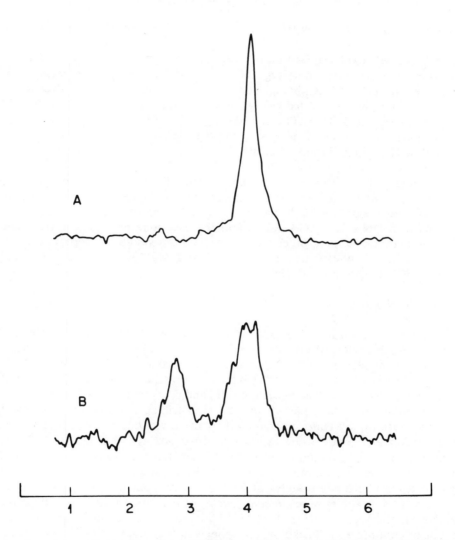

Figure 10. The 145.7 MHz Fourier transform NMR spectra (1 to 6 ppm upfield from trimethylphosphate) of 37.5 mM poly(dG-dC), $S_{20,\omega} = 6.9$, in the absence (A) and presence (B) of 4 M NaCl in 10 mM NaCl, 1 mM phosphate, 1 mM EDTA, ^2H_2O, pH 6.3 at 84.4° and 80.5°C, respectively.

normal chemical shift of $\sim$4.1 ppm as will the internucleotide phosphates in duplex regions with *gauche, gauche* orientations about the O-P bonds. By contrast, the internucleotide phosphates in the duplex region with altered orientation about the O-P bonds could resonate downfield at 2.8 ppm. We have also investigated the ^{31}P NMR spectrum of $(dG-dC)_8$ in high salt (to minimize contributions from single-stranded regions) and observe two ^{31}P resonances with approximately equal areas separated by 1.5 ppm in 4 M NaCl solution. We therefore favor a structure for the duplex regions of poly(dG-dC) in high salt in which every other phosphodiester linkage changes its conformation from the *gauche, gauche* orientation about the O-P bonds.

Conformational Aspects

The above NMR data suggest that poly(dG-dC) in high salt adopts a conformation for which the symmetry unit repeats every two base pairs. The chemical shift parameters for poly(dG-dC) in high salt demonstrate that every other glycosidic torsion angle and phosphodiester linkage adopts a different conformation from that observed in regular B-DNA. The nonexchangeable proton data suggest a somewhat altered base pair overlap geometry in the high salt form with greater overlap of pyrimidine H-5 with adjacent base pairs.

Berman and collaborators (private communication) have demonstrated from computer model building studies that oxygens on pairs of adjacent phosphates can be 4Å apart for direct coordination to metal ions when the glycosidic torsion angles alternate between DNA and RNA values and the backbone phosphate torsion angles alternate between *gauche, gauche* and *gauche, trans* values. It appears likely that pairs of adjacent phosphates generate Na^+ and Mg^{++} ion binding sites for the poly(dG-dC) conformation in high salt solution.

The Alternating B-DNA Model

Viswamitra, Kennard and their co-workers have solved the X-ray structure of the ammonium salt of the self-complementary tetranucleotide dA_1-dT_2-dA_3-dT_4[44]. They observe that the segments dA_1-dT_2 and dA_3-dT_4 on one strand form Watson-Crick hydrogen bonds with complementary sequences on separate strands. The sharp turn in the polynucleotide backbone between dT_2 and dA_3 prevented formation of a self-complementary four base-paired duplex. The striking features observed for this structure[44] are (a) C2′*endo* sugar pucker at the thymidine residues and C3′ *endo* sugar pucker at the adenosine residues in dA-dT-dA-dT in contrast to C2′ *endo* sugar pucker for all residues in B-DNA. (b) The glycosidic torsion angles were 70° for the thymidine residues but 0° for the adenosine residues in dA-dT-dA-dT in contrast to glycosidic torsion angles of 85° for all residues in B-DNA. (c) The O-P backbone torsion angles were *gauche, gauche* for the segment dA(3′-5′)dT but *trans, gauche* for the segment dT(3′-5′)dA in dA-dT-dA-dT in contrast to the *gauche, gauche* orientation at all backbone phosphates in B-DNA. (d) The X-ray data show considerable overlap between bases in the dA(3′-5′)dT segment but no stacking in the dT(3′-5′)dA segment.

Klug and Jack (personal communication) have derived an "alternating B-DNA" structure for alternating deoxypurine:deoxypyrimidine polynucleotide duplexes based on energy minimization calculations which incorporates the structural features observed in the X-ray structure of dA-dT-dA-dT.[44]

A choice amongst possible models for the "alternating B-DNA" conformation of poly(dG-dC) in high salt solution must await the resultss of selective labelling studies to determine which glycosidic torsion angle (pyrimidine or purine) and which phosphodiester linkage [dG(3′-5′)dC or dC(3′-5′)dG] adopts a conformation different from that observed in B-DNA.

Steroid Diamine • Poly(dG-dC) Complex

Introduction

The flexibility of the DNA double helix is manifested in the folding of nucleic acids around histones in chromatin and the packaging of DNA into phage heads. This flexibility can occur by either a smooth folding of the DNA double helix where the structural changes are distributed along the chain[45/46] or by an abrupt change in chain direction where the structural changes are localized at kink sites.[47/48] Sobell has suggested that the kink is a key intermediate in the process of drug intercalation into DNA and proposed further that the base pairs are kinked at the drug binding site in steroid diamine • nucleic acid complexes.[48]

We have utilized proton NMR spectroscopy to monitor drug•poly(dA-dT) complexes for ligands that intercalate between base pairs and those that bind to the sugar- phosphate backbone.[16] It was of interest to deduce the NMR parameters for the neighbor exclusion steroid diamine•poly(dG-dC) complex and compare them with those for poly(dG-dC) in high salt solution since it has been postulated that the nucleic acid conformation is β-kinked in both cases.[49] Because NMR spectroscopy monitors proton markers distributed throughout the base pairs and the sugar rings of the polynucleotide, as well as the steroid diamine, the components of the complex can be independently monitored through the temperature dependent melting transition.

The NMR studies were undertaken with dipyrandium ($3\beta,17\beta$-dipyrrolidin-1′-yl-5α-androstane) rather than irhediamine due to the stability of the former steroid diamine at high temperatures. Waring and Henley have reported on the stereochemical aspects of the interaction between steroid diamines and DNA.[50]

Hydrogen bonds

The stability of the Watson-Crick base pairs in the steroid diamine • synthetic DNA complex have been evaluated from the chemical shift and line width of the guanosine H-1 resonance as a function of temperature and pH in solution.[21/23/25] This proton exhibits a broad resonance at ~13.0 ppm in the spectrum of the P_i/drug = 5 dipyrandium•poly(dG-dC) complex in 10mM NaCl, H_2O, pH 6.9 solution at 37°C which demonstrates that the base pairs are intact in the complex. (Figure 11A).

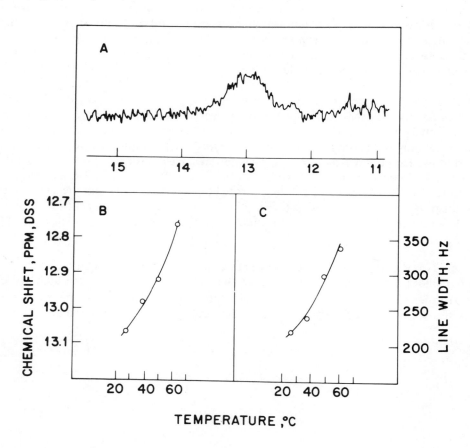

Figure 11. (A) The continuous wave 360 MHz proton NMR spectrum (11 to 15 ppm) of the dipyrandium•poly(dG-dC) complex, P_i/drug = 5, in 10 mM NaCl, 1 mM phosphate, 1 mM EDTA, H_2O, pH 6.9 at 37°C. The temperature dependence of the chemical shift (B) and line width (C) of the guanosine H-1 exchangeable proton in the complex. The concentration of poly(dG-dC) was 26.3 mM and the sedimentation coefficient ranged between 5.4 and 6.9.

The guanosine H-1 proton shifts upfield by 0.29 ppm on raising the temperature from 30° to 60°C in the neighbor exclusion steroid diamine • poly(dG-dC) complex (Fig. 11B) and broadens out in spectra recorded at $\geq$ 70°C. By contrast, a smaller upfield shift at 0.04 ppm is observed for this proton in poly(dG-dC) in the same temperature range (Fig. 1B) and the resonance can be readily monitored up to 90°C. The chemical shift data suggest a weakening of the imino Watson- Crick hydrogen bond in the steroid diamine • synthetic DNA complex with increasing temperature in the premelting transition region.

The line width of the guanosine H-1 proton in the neighbor exclusion dipyrandium • poly(dG-dC) complex (Fig. 11C) is greater by a factor of ∿3 compared to the value for this proton in the synthetic DNA alone (Fig. 1C) in solution. Further, this exchangeable resonance in the complex can be observed in spectra recorded at pH 6.9 and pH 8.0 at 29°C but broadens out in the spectrum recorded at pH 9.0 at the same temperature. The larger line widths of the guanosine H-1 resonance and its susceptibility to base catalysts suggest that the base pairs are partially exposed to solvent at the steroid diamine binding site.

Base pair unstacking

The nonexchangeable proton spectra of poly(dG-dC) and the dipyrandium • poly(dG-dC) complex, P_i/drug = 5, in 10 mM NaCl, ^2H_2O solution are compared in Figs. 12A and 12B respectively. Individual resonances in the two sets of spectra were correlated from mixing experiments since the exchange of steroid diamine between potential binding sites was fast on the NMR time scale.

The base and sugar H-1′ resonances of poly(dG-dC) shift downfield on formation of the neighbor exclusion steroid diamine complex (Fig. 13 and Table III). The relative magnitude of the downfield shifts at the various protons on complex formation (Table III) correlates with the relative magnitude of the downfield shifts associated with the duplex to strand transition of poly(dG-dC) in solution (Table I). The downfield shifts of the nucleic acid protons on formation of the steroid diamine • synthetic DNA complex must reflect the loss of ring current contributions[24/29] from adjacent base pairs(s). These chemical shift data require unstacking of the base pairs at the steroid diamine binding site.

Steroid diamine - base pair interactions

The CH_3 and NCH_3 resonances of the steroid diamine have been monitored for dipyrandium alone and in the P_i/drug = 5 dipyrandium • poly(dG-dC) complex as a function of temperature. The chemical shift data are summarized in Table IV with the CH_3 and NCH_3 groups shifting to high field on complex formation. The upfield complexation shifts suggest that the steroid diamine is located above the nucleic acid base pair planes at the binding site and its protons experience upfield ring current contributions of the guanosine and cytidine rings.[24/29]

The magnitude of the upfield complexation shifts at the CH_3 protons (0.53 ± 0.05 ppm) is much larger than at the NCH_3 protons (0.17 ± 0.04 ppm) which requires that

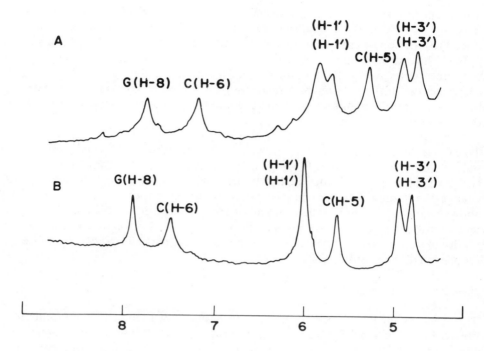

Figure 12. The 360 MHz Fourier transform proton NMR spectra (4.5 to 8.5 ppm) of (A) poly(dG-dC), pH 6.3 at 89°C and (B) the dipyrandium•poly(dG-dC) complex, P_i/drug = 5, pH 6.85 at 83°C in 10 mM NaCl, 1 mM phosphate, 1 mM EDTA, ^2H_2O solution. The concentrations of the synthetic DNA was 37.5 mM and 26.3 mM in the absence and presence of steroid diamine, respectively.

Table III

Chemical Shifts (ppm) of the Nonexchangeable Base and Sugar Protons of Poly(dG-dC) and the Dipyrandium•poly(dG-dC) Complex, P_i/drug = 5, in 10 mM NaCl, ^2H_2O at 80°C.

Protons	Poly(dG-dC)	Dipyrandium• Poly(dG-dC) Complex	Downfield Complexation Shift
G(H-8)	7.780	7.905	0.125
C(H-5)	5.280	5.620	0.340
C(H-6)	7.210	7.480	0.270
(H-1′)	5.690	5.985	0.295
(H-1′)	5.830	5.985	0.155
(H-3′)	4.755	4.800	0.045
(H-3′)	4.915	4.940	0.025

the steroid diamine CH$_3$ groups be located more directly over the base pairs than the NCH$_3$ groups (Table IV).

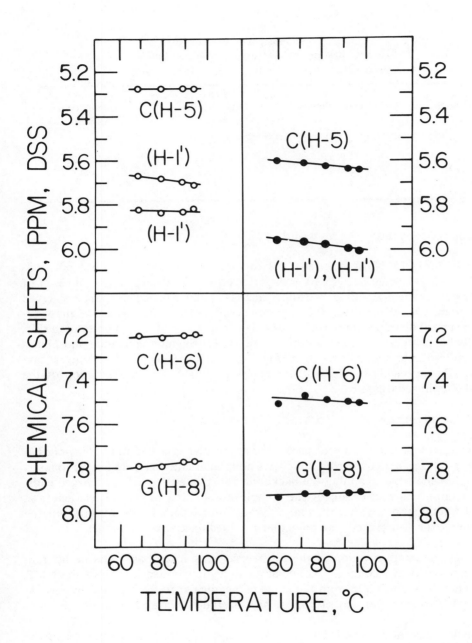

Figure 13. The temperature dependence of the base and sugar proton resonances in the duplex state of (A) poly(dG-dC) and (B) the dipyrandium•poly(dG-dC) complex, P_i/drug = 5, in 10 mM NaCl, 1 mM phosphate, 1 mM EDTA, ^2H_2O solution. The concentration of the synthetic DNA was 37.5 mM and 26.3 mM in the absence and presence of steroid diamine, respectively.

Table IV
Chemical shifts (ppm) of the CH₃ and NCH₃
groups of dipyrandium and the dipyrandium•poly(dG-dC)
complex, P_i/drug = 5, in 10 mM NaCl, ^2H_2O, at 90°

Groups	Dipyrandium	Dipyrandium• Poly(dG-dC) Complex	Upfield Complexation Shift
CH₃	0.835	0.260	0.575
CH₃	1.025	0.550	0.475
NCH₃	2.890	2.755	0.135
NCH₃	3.025	2.820	0.205

Backbone phosphates

The 145.7 MHz ^{31}P NMR spectra of poly(dG-dC) and the dipyrandium • poly(dG-dC) complex, P_i/drug = 5, in 10 mM buffer have been recorded as a function of temperature in the premelting transition range. The single broad resonance observed for the synthetic DNA shifts ∼0.1 ppm downfield on formation of the neighbor exclusion steroid diamine complex (Fig. 14). The magnitude of this complexation shift is negligible compared to the 1.5 ppm separation for the internucleotide phosphates of poly(dG-dC) in high salt solution (Fig. 10). The experimental data suggest that changes in the backbone O-P torsion angles do not occur on generation of the steroid diamine binding site.

Models for steroid diamine • synthetic DNA complex

The maximal separation between parallel base pair planes is 6.8Å and corresponds to an intercalation site. The steroid diamine is nonplanar and therefore unable to insert into such a site while spanning the backbone phosphates through its charged ends. The magnitude and direction of the complexation shifts of the nucleic acid and steroid diamine protons suggests that dipyrandium partially inserts into the binding site generated by unstacked base pairs that are tilted relative to each other.

The magnitude and direction of the tilt between unstacked base pairs in dipyrandium • poly(dG-dC) complex cannot be unambiguously deduced from the NMR parameters.[27] A tilt of 30° from the plane perpendicular to the orientation axis has been evaluated from transient electric dichroism studies of the complex.[51]

The nucleic acid and steroid diamine proton chemical shifts on formation of the neighbor exclusion dipyrandium • poly(dG-dC) complex are consistent with the general concepts put forward by Sobell for this interaction.[48] The Sobell model proposes that the Watson-Crick hydrogen bonds are intact (verified by the observation of the guanosine H-1 resonance in H₂0 solution) in the neighbor exclusion complex and that every other set of base pairs partially unstacks [verified by loss of ring current contributions from adjacent base pair(s) resulting in a downfield shift in the base and sugar protons] and the steroid diamine partially inserts between tilted base

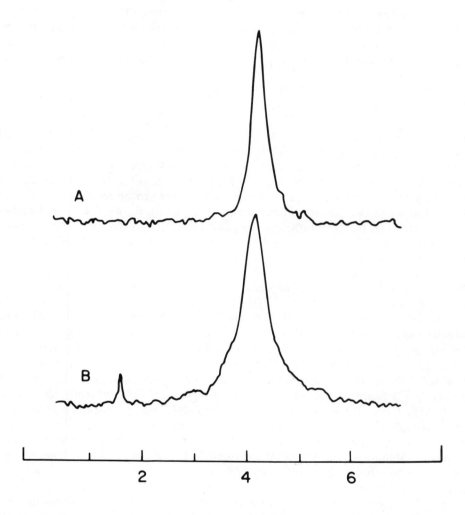

Figure 14. The 145.7 MHz Fourier transform NMR spectra (1 to 7 ppm upfield from trimethylphosphate) of (A) poly(dG-dC), pH 6.3 at 42.4°C and (B) the dipyrandium•poly(dG-dC) complex, P_i/drug = 5, pH 6.85 at 45.5°C in 10 mM NaCl, 1 mM phosphate, 1 mM EDTA, ^2H_2O solution. The concentration of the synthetic DNA was 37.5 mM and 26.3 mM in the absence and presence of steroid diamine, respectively.

pairs at this site (verified by upfield NMR ring current shifts of steroid diamine protons from base pairs at the binding site). The Sobell model generates the steroid diamine binding site through a change in the sugar pucker and glycosidic torsion angle with a minimum perturbation in the O-P backbone torsion angles.[48] We are unable to estimate the sugar puckers in the steroid diamine complex at the synthetic DNA level but have demonstrated that the conformation about the O-P bonds remains unchanged on complex formation.

Our NMR results demonstrate that poly(dG-dC) in high salt is a stacked base paired duplex structure in contrast to the steroid diamine • poly(dG-dC) complex in which every other base pair unstacks to generate a ligand binding site. This rules out proposals[49] that suggest that the high salt form of poly(dG-dC) adopt a β-kinked B-DNA structure similar to the conformation postulated for steroid diamine • nucleic acid interactions.

Summary

The helix:coil transition of poly(dG-dC) in low salt (10 mM NaCl) has been followed by NMR spectroscopy between 40°C and 100°C in aqueous solution. The hydrogen bonds are monitored at the exchangeable resonances, the base pair overlap geometries estimated from the upfield base proton shifts on duplex formation, the glycosidic torsion angle changes followed at the sugar proton resonances and the O-P torsion angle variations monitored at the chemical shifts of the backbone phosphates. The NMR parameters for poly(dA-dT),[18] poly(dI-dC)[17] and poly(dG-dC) exhibit similar trends consistent with formation of sequence independent double helical structures for alternating purine- pyrimidine synthetic DNA's in solution.

We have investigated the salt-induced conformational transition of poly(dG-dC)[20] by NMR Spectroscopy and observe selective and dramatic perturbations on addition of 4M NaCl to the synthetic DNA in solution. The Watson-Crick hydrogen bonds are detected in the low and high salt forms of poly(dG-dC) confirming the double helical nature of both states. Two well resolved ^{31}P resonances separated by 1.25 ppm are detected for poly(dG-dC) in high salt in contrast to a single resonance in low salt. These results suggest that every other phosphodiester linkage in the poly(dG-dC) duplex in high salt adopts a conformation different from that observed in B-DNA. The cytidine H-5 resonance in the poly(dG-dC) duplex in high salt is 0.2 ppm to higher field than in low salt form indicative of a greater overlap of the pyrimidine 5 position with adjacent base pairs in the high salt conformation of the synthetic DNA in solution. One of the sugar H-1′ protons exhibits a large chemical shift difference between the low and high salt forms of poly(dG-dC) indicative of a change of either the guanosine or cytidine glycosidic torsion angles between the two forms of the synthetic DNA. The experimental results are discussed in relation to the recent X-ray structure of pdA-dT-dA-dT[44] and the incorporation of this crystallographic information into a model designated the "alternating B-DNA" conformation for synthetic DNA with an alternating purine-pyrimidine sequence. (Klug and Jack, private communicaton).

We have studied the neighbor exclusion complex formed between the steroid diamine dipyrandium iodide and poly(dG-dC) in 10 mM buffer solution in order to evaluate the structural and kinetic aspects of the binding of a nonintercalative drug to a synthetic DNA in solution. The nonexchangeable proton chemical shift parameters for the dipyrandium • poly(dG-dC) complex demontrates unstacking of base pairs and partial insertion of the steroid diamine at the complexation site. The chemical shifts and line-widths of the exchangeable protons as a function of pH demonstrate that the base pairs are intact but partially exposed to solvent at the steroid diamine binding site. The phosphorus chemical shifts suggest that the base pairs unstack upon complex formation without changes in the O-P polynucleotide backbone torsion angles. The NMR line shape parameters require rapid exchange of the steroid diamine among potential binding sites and are consistent with greater segmental flexibility in the complex compared to the synthetic DNA in solution. The NMR experiments are discussed in relation to Sobell's proposed model for the steroid diamine • DNA complex.[48]

The ^{1}H and ^{31}P NMR parameters for poly(dG-dC) in high salt and for the neighbor exclusion steroid diamine • poly(dG-dC) complex are very different and rule out models[49] that propose a common conformation for the nucleic acid in these two ligand - DNA complexes.

References and Footnotes

1. Arnott, S., Chandrasekaran, R., and Selsing, E., in *Structure and Conformation of Nucleic Acids and Protein-Nucleic Acid Interactions*, Eds. Sundaralingam, M., and Rao., S. T., University Park Press, Baltimore, 577-596 (1975).
2. Brunner, W. C., and Maestre, M. F., *Biopolymers, 13*, 345-357 (1975).
3. Pilet, J., Blicharski, J., and Brahms, J., *Biochemistry, 14*, 1869-1876 (1975).
4. Wells, R. D., Larson, J. E., Grant, R. C., Shortle, B. E., and Cantor, C. R., *J. Mol. Biol., 54*, 465-497 (1970).
5. Ivanov, V. I., Minchenkova, L. E., Schyolkina, A. K., and Poletayer, A. I., *Biopolymers, 12*, 89-110 (1973).
6. Cross, A. D., and Crothers, D. M., *Biochemistry, 10*, 4015-4023 (1971).
7. Patel, D. J., *Biochemistry, 13*, 2396-2402 (1974).
8. Borer, P. N., Kan, L. S., and T'so, P. O. P., *Biochemistry, 14*, 4847-4863 (1975).
9. Kallenbach, N. R., Daniel, W. E., Jr., and Kaminker, M. A., *Biochemistry, 15*, 1218-1224 (1976).
10. Early, T. A., Kearns, D. R., Burd, J. F., Larson, J. E., and Wells, R. D., *Biochemistry, 16*, 541-551 (1976).
11. Patel, D. J., *Biopolymers, 15*, 533-558 (1976).
12. Baldwin, R. L., *Acc. Chem. Res., 4*, 265-272 (1971).
13. Spatz, H. Ch., and Baldwin, R. L., *J. Mol. Biol., 11*, 213-222 (1965).
14. Patel, D. J., and Canuel, L. L., *Proc. Natl. Acad. Scs., USA, 73*, 674-678 (1976).
15. Kearns, D. R., *Prog. Nucleic Acids Res. and Mol. Biol., 6*, 477-523 (1977).
16. Patel, D. J., *Acc. Chem. Res., 12*, 118-125 (1979).
17. Patel, D. J., *Eur. J. Biochem., 83*, 453-464 (1978).
18. Patel, D. J., *J. Polymer Sci., Polymer Symposium, 62*, 117-141 (1978).
19. Early, T. A., Olmstead, J., Kearns, D. R., and Lewis, A. G., *Nucleic Acids Res., 5*, 1955-1970 (1978).
20. Pohl, F. M., and Jovin, T. M., *J. Mol. Biol., 67*, 375-396 (1972).
21. Kearns, D. R., Patel, D. J., and Shulman, R. G., *Nature. 229*, 338-339 (1971).
22. Patel, D. J., *Biopolymers, 18*, 553-569 (1979).
23. Patel, D. J., and Hilbers, C. W., *Biochemistry, 14*, 2656-2660 (1975).
24. Giessner-Prettre, C., Pullman, B., Borer, P. N., Kan, L. S., and T'so, P. O. P., *Biopolymers, 15*, 2277-2286 (1976).
25. Crothers, D. M., Hilbers, C. W., and Shulman, R. G., *Proc. Natl. Acad. Scs., USA, 70*, 2899-2901 (1973).
26. Patel, D. J., *Biopolymers, 16*, 1635-1656 (1977).
27. Patel, D. J., and Canuel, L. L., *Proc. Natl. Acad. Scs., USA, 76*, 24-28 (1979).
28. Bovey, F. A., *NMR Spectroscopy*, Academic Press, pg. 187 (1969).
29. Arter, D. B., and Schmidt, P. G., *Nucleic Acids Research, 3*, 1437-1447 (1976).
30. Giessner-Prettre, C., and Pullman, B., *Biochem. Biophys. Res. Commun., 70*, 578-581 (1976).

31. Arnott, S., and Hukins, D. W. L., *Biochem. Biophys. Res. Commun.*, *47*, 1504-1508 (1972).
32. Giessner-Prettre, C., Pullman, B., and Caillet, J., (1977), *Nucleic Acids Research*, *4*, 99-116 (1977).
33. Giessner-Prettre, C., and Pullman, B., *J. Theor. Biol.*, *65*, 171-188, 189-201 (1977).
34. Gueron, M., and Shulman, R. G., *Proc. Natl. Acad. Scs.*, *USA*, *72*, 3483-3485 (1975).
35. Tewari, R., Nanda, R. K., and Govil, G., *Biopolymers*, *13*, 2015-2035 (1974).
36. Olson, W. K., *Biopolymers*, *14*, 1797- 1810 (1975).
37. Yathindra, N., and Sundaralingam, M., *Proc. Natl. Acad. Scs.*, *USA*, *71*, 3325- 3328 (1974).
38. Pohl, F. M., *Nature*, *260*, 365-366 (1976).
39. Mercardo, C. M., and Tomasz, M., *Biochemistry*, *16*, 2040-2046 (1977).
40. Pohl, F. M., Jovin, T. M., Baehr, W., and Holbrook, J. J., *Proc. Natl. Acad. Scs.*, *USA*, *69*, 3805-3809 (1972).
41. Drew, H. R., Dickerson, R. E., and Itakura, K., *J. Mol. Biol.*, *25*, 535-543 (1978).
42. Olson, W. K., *Proc. Natl. Acad. Scs.*, *USA*, *74*, 1775-1779 (1977).
43. Sobell, H., Tsai, C. C., Jain, S., and Gilbert, S., *J. Mol. Biol.*, *114*, 333-365 (1977).
44. Viswamitra, M. A., Kennard, O., Jones, P. G., Sheldrick, G. M., Salisbury, S., Falvello, L., and Shakked, Z., *Nature*, *273*, 687-688 (1978).
45. Sussman, J. L., and Trifonov, E. N., *Proc. Natl. Acad. Scs.*, *USA*, *75*, 103-107 (1978).
46. Levitt, M., *Proc. Natl. Acad. Scs.*, *USA*, *75*, 1775-1779 (1978).
47. Crick, F. H. C., and Klug, A., *Nature*, *255*, 530-533 (1975).
48. Sobell, H. M., Tsai, C. C., Gilbert, S. G., Jain, S. C., and Sakore, T. D., *Proc. Natl. Acad. Scs.*, *USA*, *73*, 3068-3072 (1976).
49. Sobell, H. M., Reddy, B. S., Bhandary, K. K., Jain, S. C., Sakore, T. D., and Seshadri, T. P., *Cold Spring Harbor Symposium. Quant. Biol.*, *42*, 87-101 (1978).
50. Waring, M.J., and Henley, S. M., *Nucleic Acids Research*, *2*, 567-585 (1975).
51. Dattagupta, N., Hogan, M., and Crothers, D. M., *Proc. Natl. Acad. Sci. USA*, *75*, 4286-4290 (1978).

Spectroscopic Studies of Drug Nucleic Acid Complexes

Thomas R. Krugh, John W. Hook, III, and S. Lin
Department of Chemistry
University of Rochester
Rochester, New York 14627

and

Fu-Ming Chen
Department of Chemistry
Tennessee State University
Nashville, Tennessee 37203

The structure, pharmacological activity, and the interaction of the actinomycins with double-stranded DNA has been the subject of an extensive number of studies (for example, see the review by Meienhofer and Atherton,[1] and the reviews and articles cited therein). The chemical structure of actinomycin D (just one of a large class of related molecules) is shown in Figure 1. Note that the (relatively) planar chromophore, commonly called the phenoxazone ring, has two cyclic pentapeptide rings attached to it. The combination of the intercalating chromophore and the cyclic pentapeptides is responsible for some of the unusual properties observed for the binding of actinomycin to DNA. For example, in 1968 Muller and Crothers[2] showed that the association of actinomycin to DNA is characterized by five separate rate constants. Even more interesting were Muller and Crothers' observations that the actinomycins dissociate very slowly from DNA and that the slow dissociation is apparently correlated with the pharmacological activity of the drugs (i.e. the inhibition of RNA polymerase). The dissociation relaxation curve (from calf thymus DNA) required a minimum of three separate exponential curves to fit the decay curve mathematically.[2] The actinomycins also exhibit a general requirement for the presence of guanine at the binding site (e.g., see Wells and Larson (1970)[3]; Kersten (1961)[4]; and Goldberg, *et al.* (1962)[5]). This unusual requirement has been frequently illustrated by citing the observation[5] that guanine-free DNA, such as poly-(dA-dT)•poly(dA-dT), does not bind actinomycin. An important exception to this observation will be presented below.

Ethidium bromide (Figure 1) is an intercalating agent with mutagenic properties. References 6-11 provide an introduction to the extensive studies on ethidium-DNA complexes as well as the important role of ethidium as a fluorescent probe. Our studies on ethidium were initiated in 1972 to determine if this relatively simple drug molecule would exhibit preferential binding to oligonucleotides. A second goal of the initial ethidium experiments was to provide another test of the use of oligonucleotides as model systems for studying drug-nucleic acid complexes. In the initial experiments we showed that ethidium does exhibit a pronounced preference for binding to certain sequences with both the ribo- and deoxyribodinucleoside monophosphates.[11-13] The most striking observation was that ethidium binds more strongly to the self-complementary pyrimidine (3'-5') purine dinucleosides (or dinucleotides) than to their sequence isomers, the purine (3'-5') pyrimidine dinucleotides.[11-13] Sobell and coworkers determined the x-ray structures of cocrystalline complexes of ethidium with the self-complementary pyrimidine (3'-5') purine dinucleoside monophosphates.[14] The crystal structures of several other drug-nucleic acid complexes have also been reported during the last four years.[15-18]

Two drugs which in our studies have exhibited no marked preference between pyrimidine (3'-5') purine and purine (3'-5') pyrimidine deoxydinucleotides are daunorubicin and actinomine (Figure 1).[19] Daunorubicin (frequently called

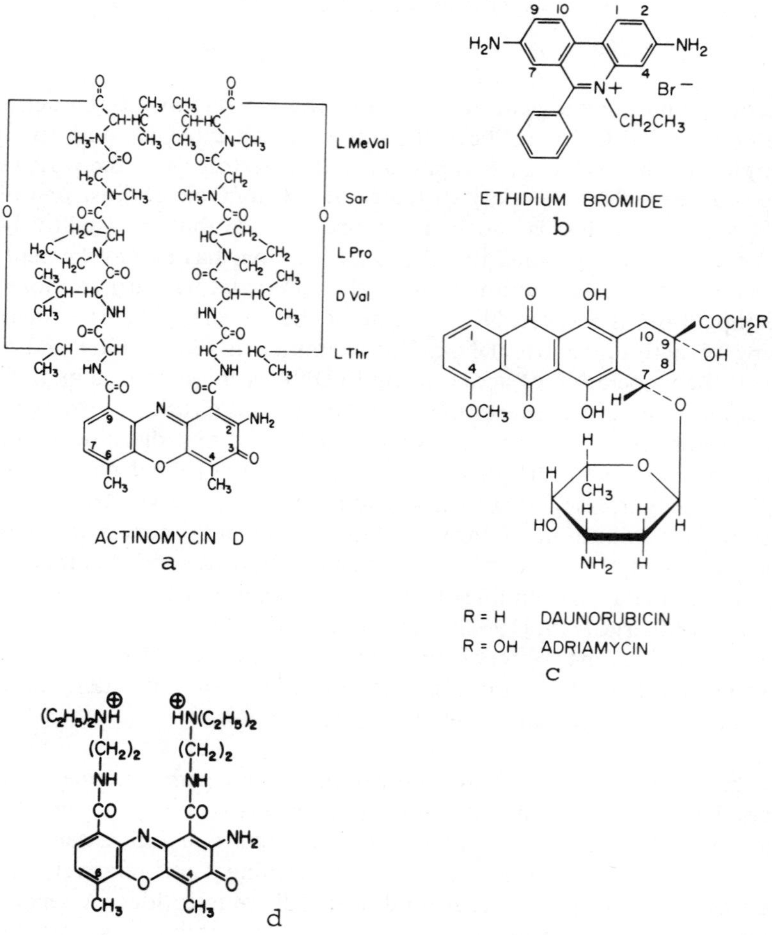

Figure 1. The chemical structures of (a), actinomycin D, (b) ethidium bromide, (c), daunorubicin and adriamycin, and (d), actinomine.

daunomycin) and the related antibiotic adriamycin are intercalating drugs which are important drugs for cancer chemotherapy.[20-22] Actinomine is an actinomycin related derivative which has been used to model the interaction of the actinomycin chromophore in binding to DNA.[2] It should be noted, however, that actinomine has two positively charged amino groups at neutral pH,[2] whereas the actinomycins are uncharged, and thus actinomine is as much different from actinomycin (two positively charged side chains versus two uncharged cyclic pentapeptide groups) as it is similar (identical phenoxazone rings).

Ethidium and Actinomycin Bind Non-competitively to Calf Thymus DNA at Low r Values

In the model oligonucleotide binding studies, ethidium bromide has exhibited a marked preference for binding to pyrimidine (3'-5') purine binding sites when compared to the purine (3'-5') pyrimidine binding sites;[11/12/13/23/24] the (dG-dG)•(dC-dC) site has also been shown to be a favorable binding site in the oligonucleotide binding studies.[25] On the other hand, actinomycin exhibits a general preference for binding to guanine (3'-5') pyrimidine sequences,[3/26-29] although (dG-dG)•(dC-dC) is also a favorable binding site.[2/3] In considering the simultaneous binding of actinomycin D and ethidium bromide to calf-thymus DNA, one expects that if the sequence preferences discussed above hold true when these molecules bind to DNA, then ethidium and actinomycin D should bind primarily in a non-competitive fashion, at least at low levels of total bound drug. This expectation has been confirmed by Reinhardt and Krugh,[7] as illustrated by the data in Figure 2a. The roughly parallel lines (at least for low r values) of the ethidium bromide titrations (with and without actinomycin D) are indicative of non-competitive binding.[30/31] On the other hand,

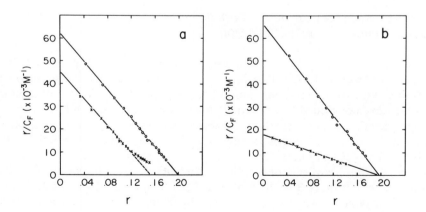

Figure 2.(a) Competition of actinomycin D with ethidium bromide for binding sites on DNA. Fluorescence Scatchard plots of EthBr (concentration 1.0-29.1 μM) in 50 mM Tris-HCl (pH 7.5), 0.2 M NaCl buffer (o) and in the presence of Act D, [DNA•P]/[Act D] = 5 (x).; (b) Competition of actinomine with ethidium bromide for binding sites on DNA. Fluorescence Scatchard plots of ethidium bromide (concentration 1.0-28.3 μM) bound to calf thymus DNA ([DNA•P] = 3.5 μM) in 50 mM Tris-HCl (pH 7.5), 0.2 M NaCl buffer (o) and in the presence of actinomine, ([DNA•P]/[actinomine]) = 0.5 (x). Reproduced with permission from Ref. 7.

the binding of ethidium bromide to calf thymus DNA in the presence of actinomine is characterized by a competition for binding sites, as illustrated by the data in Figure 2b. This competition is consistent with the oligonucleotide binding studies with actinomine, which shows no strong preference for purine (3'-5') pyrimidine or pyrimidine (3'-5') purine sequences.[7] These types of approaches, at both the oligonucleotide and polynucleotide levels, will lead to a more complete understanding of the sequence preferences that drugs exhibit, and more importantly, a molecular understanding of the origin of these sequence preferences for various classes of drugs.

A further test of the relative preferences for ethidium binding to the (dC-dG)• (dC-dG) sites compared to the (dG-dC)•(dG-dC) sites was done using the method of continuous variation to study complex formation of ethidium to the deoxyhexanucleotides (dG-dC)₃ and (dC-dG)₃. The data in Figure 3 show that the most favorable complex formed in the ethidium experiment with (dG-dC)₃ involves the complex formation between two ethidiums and two (dG-dC)₃ strands (i.e., the lines intersect at the 0.5 mole fraction point). With (dC-dG)₃ the stoichiometry of complex formation is 3 ethidiums: 2 strands (or 3 ethidiums per helix), since the maximal point of complex formation is at the 0.6 ethidium mole fraction point. These stoichiometries of complex formation have been independently verified by circular dichroism and fluorescence titrations [M. S. Balakrishnan and T. R. Krugh, to be published]. The 3:2 complex formation with (dC-dG)₃ is consistent with intercalation at the three dC-dG sequences of the double helix while the 2:2 stoichiometry for the binding of ethidium with (dG-dC)₃, which has two dC-dG sequences, is evidence that ethidium has a strong preference for binding at the dC-dG sequences (as opposed to the three (dG-dC) sequences).

Fluorescence Lifetime Measurements of Ethidium Complexes in Solution and in the Solid State

The cocrystalline complexes of ethidium bromide with the dinucleoside monophosphates UpA and CpG used for the x-ray analysis had a stoichiometry of two ethidiums and two dinucleoside monophosphates.[14/15/32] From the solution experiments we have concluded that when UpA or CpG are present in large excesses compared to the ethidium concentration, the predominant complex formed is one in which essentially all the ethidium is intercalated into a miniature double helix.[6] Reinhardt and Krugh[7] have performed fluorescence lifetime measurements on two separate cocrystalline complexes of ethidium with CpG (2:2 stoichiometry) as well as on a solution of ethidium (1.3×10^{-4} M) with CpG (6.9×10^{-3} M). In solution the experimental decay curve was well represented by a single exponential decay function with a lifetime of 23 nsec, a value which is in close agreement with the previously measured value for ethidium intercalated into double-stranded RNA.[33] In the cocrystalline complexes two distinct fluorescence decays were observed with lifetimes of approximately 4.5 and 11 nsec. The two distinct lifetimes observed in the crystalline state are consistent with the two environmentally distinct ethidiums observed in the x-ray structures and thus the fluorescence experiments provide a link between the solid state studies and the solution studies.

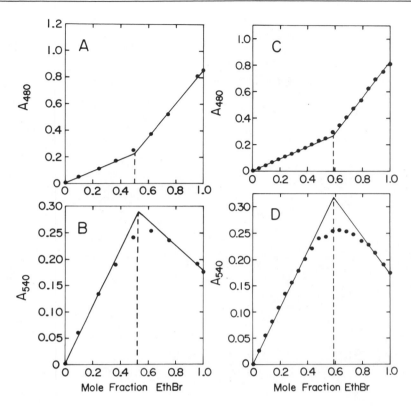

Figure 3. Absorbance vs. mole fraction EthBr for the continuous variation experiment involving EthBr plus (dG-dC)₃ or (dC-dG)₃. The concentration of EthBr plus hexanucleotide strand was kept constant at 150 μM throughout the experiment. (dG-dC)₃ absorbance monitored at 480 nm (A) and 540 nm (B). (dC-dG)₃ absorbance monitored at 480 nm (C) and 540 nm (D). Figures A and B reproduced with permission from reference 25. Figures C and D from M. S. Balakrishnan and T. R. Krugh, to be published.

Daunorubicin and Adriamycin Facilitate Actinomycin Binding to poly(dA-dT)•poly(dA-dT)

As noted above, the classic example of the actinomycin requirement for guanine at the intercalation site is the observation that actinomycin D does not bind to double-stranded poly(dA-dT)•poly(dA-dT). However, the circular dichroism spectra of solutions of poly(dA-dT)•poly(dA-dT) with various combinations of either daunorubicin or adriamycin clearly show that these drugs facilitate the binding of actinomycin to this polynucleotide (Figure 4). We believe that this is the first example of cooperative binding of two intercalating drugs. Neither ethidium bromide nor acridine orange facilitates the binding of actinomycin D to poly-(dA-dT)•poly(dA-dT) (although both of these molecules intercalate into the polynucleotide) which illustrates the specificity of the cooperative interaction of actinomycin and daunorubicin when both are bound to poly(dA-dT)•poly(dA-dT). It will be interesting to explore the nature of the structural changes in the conformation of poly(dA-dT)•poly(dA-dT) that result from the binding of either

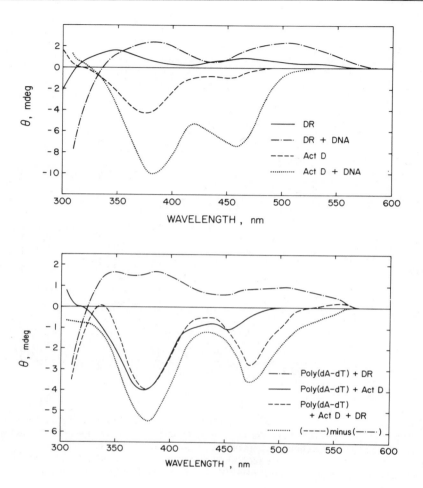

Figure 4. Circular dichroism spectra of solutions of: *a*, 8.5 ×10⁻⁶M daunorubicin (DR) and ac-tinomycin D (Act D) alone and in the presence of 8.5 × 10⁻⁵M DNA; *b*, 8.5 × 10⁻⁶ M daunorubicin plus 8.5 × 10⁻⁵M poly(dA-dT)•poly(dA-dT); 8.5 ×10⁻⁶M actinomycin D plus 8.5 ×10⁻⁵M poly(dA-dT)•poly(dA-T); 8.5 ×10⁻⁶M actinomycin D plus 8.5 ×10⁻⁶ daunorubicin plus 8.5 ×10⁻⁵M poly(dA-dT)•poly(dA-dT). The curve ...... was calculated by subtracting the poly(dA-dT)•poly(dA-dT) + DR spectrum from the poly(dA-dT)•poly(dA-dT) + DR + Act D spectrum which, to a first approxima-tion, is an estimate of the circular dichroism spectrum of actinomycin D when bound to poly(dA-dT)•poly(dA-dT). All spectra were recorded on a Jasco J-40 circular dichroism instrument in a 4-cm path length cell at 20°C. All solutions contained 10 mM potassium phosphate buffer, pH 7.0. Essentially similar results were obtained at lower drug-to-phosphate ratios, as well as in the presence of 0.1 M NaCl. The spectra in which adriamycin were used in place of daunorubicin gave qualitatively similar results. The circular dichroism spectra in which ethidium bromide is used in combination with actinomycin D and poly(dA-dT)•poly(dA-dT) do not show the appearance of the large negative band in the 440-480 nm region. Reproduced with permission from reference 22.

daunorubicin or adriamycin because these conformational changes in the double helical geometry of poly(dA-dT)•poly(dA-dT) are presumably responsible for the facilitation of actinomycin D binding to this polynucleotide.

Enzyme Inhibition Studies

Actinomycin D is a very effective inhibitor of RNA polymerase when native DNA is used as a template (e.g., see references 1, 2, 34). However, when poly-(dA-dT)•poly(dA-dT) is used as a template there is no inhibition of *E. coli* RNA polymerase (Figure 5), an observation which has also been taken as evidence for the non-binding of actinomycin D to this guanine-free polynucleotide. If sufficient daunorubicin is added to the solutions to facilitate actinomycin D binding to the template, one observes that actinomycin D does not enhance the inhibition of *E. coli* RNA polymerase (Figure 5, from A. H. McHale, S. S. Holcomb, R. Josephson, and T. R. Krugh, unpublished data). In other words, even though actinomycin D binds to poly(dA-dT)•poly(dA-dT) in the presence of daunorubicin, actinomycin D does not appear to interfere with RNA polymerase activity. These experiments suggested a direct test of the hypothesis that the inhibition of RNA polymerase is closely coupled to the extremely slow dissociation of actinomycin from calf thymus DNA[2] since the failure of actinomycin to inhibit RNA polymerase predicts a rapid dissociation rate for the actinomycin D-poly(dA-dT)•poly(dA-dT) daunorubicin complex. We have experimentally verified that actinomycin D dissociates much faster from poly(dA-dT)•poly(dA-dT) than from calf thymus DNA.

Actinomycin Dissociation Kinetics

Absorption, CD and fluorescence studies in our laboratory have revealed that actinomycin D and its fluorescent derivative, 7-amino-actinomycin D, bind to poly(dG-dC)•poly(dG-dC) roughly an order of magnitude more strongly than to calf thymus DNA. The saturation binding appears to occur at a phosphate-to-drug

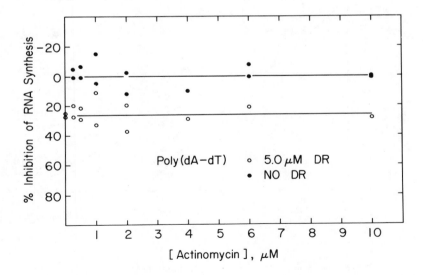

Figure 5. Drug induced inhibition of RNA synthesis using DNA dependent RNA polymerase from *E. coli* with poly(dA-dT)•poly(dA-dT) as the template as a function of actinomycin D concentration. [A. H. McHale, S. S. Holcomb, R. Josephson and T. R. Krugh, unpublished data.]

ratio (P/D) of approximately 8:1. There is some uncertainty in this ratio, however, due to conflicting values for the molar extinction coefficient of poly(dG-dC)•poly(dG-dC).[35-37]

Three representative dissociation curves for actinomycin D-poly(dG-dC)•poly(dG-dC) complexes are shown in Figure 6. Two characteristics are evident in these dissociations: (1) The time dependence of the absorbance is well described by a single exponential; and (2) The apparent dissociation lifetime, τ, depends on the phosphate/drug (P/D) ratio (Figure 7). The obvious linearity of the dissociation curves throughout the whole range of observed τ's precludes explaining the P/D dependence of τ with two exponential processes in which the relative magnitudes of the individual ΔA's vary. This would be the case, for example, if there were two classes of binding sites whose relative populations vary with P/D. Any two exponential model in which the apparent τ varies over a factor of two, as is observed here, would predict that at least one of the three curves in Figure 6 should exhibit a significant concave curvature. Such curvature is not observed

It is intriguing to note that the dissociating complex's behavior is dependent on its history. For instance, when the 7.1:1 (P/D) complex is 60 percent dissociated, there is approximately one drug molecule bound for every 18 phosphates, as at the beginning of the 17.7:1 dissociation shown in the middle curve of Figure 6. However, though the phosphate/bound drug ratio is the same in these two cases, the behavior is quite different. One possible explanation for this behavior would be that the polynucleotide structure might be locally distorted by the drug molecule in some manner which slowed the dissociation of the other drug molecules bound to the

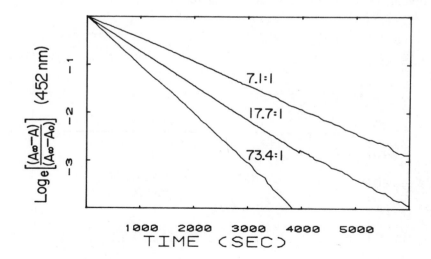

Figure 6. Absorbance changes due to dissociation of actinomycin D from poly(dG-dC)•poly(dG-dC) in 0.1 M NaCl, 10^{-4} M EDTA, 0.01 M sodium phosphate buffer at pH 7.0 and 22°C. The three curves are for phosphate-to-drug ratios of 7.1:1, 17.7:1, and 73.4:1 as marked. In these experiments the initial drug concentration ranged from 7 to 10 × 10^{-5}M. Dissociation was induced by mixing with a 5 percent solution of sodium dodecyl sulfate (SDS).[2]

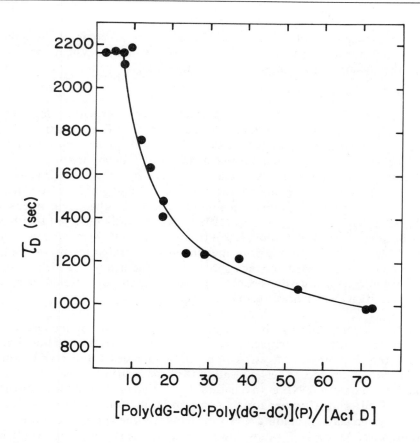

Figure 7. Dissociation lifetimes for the poly(dG-dC)•poly(dG-dC)-actinomycin D complex plotted as a function of phosphate/drug ratio. The conditions are the same as in Figure 6.

polynucleotide. If relaxation from such a perturbed structure was long with respect to the dissociation time of the drug, the dissociation kinetics of the drug would appear as a single exponential with τ dependent upon the initial P/D ratio. Such a structural distortion might be expected to alter the circular dichroism (CD) spectrum of the polynucleotide. A 7.1:1 (P/D) mixture of poly(dG-dC)•poly(dG-dC) and actinomycin D was dissociated with SDS, and the 240-500 nm CD spectrum was observed periodically following the initial mixing. The change in θ at all wavelengths qualitatively followed the dissociation curve observed in the visible absorbance band at 452 nm. No significant changes in the CD spectrum were observed past 7500 seconds (the dissociation was monitored for 22,000 seconds). Experiments in which the dissociation was monitored with CD at single wavelengths of 250, 290, and 460 nm all gave the same observed lifetimes (within experimental error) as the absorbance experiments.

Preliminary experiments on the temperature dependence of the dissociation time constants for an actinomycin D - poly(dG-dC)•poly(dG-dC) complex at a 15:1 P/D initial ratio indicate that the activation energy is ~22 kcal/mole. The activation energy for the slowest dissociating component of the actinomycin D - calf thymus

DNA complex was ~19 kcal/mole. It will be interesting to compare the activation enthalpies and entropies as a function of the P/D ratio in these systems, as well as to compare these values with other native and synthetic DNAs with varying lengths and compositions.

The transmission of the distortions of the nucleic acid conformation along the double helix involved in the saturation dependence of the actinomycin D poly(dG-dC)•poly(dG-dC) dissociation and in facilitation of actinomycin binding to poly(dA-dT)•poly(dA-dT) by daunorubicin may also be an important component in the selective recognition of nucleic acid sequences (as for example, in promoter-operator complexes), and in the transmission of thermal stability from the G-C region to the A-T region which has been observed in block oligonucleotides such as $d(C_{15}A_{15})•d(T_{15}G_{15})$.[38-40] Hogan et al. (1979)[41] have recently observed that the binding of distamycin induces a cooperative transition of calf thymus DNA to a new form with higher affinity for the drug and altered structural properties of the DNA. Several other DNA's tested did not show this phenomenon,[41] which further illustrates the interesting questions concerning allosteric conformational changes and sequence specificity which have yet to be answered.

The single-exponential decay observed in of the dissociation of the actinomycin-poly(dG-dC)•poly(dG-dC) complex suggests that the multiple dissociation time constants for the actinomycin D-calf thymus DNA complex (Muller and Crothers, 1968;[2] and below) are most likely a result of the heterogeneity of the actinomycin D binding sites on natural DNA's. In a natural DNA containing several classes of binding sites of differing affinities for actinomycin D, the fraction of drug molecules bound to a given class of binding sites will change as one approaches saturation of all binding sites. At low D/P ratios (low saturation), the stronger sites would be expected to have a larger fraction of the bound actinomycin D. As more drug is added and saturation is approached, a larger fraction of the bound drug will be located at the weaker sites. If dissociation from each class of binding sites is a single exponential with τ dependent on the class of binding sites, then the relative contribution of each exponential will vary as a function of the relative population of drugs bound to that class of sites. Several dissociation curves for actinomycin from calf thymus DNA at different D/P ratios are shown in Figure 8. The dissociation of the complex can be mathematically described as a sum of three exponential terms:

$$A_0 - A = \Delta A_1 e^{-t/\tau_1} + \Delta A_2 e^{-t/\tau_2} + \Delta A_3 e^{-t/\tau_3} \tag{1}$$

The data in Figure 8 were analyzed using equation 1 and a non-linear least squares fitting routine. In each dissociation experiment, there is a long-lived component, τ_s, with a lifetime of about 1000 seconds. The relative magnitude of this slow component is a function of the D/P ratio as illustrated in Figure 9. At a D/P ratio equal to 1:200 the slow component is the dominant feature, accounting for almost 80 percent of the observed absorbance change. As the D/P ratio increases, the faster components become relatively more important. By a D/P ratio equal to 1:40, ΔA_s represents less than half of the observed change. The observed dependence of ΔA_s

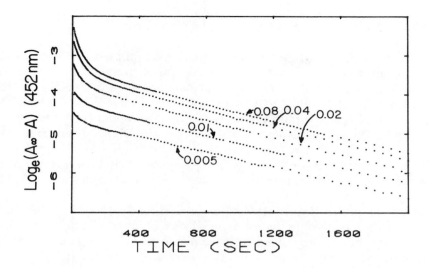

Figure 8. Absorbance changes due to dissociation of actinomycin D from calf thymus DNA in BPES buffer (pH 7.0) at 24°C. The initial concentration of DNA was 3.1×10^{-3}M in each experiment.

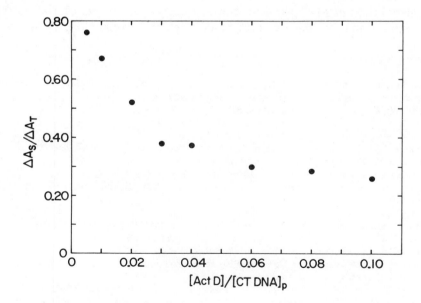

Figure 9. Dependence of the fractional absorbance change due to the slow component on the drug-to-phosphate ratio in the dissociation of the actinomycin D-calf thymus DNA complex. Conditions are as in Figure 8.

on D/P fits well with the site preference model discussed above.

Since actinomycin apparently has a general requirement for gaunine in binding to DNA, it is tempting to speculate that the population of strong binding sites would depend on the GC content of the DNA. Preliminary results on *Cl. perfringens* DNA (30 percent $G + C$), calf thymus DNA (42 percent $G + C$), and *M. lysodeikticus* DNA (72 percent $G + C$) at a P/D ratio equal to 55:1 (low saturation) are consistent with this hypothesis. The observed values of τ_s in each of these DNA's is very close to the low saturation lifetime observed for poly(dG-dC)•poly(dG-dC). While it may be interesting to speculate that a (dG-dC)•(dG-dC) sequence may be a member of the strong binding class, more data are needed to make any firm assignment.

In summary, the present experiments have demonstrated several new phenomena concerning the influence of the conformational state of the nucleic acids upon their interaction with drugs. These types of experiments may also serve as models for understanding the concepts of protein-nucleic acid recognition and (allosteric) control, as well as for providing an ever increasing rational basis for new drug design.

Acknowledgments

This research was supported by NIH Grants CA-17865, CA-14103, as well as by a Research Career Development Award (CA-00257), and an Alfred P. Sloan Fellowship (to TRK), and a MARC Fellowship (to FMC). The authors acknowledge the collaboration of M. Petersheim during a portion of the kinetic studies, and Dr. M. S. Balakrishnan for the use of his data (shown on the right in Figure 3).

References and Notes

1. Meienhofer, J., and Atherton, E. in *Structure-Activity Relationships Among the Semisynthetic Antibiotics* (Perlman, D., Ed.), pp. 427-529, Academic Press, New York (1977).
2. Muller, W., and Crothers, D. M., *J. Mol. Biol. 35*, 251-290 (1968).
3. Wells, R. D., and Larson, J. E., *J. Mol. Biol. 49*, 319-342 (1970).
4. Kersten, W., *Biochim. Biophys. Acta 47*, 610 (1961).
5. Goldberg, I. H., Rabinowitz, M. and Reich, E., *Proc. Natl. Acad. Sci. USA 48*, 2094 (1962).
6. Reinhardt, C. G. and Krugh, T. R., *Biochemistry 16*, 2890-2895 (1977).
7. Reinhardt, C. G., and Krugh, T. R., *Biochemistry 17*, 4845-4854 (1978).
8. Le Pecq, J.-B. in *Methods of Biochemical Analysis 20*, (Glick, D., Ed.) pp. 41-86, John Wiley and Sons, New York (1971).
9. Lee, C.-H. and Tinoco, I., Jr., *Nature 274*, 609-610 (1978).
10. Patel, D. J. and Shen, C., *Proc. Natl. Acad. Sci. USA 75*, 2553-2557 (1978).
11. Krugh, T. R., Wittlin, F. H., and Cramer, S. P., *Biopolymers 14*, 197-210 (1975).
12. Krugh, T. R., and Reinhardt, C. G., *J. Mol. Biol. 97*, 133-162 (1975).
13. Krugh, T. R. in *Molecular and Quantum Pharmacology* (Bergmann, E. D. an Pullman, B., Eds.), pp. 465-471, Reidel Pub. Co., Dordrecht, Holland (1974).
14. Tsai, C.-C., Jain, S. C., and Sobell, H. M., *Proc. Natl. Acad. Sci. USA 72*, 628-632 (1975).
15. Sobell, H. M., Jain, S. C., Sakore, T. D., Reddy, B. S., Bhandary, K. K., and Seshadri, T. P. in *International Symposium on Biomolecular Structure, Conformation, Function, and Evolution, Madras, India, January 4-7, 1978*, (Srinivasan, R., Ed.) Pergamon Press, Inc., New York (in press).
16. Neidle, S., Archai, A., Taylor, G. L., Berman, H. L., Carrell, H. L., Glusker, J. P., and Stallings, W. C., *Nature 269*, 304-307 (1977).
17. Seeman, N. C., Day, R. O., and Rich, A., *Nature 253*, 324-326 (1975).

18. The reader is also referred to the articles in this volume by Drs. H. M. Berman, H. M. Sobell, A. Rich, and their coworkers, and the references therein.

19. Krugh, T. R., Holcomb, S., Moehle, W. E., unpublished results.

20. Henry, D. W., in *Cancer Chemotherapy*, (Sartorelli, A. C., Ed.) Amer. Chem. Soc. Symp. *30*, Amer. Chem. Soc., Washington, D.C., pp. 15-17 (1976).

21. Arcamone, F., Cassinelli, G., Franceschi, G., Penco, S., Pol, C., Redaelli, S., and Selva, A., in *International Symposium on Adriamycin* (Eds., Carter, S. K., Di Marco, A., Ghione, M., Krakoff, I. H., and Mathe, G.) 9-22, Springer-Verlag, New York, New York (1972).

22. Krugh, T. R., and Young, M. A., *Nature 269*, 627-628 (1977).

23. Patel, D. J., *Biopolymers 15*, 533-558 (1976).

24. Patel, D. J., *Biochim. Biophys. Acta 442*, 98-108 (1976).

25. Kastrup, R. V., Young, M. A., and Krugh, T. R., *Biochemistry 17*, 4855-4865 (1978).

26. Krugh, T. R., *Proc. Natl. Acad. Sci. USA 69*, 1911-1914 (1972).

27. Patel, D. J., *Biochemistry 13*, 2396-2402 (1974).

28. Krugh, T. R., Mooberry, E. S., and Chiao, Y.-C. C., *Biochemistry 16*, 740-747 (1977).

29. Sobell, H. M., *Prog. Nuc. Acid Res. and Mol. Biol. 13*, 153-190 (1973).

30. Le Pecq, J.-B., and Paoletti, C., *J. Mol. Biol. 27*, 87-106 (1967).

31. A more rigorous analysis of the binding data is actually required to accurately characterize the complex equilibria involved, but the qualitative conclusion that these two drugs bind in a non-competitive manner at low r values appears to be justified (see the discussion in Reinhardt and Krugh, reference 7).

32. Jain, S. C., Tsai, C.-C., and Sobell, H. M., *J. Mol. Biol. 114*, 317-331 (1977).

33. Burns, V. W. F., *Arch. Biochem. Biophys. 145*, 248 (1971).

34. Wells, R. D., in *Progress in Molecular and Subcellular Biology 2*, (Hahn, F. E., Ed.) 21-32, Springer-Verlag, New York (1971).

35. Wells, R. D., Larson, J. E., Grant, R. C., Shortle, B. E., and Cantor, C. R., *J. Mol. Biol. 54*, 465-497 (1970).

36. Pohl, F. M., Jovin, T. M., Baehr, W., and Holbrook, J. J., *Proc. Natl. Acad. Sci. USA 69*, 3805-3809 (1972).

37. Lee, K.-R., this laboratory

38. Burd, J. F., Wartell, R. M., Dodgson, J. B., and Wells, R. D., *J. Biol. Chem. 250*, 5109-5113 (1975).

39. Burd, J. F., Larson, J. E., and Wells, R. D., *J. Biol. Chem. 250*, 6002-6007 (1975).

40. Early, T. E., Kearns, D. R., Burd, J. F., Larson, J. E., and Wells, R. D., *Biochemistry 16*, 541-551 (1977).

41. Hogan, M., Dattagupta, N., and Crothers, D. M., *Nature 278*, 521-524 (1979).

Part V
Intact Biological Systems

Carbon-13 Cross-Polarization Magic-Angle NMR Studies of Biological Systems[1]

Steven S. Danyluk and Herbert M. Schwartz
Division of Biological and Medical Research
Argonne National Laboratory
Argonne, Illinois 60439

Introduction

High-resolution nuclear magnetic resonance spectroscopy (continuous wave and Fourier transform) has yielded an enormous wealth of information about structures conformational and dynamical properties, and intermolecular interactions of naturally occurring biological molecules and their simpler constituent fragments in solution.[2-4] Such knowledge provides an important base-line for interpretation of functional mechanisms for these molecules. There is nevertheless a compelling need to extend these "model system" studies to condensed states more typical of those existing in intact biological systems, i.e., whole cells, tissues, organs. However, attainment of high-resolution NMR in these systems has been hampered by several major constraints. Spectra for most nuclei are usually characterized by a marked line-broadening, particularly for large, immobilized biomolecules. An additional complication arises from the extremely low sensitivity of several biologically important nuclei, ^{13}C, ^{15}N. Further exacerbating these problems is extensive sample heterogeneity, with tissues generally presenting a mosaic of molecular types and dynamical properties.

In condensed states of diamagnetic materials, line-broadening is traceable to two principal sources, magnetic dipole-dipole interactions and chemical shift anisotropies. Neither of these presents a problem in liquids where rapid molecular tumbling averages dipolar interactions to zero, and anisotropic chemical shifts to the isotropic values observed in high-resolution spectra. In contrast, the resonance widths in condensed states are dispersed over spectral widths up to several kiloHertz. It is apparent that removal of dipolar and shift anisotropy broadenings in biological systems would open up possibilities for high-resolution NMR measurements on a scale approaching solution conditions. Realization of this goal has been advanced greatly by the major contributions of Waugh and coworkers,[5-8] Schaefer and Stejskal,[9-12] and Torchia.[13-15]

Cross-Polarization Magic-Angle Techniques

The theoretical and experimental basis for amelioration of line-broadening and sensitivity problems in crystalline and amorphous solids has been developed in detail by

Pines, Gibby, and Waugh.[5] For randomly oriented rigid systems, the spatially dependent static dipolar interactions between ^{13}C and ^{1}H spins can be removed by strong proton resonant decoupling (dipolar decoupling), analogous to the usual scalar J decoupling except for the much higher power levels ($\gamma_1 H_2/2\pi \sim$ 10-50 kHz) required for the former. The consequent improvements in resolution can be further augmented by taking advantage of the cross-polarization (CP) phenomenon. This effect, originally described by Hartmann and Hahn,[16] provides a mechanism for transfer of magnetic polarization from an abundant magnetic nucleus, ^{1}H, to a less abundant nucleus, ^{13}C, via the thermal reservoir. Under appropriate conditions, magnetization transfer can perturb the Boltzmann distribution of ^{13}C so as to favor enhancement of integrated signal intensity for the latter. One relatively straightforward method for achieving contact between the two spin systems involves simultaneous high rf power irradiation at ^{13}C and ^{1}H resonance frequencies such that the condition $\gamma_{1_H} H_{1(^1H)} = \gamma_{^{13}C} H_{1(^{13}C)}$ is satisfied.[16] By a correct choice of pulse cycling times for the two spins, significant enhancement of ^{13}C intensity is attained over the normal free induction decay. As noted recently by Torchia,[17] direct information about relative dynamical properties is derivable from measurements of scalar decoupled, dipolar decoupled and CP proton enhanced spectra. This is so because signal intensities of scalar decoupled spectra reflect only mobile reorienting molecules and/or groups, while dipolar decoupled spectra comprise both mobile and rigid states.

While the combination of proton dipolar decoupling and cross-polarization produces significant line narrowing and signal enhancement in solid state ^{13}C spectra, the full measure of these techniques is very often obscured by shielding effects. Anisotropy broadenings are particularly noticeable for ^{13}C nuclei involved in unsaturated bonds, and can produce line-broadenings in the kHz range. Fortunately, this complication can be removed by making use of magic-angle [$(1-3 \cos^2\theta) = 0$] rapid spinning techniques originally proposed by Andrews and coworkers[18] and Lowe[19] as a means for reducing anisotropic shift and dipolar coupling effects. Since the spatial dependence of chemical shift and dipolar tensors is the same, rapid spinning, $\omega_r > \Delta\delta$, at the magic-angle $54°, 44'$ will effectively average these interactions to their isotropic values. Thus by supplementing cross-polarization and dipolar decoupling with magic-angle spinning, dramatic increases in spectral resolution have been achieved for a variety of systems including solid polymers,[10/11] polycrystalline organic solids,[20/21] biomembranes,[22-24] and certain biological materials.[25] An especially attractive area opened up by these techniques is the investigation of molecular dynamical behavior in intact biological samples under essentially *in vivo* conditions. This possibility is explored in the following sections.

^{13}C Spectra For Selected Tissues

FT scalar decoupled spectra.

Conventional ^{13}C FT (scalar decoupling) NMR has been employed with considerable success in structural studies of peptides, proteins, and other relatively large biological molecules[26/27] in solution. High-resolution ^{13}C spectra were recently reported for rapidly reorienting membrane constituents of bacterial cells,[22/28] both at natural abundance and using ^{13}C enrichment. Under favorable conditions, where translational and/or diffusional motion exceeds rates of $\sim 10^7$ sec^{-1}, well-resolved FT spectra should be detectable for molecules resident within an otherwise rigid matrix typified by intact tissue sections. This possibility was tested for freshly excised tendon from three mammalian species, canine Achilles tendon, bovine Achilles tendon, and mouse tail. Tendon is an obvious choice because of its high biomolecular homogeneity as distinct from other biological materials, e.g., bacterial

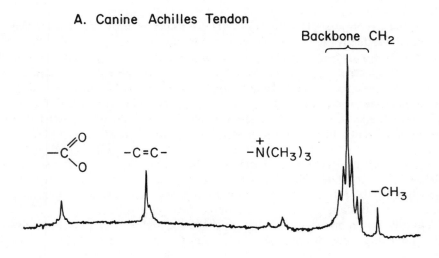

A. Canine Achilles Tendon

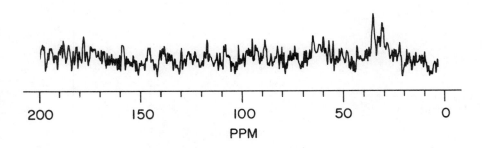

B. Treated Tendon

Figure 1. 15.089 MHz natural abundance ^{13}C FT scalar decoupled spectra of oriented canine Achilles tendon at 30°C. Spectra were measured with a Nicolet-modified spectrometer. Chemical shifts are given in ppm relative to external TMS. (A) 13,840 FID's with a 2.7 sec acquisition time. (B) Canine Achilles tendon after extraction with 1:1 CHCl$_3$:CH$_3$OH; the spectrum represents 5,000 FID's.

or mammalian cells. Two-dimensional gel electrophoresis on an ISO-DALT system[29] confirmed collagen as the main constituent in each tissue, comprising 65 percent dry weight of canine Achilles tendon and 85 percent of dry weight of bovine tendon, with the remainder (10-15 percent) made up of connective molecules (proteoglycans, keratin, chondroitin sulfate) and small amounts of lipoprotein ~ 2 percent, actin < 1 percent, and serum albumin $\sim$ 2-3 percent. Canine tendon showed a somewhat greater heterogeneity primarily because of the presence of vasculature and residual collagen secreting cells in the immature tendon.

^{13}C scalar decoupled (^1H field $\sim$ 2-3 kHz) FT spectra measured for tendon samples oriented vertically w.r.t. H_o show a surprising amount of resolvable near high-resolution signals for canine, Figure 1A, and mouse tail, Figure 2A, tendon. By contrast, the spectrum of bovine Achilles tendon (two-year-old animal), Figure 1B, shows no evidence of fine structure apart from a broad component at $\sim$ 20 ppm. Also, prolonged soaking (36 hours) of canine tendon in 1:1 $CH_3OH/CHCl_3$ leads to a disappearance of the high-resolution signals as would be expected if extraction with this solvent removes mobile and accessible lipid constituents.

Both the multiplet patterns and chemical shift ranges for the tendon signals are very similar to spectral characteristics for cellular membrane constituents[30-32] (phospholipids, fatty acids), as is apparent from comparison of spectra for canine tendon with unilamellar vesicles prepared from egg yolk lecithin (EYLV), Figures 3A and B. Based on this close identity, spectral assignments can be made as in Figure 1A and Table I. Comparison of shifts from the major assignable carbons, Table I, shows only minor variations, < 1 ppm, in specific lipid carbon resonances across

Table I
Chemical Shift† Assignments
For Lipids From Various Sources

Assignment	Achilles Tendon Canine	Mouse Tail	EYLV[31/33]
Carboxyl	171.8	173.5	173.0
	--	172.5	--
C=C-	130.0	130.0	129.2
	128.1	128.4	127.6
Choline CH_2-N^+	69.4	69.4	65.8
Choline $(CH_3)_3N^+$	62.7	62.4	54.4
CH_2-CO_2	34.1	34.2	33.8
CH_2-CH_2-CH_3	32.1	32.3	31.7
Main Chain CH_2	30.0	30.2	29.8
		29.7	
CH_2-CH_2-CH=CH	27.7	27.5	27.2
-CH=CH-CH_2-CH=CH	--	25.8	25.6
-CH_2-CH_2CO_2	25.4	25.2	--
CH_2-CH_3	23.1	23.1	22.9
-CH_3	14.0	14.3	13.9

†Shifts reported as ppm from external TMS with an estimated error of $\pm$ 0.5 ppm.

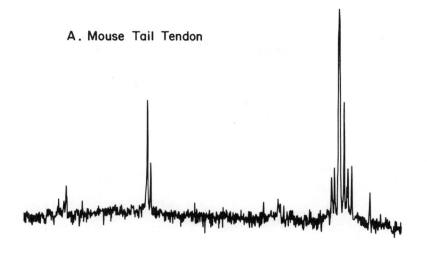

A. Mouse Tail Tendon

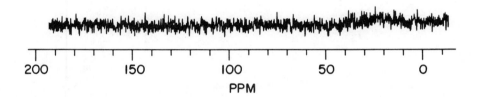

B. Bovine Achilles Tendon

| 200 | 150 | 100 | 50 | 0 |

PPM

Figure 2. 15.089 MHz ^{13}C FT scalar decoupled spectra for oriented mouse tail tendon (A) and bovine Achilles tendon (B) at 30°C. Shifts are given in ppm relative to external TMS. Spectrum A was acquired after 1,000 FID's, 0.7 sec acquisition time; spectrum B was obtained after 4,000 FID's.

diverse biological states, *except* for the $(CH_3)_3N^+$-cholyl head group signal which is deshielded by ~ 8.0 ppm in canine tendon relative to EYLV. Since the phosphoryl cholyl moiety of phospholipids in bilayer vesicles exists in contact with water molecules, whereas the apolar terminus is buried inside the bilayer, it is not surprising that a change in state to the environment within tissues produces detectable shift changes for cholyl signals but not for other carbon resonances.

Several factors could account for the above observation, the most likely being deshielding by negatively charged amino acid side-chains, e.g., Glu, of collagen, and/or charged glucuronyl groups of chondroitin sulfates. It is noteworthy that while lipid components of an intact, rigid, tissue sample are clearly resolvable, the endogenous proteins (collagen and minor constituents) are not detectable, a behavior readily explainable by a much greater rate of rotational and diffusional mobility for the former. The origin of the lipid signals for canine and mouse tendon is presently

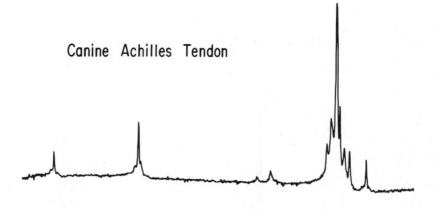

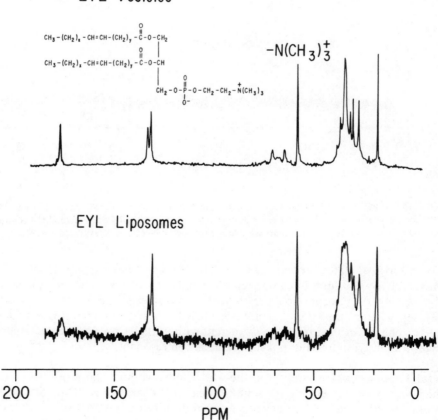

Figure 3. Comparison of canine Achilles tendon (see Figure 1A for details) with model lipid vesicles. ^{13}C spectra in all three traces were recorded at 15.089 MHz in the FT scalar decoupled mode. The middle spectrum (3,000 FID, 2.7 sec) was obtained for bilayer vesicles, $\sim$ 300 Å diameter, prepared by sonication of EYL in D_2O at pD $\sim$7.0 under a nitrogen atmosphere. The spectrum in the lower trace (1,000 FID, 2.7 sec acq.) was recorded for multilamellar spherical liposomes, $\sim$ 900 Å, in D_2O.

uncertain,[34] but the likelihood is that these are attributable to cellular membranes of vascular and collagen secreting cells.

A further insight into lipid mobilities in tissue states can be gained from spin lattice relaxation (T_1) data for the individual carbon resonances. Using the standard $180°-\tau-90°$ inversion recovery method,[35] T_1 values were measured for canine and mouse tendon, and, for comparative purposes, in unilamellar (bilayer) and multilamellamer (liposome) vesicles of egg yolk lecithin. Not surprisingly, the relaxation times are much shorter for tissue lipids, Table II, the change being most dramatic for the cholyl $-N^+(CH_3)_3$ group. Thus, while the lipids are clearly less mobile in tissue versus liposomes, the extent of relative chain immobilization would appear to be greatest for the exposed, polar head group[36], i.e., there is less perturbation of apolar fatty acid side-chains located in the interior of the membrane. It is tempting to speculate that the $-N(CH_3)_3$ chemical shift and T_1 changes are traceable to a common origin, namely the interaction with oppositely charged groups on connective tissue biomolecules, e.g., chondroitin sulfate, proteoglycans, etc. Further work is needed to confirm this point. From Table II, it can also be concluded that motional restriction of the lipids increases quantitatively in the order EYLV < EYLL << tendon, i.e., phospholipids are freer to reorient in tightly packed liposomes than in tissue. The generality of this observation will require further measurements with a range of tissue types.

^{13}C Magic-angle spectra for tendon.

As noted previously, dipolar and shift anisotropy effects preclude observation of high-resolution spectra for biological macromolecules in tissue by conventional FT spectroscopy. Both sources of line- broadening can be removed by combinations of CP, dipolar decoupling, and magic-angle spinning as has been so strikingly demonstrated for collagen fibrils,[13] bovine nasal cartilage,[14] and collagen in lyophilized bovine tendon, calf skin, and rat skin.[25]

In the present work, the natural abundance ^{13}C spectrum for canine Achilles tendon was measured at 37.7 MHz with magic-angle spinning at a rate of 1.75 kHz (CP was not employed in this instance, but the ^{1}H decoupling field was relatively strong,

Table II
^{13}C Spin Lattice Relaxation Times† (T_1) For Lipid Groups.

Assignment	Achilles Tendon	Mouse Tail	EYL (liposomes)	EYL (vesicles)	EYL (vesicles) Ref. 33
C=O	0.65	0.80	1.7	1.7	1.8
C=C	0.25	0.28	0.7	--	0.57-0.75
Choline $(CH_3)_3$	0.04	0.07	0.50	0.71	0.62
Chain CH_2	0.14	0.13	0.37	0.39	0.40
CH_2-CH_3	--	0.87	0.73	1.09	1.4
CH_3	0.43	1.01	1.7	2.0	2.8

†Measured by T_1 inversion recovery method; estimated error, ± 15 percent for canine tendon and mouse tails, and ± 10 percent for EYL vesicles and liposomes.

$\gamma H_1/2\pi = 35$ kHz). A striking number of new, well-resolved signals (line-widths 5-10 Hz) become evident, Figure 4, as compared with the FT spectrum, Figure 1A. Since collagen, the predominant biological macromolecule in tendon, is mainly composed (> 75 percent) of Pro, Gly, Ala, Glu, and hydroxy Pro with smaller contributions (3 percent) from Val, Leu, Arg, and aromatic amino acids, the signals grouped at 170-185 ppm, 45-75 ppm, and 15-30 ppm can be attributed[3/26] to peptide carbonyls, αCH, and methyl carbons respectively, Table III. Two particularly distinct, narrow signals centered at 159.0 ppm and 44.0 ppm are unmistakably due to C_ξ atoms of arginine/tyrosine and glycine α-carbon atoms, respectively.[3/26] The magic-angle spectrum for tendon has a close resemblance to the FT spectrum for fibrous aortic elastin,[37] and to denatured calf Achilles tendon,[38] a not unexpected result since the amino acid contents of the major proteins are similar.

An indication of magic-angle spinning enhancement is given in Figure 5 where spectra are compared under rapid spinning conditions for two orientations of the spinning axis, $\beta = 0°$ and magic angle, with respect to the static polarizing field, H_o. Significant broadenings occur at $\beta = 0°$ for signals across the entire spectrum, with the change being most striking for the C_ξ (Arg, Tyr) signals at 159 ppm, an effect in accord with large chemical shift anisotropies for the latter nuclei.

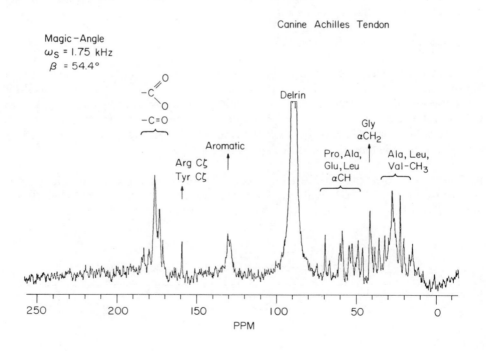

Figure 4. ^{13}C natural abundance spectrum for canine Achilles tendon measured with rapid magic-angle spinning. Tissue samples were tightly packed with the "long-axis" of the tissue vertically oriented in the rotor. The spectrum was recorded at ambient temperature on a Nicolet NT-150 CP spectrometer at a ^{13}C resonance frequency of 37.735 MHz and a constant spinning rate of $\omega_s = 1.75$ kHz. A proton dipolar decoupling field of $\gamma H_2/2\pi \sim 35$ kHz was used and the spectrum represents 2,000 accumulated FID's. No cross-polarization contact was employed in obtaining the tendon spectrum.

Table III
^{13}C Chemical Shifts† for Canine Tendon Protein Constituents

Carbon	Group, Amino acid	Tendon	Elastin[37]
C=O	Peptide C=O	172-184	172-178
Cξ	Arg, Tyr	159.0	--
C	Aromatic	138-140	132-134
αCH	Pro, Ala, Val, Leu,Glu	48-72	44-64
αCH$_2$	Gly	44.0	45.0
β,γCH$_2$	Pro	35-42	24-32
-CH$_3$	Ala, Val, Leu	18-30	19-24

†Shifts are in ppm relative to external TMS, and are accurate to ± 0.5 ppm.

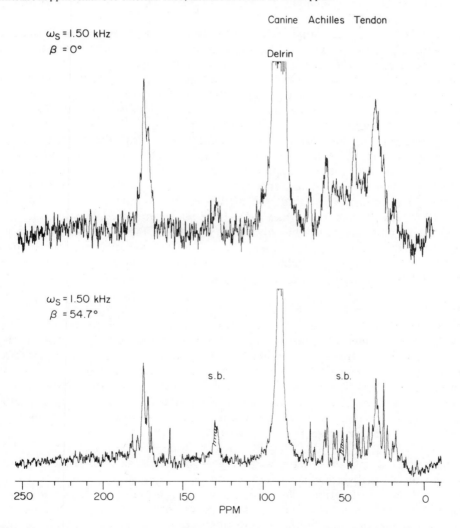

Figure 5. Effect of magic-angle variation on the ^{13}C ^{1}H dipolar decoupled spectrum of canine tendon. Spectrometer operating conditions were identical to those in Figure 4, except for the change in β, the angle between the rotor axis and H$_o$.

The dependence of line-widths on spinning rate and angular orientation for canine tendon collagen signals is consistent with considerable immobilization of the collagen molecules as compared with tissue lipids. When compared with results for bovine Achilles tendon[13] and rat skin,[25] however, the collagen in canine tendon is appreciably more mobile than in either of the other cases where detection of the protein signals requires CP, dipolar decoupling conditions. A partial explanation may lie in the relatively immature nature of the canine tendon obtained from an animal not yet fully developed. Extensive cross-linking typical of "aged" tendon is generally not present in newly formed collagenous tissue. Further, more quantitative knowledge of molecular dynamics of collagen in different tissues will require measurements of T_1 and T_2 relaxation times.

Cross-polarization magic-angle spectra.

Cross-polarization magic-angle spinning techniques have been applied in this study to the measurement of ^{13}C natural abundance spectra for several polycrystalline and amorphous biomolecular materials. One of the immediate objectives of such measurements is the detection of structural and conformational differences for biomolecules in crystalline and solution states. If the existence of such conformational divergences proves to be commonplace then obvious concerns would be raised about the validity of structural extrapolations from crystallographic data. A direct test can be made by comparing isotropic dipolar decoupled shift anisotropy averaged resonances in the solid state with values measured in solution. Griffin and coworkers[24] recently reported an unusual doubling of the center-band and side-band signals in the CP magic-angle spectrum for the ^{13}C-enriched carboxyl group of glycine. This unusual splitting was attributed to co-crystallization of different crystallographic forms from ethanol. We have remeasured the CP magic-angle spinning spectrum of polycrystalline glycine at 37.7 MHz, Figure 6A, and, in contrast, detect no signal doubling. Moreover, the solid state resonance frequencies for the carboxyl and α-carbons coincide with solution values, Figure 6B. In this instance, the shielding environments are clearly very similar in the two states.

An extension of CP, dipolar decoupling magic-angle spinning methods to the measurement of ^{13}C resonances for polycrystalline nucleic acids and their constituents has so far met with limited success. For reasons that are as yet unclear, ^{1}H-^{13}C cross-polarization appears to be relatively inefficient for these molecules, although weak signals are detected for polycrystalline yeast m-RNA in regions corresponding to base ring, 120-150 ppm, and furanose ring, 70-90 ppm, carbon resonances. No ^{13}C spectra were detectable for monomeric ribonucleosides/tides even after prolonged accumulations. Further exploratory work is needed to sort out the reasons for the lack of CP on the monomers.

Although coal is not strictly a biological material, this amorphous solid is nonetheless of early biotic origin and represents an interesting system for CP magic-angle NMR investigation. The molecular constitution of coal is only poorly understood with most, if not all, parameters for characterizing its properties being of a classical nature, i.e., hardness, reflectivity, rank, etc. Fundamental information about internal chemical structure, of critical importance for projection of combustion and conversion characteristics, e.g., ratio of aromatic to aliphatic carbons,[39] relative mobility of various groups and side-chains, is currently not obtainable by direct, non-destructive means. Such data can be derived in principle from solid state spectra for coal.

An example of a CP, dipolar decoupled magic-angle ^{13}C spectrum for a bituminous coal sample (Sewickley, mesh -12 to 100) is illustrated in Figure 7. This spectrum, recorded at 37.7 MHz for a polycrystalline plug packed into the spinning rotor,

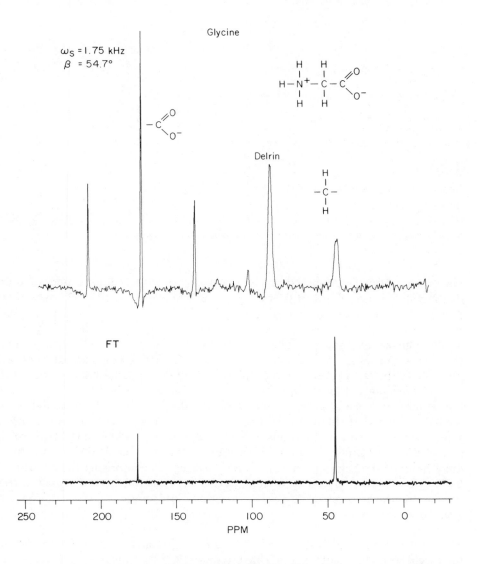

Figure 6. 37.73 MHz ^{13}C natural abundance dipolar decoupled, CP magic-angle spectrum for polycrystalline glycine, upper spectrum. Spectrometer settings are the same as in Figure 4, except for inclusion of a cross-polarization contact time of 1.0 ms. The lower spectrum shows the FT noise-decoupled spectrum for glycine in D_2O solution, pD $\sim$ 7.

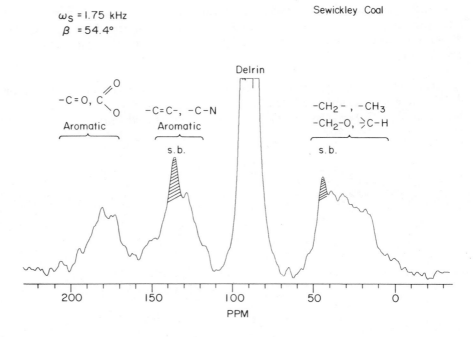

Figure 7. CP magic-angle ^{13}C spectrum for coal. The spectrum was measured at 37.73 MHz with a ^{1}H dipolar decoupling field of 50 kHz and cross-polarization contact time of 1.0 ms.

represents 1100 FID accumulations in $\sim$ 10 minutes of spectrometer operating time. In comparison, no ^{13}C spectrum was observable above noise after 12 hours of standard FT proton noise-decoupled operation.

Several clusters of signals centered at 180, 130, and 30 ppm are clearly discernable, together with a superimposed relatively intense ^{13}C signal with side-bands at $\pm$ 1.7 kHz due to the Delrin rotor. Assignment of the coal resonance bands is straightforward; with signals at low field, 180 ppm, attributable to carbonyl and aromatic carbon atoms, vinylic and aromatic atoms at $\sim$ 130 ppm and methine methylene, and methyl carbons at high field. From the integrated areas the ratio of unsaturated/saturated carbon atoms is estimated as 40:60 for the Sewickley coal sample. The observation of CP dipolar decoupled spectra for coal confirms the relative immobility of the heterogeneous molecular aggregates comprising coal. Further ^{13}C CP magic-angle measurements of coal, particularly as a function of temperature and solvent treatment promise to reveal important information about side-chain and group mobilities.

The above essentially "range-finding" spectroscopic measurements, together with the recent exciting results of other investigators,[13/14/24/25] suggest an extremely promising future for cross-polarization magic-angle NMR spectroscopy in structural and conformational studies of biological molecules in intact biological samples. Although work reported so far has focused on ^{13}C, the methods developed are readily adaptable for studies of other low abundance nuclei, e.g., ^{15}N, ^{17}O with the proviso that relaxation times are compatible with conditions for cross-polarization.

Acknowledgments

We are indebted to Nicolet Technology for providing access to their NT 150 CP spectrometer and to Charles Peters of Nicolet Technology for his careful CP magic-angle measurements at 37.7 MHz. Acknowledgment is also made to Bruce Hammer for the ^{13}C FT vesicle data. Dr. Thomas E. Fritz and Ms. Sandra Tollaksen are gratefully acknowledged for providing canine tendon material and carrying out the ISO-DALT analyses respectively.

References and Notes

1. This work was supported by the U.S. Department of Energy.
2. Nuclear Magnetic Resonance Spectroscopy in Molecular Biology, Vol. 11, The Jerusalem Symposia on Quantum Chemistry and Biochemistry, ed. B. Pullman; D. Reidel Publishing Co., Holland, 1978.
3. Wuthrich, K. Nuclear Magnetic Resonance in Biological Research - Peptides And Proteins, North-Holland Publ., Amsterdam, 1976.
4. Knowles, P. F., Marsh, D., and Rattle, H. W. E. Magnetic Resonance of Biomolecules: An Introduction to Theory and Practice of NMR and ESR in Biological Systems, Wiley-Interscience, New York, 1976.
5. Gibby, M. G., Pines, A., and Waugh, J. S. *Chem. Phys. Letters 16*, 296 (1972).
6. Pines, A., Gibby, M. G., and Waugh, J. S. *J. Chem. Phys. 59*, 569 (1973).
7. Gibby, M. G., Griffin, R. G., Pines, A., and Waugh, J. S. *Chem. Phys. Letters 17*, 80 (1972).
8. Gibby, M. G., Pines, A., and Waugh, J. S. *J. Amer. Chem. Soc. 94*, 6231 (1972).
9. Stejskal, E. O. and Schaefer, J. *J. Mag. Resonance 18*, 560 (1975).
10. Schaefer, J., Stejskal, E. O., and Buchdahl, R. *Macromolecules 8*, 291 (1975).
11. Schaefer, J. and Stejskal, E. O. *J. Amer. Chem. Soc. 98*, 1031 (1976).
12. Stejskal, E. O., Schaefer, J., and Waugh, J. S. *J. Mag. Resonance 28*, 105 (1977).
13. Torchia, D. A. and VanderHart, D. L. *J. Mol. Biol. 104*, 315 (1976).
14. Torchia, D. A., Hanson, M. A., and Hascall, V. C. *J. Biol. Chem. 252*, 3617 (1977).
15. Torchia, D. A. *J. Mag. Resonance 30*, 613 (1978).
16. Hartmann, S. and Hahn, E. L. *Phys. Rev. 128*, 2042 (1962).
17. Torchia, D. (private communication).
18. Andrew, E. R., Bradbury, A., and Eades, R. G. *Nature 182*, 1659 (1958).
19. Lowe, I. J. *Phys. Rev. Letters 2*, 285 (1959).
20. Fyfe, C. A., Lyerla, J. R., and Yannoni, C. S. *J. Amer. Chem. Soc. 100*, 5635 (1978).
21. Steger, T. R., Stejskal, E. O., McKay, R. A., Stults, B. R., and Schaefer, J. *Tetrahedron Letters 122*, 295 (1979).
22. Urbina, J. and Waugh, J. S. *Proc. Nat. Acad. Sci. USA 71*, 5062 (1974).
23. Kohler, S. J. and Klein, M. P. *Biochemistry 15*, 967 (1976).
24. Haberkorn, R. A., Herzfeld, J., and Griffin, R. G. *J. Amer. Chem. Soc. 100*, 1296 (1978).
25. Schaefer, J., Stejskal, E. O., Brewer, C. F., Keiser, H. D., and Sternlicht, H. *Arch. Biochem. and Biophys. 190*, 657 (1978).
26. Howarth, O. W. and Lilley, D. M. J. Progress in Nuclear Magnetic Resonance Spectroscopy, ed. J. W. Emsley, J. Feeney, and L. H. Sutcliffe, Pergamon Press, Oxford, Vol. 12, 1-40 (1978).
27. Allerhand, A., *Accts. Chem. Research 11*, 469 (1978).
28. Smith, I. C. P., Tulloch, A. P., Stockton, G. W., Schrier, S., Joyce, A., Butler, K. W., Boulanger, Y., Blackwell, B., and Bennett, L. G. *Ann. N. Y. Acad. Sci. 308*, 8 (1978).
29. Anderson, L. and Anderson, N. G. *Proc. Natl. Acad. Sci. USA 74*, 5421 (1977).
30. Levine, Y. K., Birdsall, N. J. M., Lee, A. G., and Metcalfe, J. C. *Biochemistry 11*, 1416 (1972).
31. a. Lee, A. G., Birdsall, N. J. M., and Metcalfe, J. C. Methods of Membrane Biology, Vol. II, ed. E. D. Korn, Plenum Press, New York, N. Y., 1974, pp. 65-75. b. Seelig, J. *Quart. Rev. Biophysics 10*, 83 (1977).

32. Smith, I. C. P. *Can. J. Chem. 57*, 1, 1979.
33. Godici, P. E. and Landsberger, F. R. *Biochemistry 13*, 362 (1974).
34. Morphological evaluation of tissue sections by scanning and transmission electron microscopy is currently underway (T. Seed, H. M. Schwartz, and S. S. Danyluk, unpublished work).
35. Vold, R. T., Waugh, J. S., Klein, M. P., and Phelps, D. E. *J. Chem. Phys. 48*, 3831 (1968).
36. Line-broadening and T_1 changes arising from residual paramagnetic metal ions are not expected to be a major factor because of their low concentrations in collagenous material and the unlikelihood of their interaction with a positively charged trimethyl amino group.
37. Urry, D. W. and Mitchell, L. *Biochem. Biophys. Res. Commun. 68*, 1153 (1976).
38. Torchia, D. A. and Piez, K. A. *J. Mol. Biol. 76*, 419 (1973).
39. Retcofsky, H. L., Schweighardt, F. K., and Hough, M. *Anal. Chem. 49*, 575 (1977).

NMR Zeugmatographic Imaging of Organs and Organisms

Paul C. Lauterbur,*⁺ Waylon V. House, Jr.,*¹
Howard E. Simon,* M. Helena Mendonca Dias,*²
Myron J. Jacobson, ≠ † Ching-Ming Lai,* and Peter Bendel*
Department of Chemistry,* Department of Radiology⁺
and
Department of Surgery ≠
State University of New York at Stony Brook
Stony Brook, New York 11794

and

Andrew M. Rudin
Northport Veterans Administration Medical Center†
Northport, New York 11768

Imaging by nuclear magnetic resonance depends upon the use of a novel principle.[3] Unlike conventional optical imaging, in which the resolution is limited by the wavelength of the radiation used, zeugmatographic imaging depends upon the application of a second field to confine the interaction between the object and the imaging radiation to a limited region smaller than the wavelength. With inhomogenous magnetic fields, it is easily possible to cause the NMR resonance frequency within an object to vary by much more than the intrinsic line width, and the resolution is in principle limited only by the sensitivity with which signals from small volumes may be detected. It can be estimated that, even for the most sensitive spectrometers now available, a practical resolution of better than 5 to 10 micrometers will be very difficult to achieve.[4]

Spatially-periodic objects, such as crystals, may in principle be imaged at atomic resolution if very large magnetic field gradients are employed, and if the required sensitivity can be achieved by coherently combining the signals from all of the unit cells, but the experimental problems have not been solved in practice.[5-7] At the present time, therefore, the resolution of NMR zeugmatographic imaging techniques is better than that of the naked eye but poorer than that of an optical microscope.

The situation is more favorable for large objects. If the linear dimensions of a region are doubled, its volume, and hence the strength of the NMR signals, increases by a factor of 8. Other things being equal, the larger an object the better the fractional resolution that can be achieved if sensitivity is the limiting factor. For example, entire animals, isolated organs, or even whole human beings may be examined, and those properties of the tissues measurable by NMR, such as nuclear concentrations, relaxation times, and spectra, may be determined.

Magnets rather different from those used in conventional NMR experiments are required for very large objects. The most common type for human whole body measurements[8/9] is a four-coil approximation to a spherical magnet, based upon the designs described some years ago by Garrett[10] and refined by Franzen.[11] Although most of those already built or under construction are made with water cooled normal conductors, one low-field superconducting magnet has been built for whole-body imaging.[12]

The four-coil magnets used in our laboratory[8] operate at 938 gauss (93.8 mT) so as to give a proton resonance frequency of 4 MHz. The first of our magnets, with a minimum bore of 42 cm, has been used extensively for imaging within a radio-frequency transmitter-receiver coil wound on a form 15 cm in diameter. The magnetic field, after shimming with a ferromagnetic plate and electrical compensation coils, has been made homogeneous to about 10 ppm (10 nT) within an oblate ellipsoid approximately 18 cm in diameter and 15 cm in length. To obtain images, linear magnetic field gradients of about 5 μT cm^{-1} are generated by additional coils wound around and through the magnet structure.[8] Any desired gradient direction may be produced by the vectorial combination of the linear orthogonal gradients generated by the individual coils.[13] The free induction decays following a radio-frequency pulse may be Fourier-transformed to give a "spectrum" that is actually a one-dimensional representation of the spatial distribution of the spins. Each different gradient direction generates a different one-dimensional projection of the NMR signal, and many such projections may be combined to "reconstruct" the distribution of signal sources in two or three dimensions.[3/8/13-15] A similar magnet, with a 62 cm bore, has been adjusted to give a homogeneous region at least 35 cm in diameter, suitable for the application of the reconstruction method to imaging of the human head and regions of the torso.

Many other methods of obtaining NMR zeugmatographic images have been developed. A bibliography may be found in a recent paper,[16] and a number of reviews are available.[17/18]

The usefulness of NMR zeugmatographic images may often be considerably increased if they contain information about relaxation times as well as spin distributions.[3/19-21] Relaxation times may be measured in images by suitable variations of any of the standard techniques used in NMR spectroscopy, such as partial saturation in CW experiments,[3/19] the pulsed inversion-recovery method,[20] or by steady-state pulsed methods.[21]

Although the natural relaxation time differences are often large enough to give good contrast between different organs and between normal and damaged or diseased tissues,[19/22/23] there are circumstances in which the differences are more subtle.[23] Because it is known that metal ions and their complexes often concentrate selectively in abnormal tissues, and that paramagnetic ions can shorten the relaxation times of water protons at low concentrations, we have carried out several series of experiments in which rats and dogs were injected with dilute solutions of manganous chloride in suitable physiological media, and the relaxation times and manganese concentrations measured *in vitro* in tissue samples. The most extensive experiments have been carried out on experimental canine myocardial infarctions. The results of one such study have been published.[24] Manganese was found to concentrate in the uninvolved regions of the heart, but to be present in much lower concentrations in the region to which blood flow had been reduced, giving very large relaxation time ratios. Such experiments, with manganous ion alone, or as various complexes, as well as with other paramagnetic species, offer the possibility of selectively modifying relaxation properties *in vivo* in ways that depend upon the physiological state of the tissues. The ultimate medical usefulness of the technique will, of course, depend upon the toxicities of the relaxation contrast agents, as well as upon their efficacies.

Another approach to increasing the specificity of biological and medical NMR zeugmatographic imaging is to make use of spectroscopic information to distinguish among chemical species. Spatially-selective excitation of the spin system in a magnetic field gradient, followed by observation in a homogeneous field of the spectrum associated with each selected region, makes possible the reconstruction of spectroscopic zeugmatograms.[25] If the chemical shifts are large enough relative to the natural line widths, so that field gradients can significantly broaden the resonances without causing complete overlap of the spectral peaks, a more direct approach is possible. These conditions may be satisfied for the ^{31}P spectra of the major metabolites in tissue, for example. Images showing the individual distributions of inorganic phosphate, ATP and phosphocreatine in test objects consisting of assemblages of glass tubes containing the separate compounds have been reconstructed from phosphorus NMR data taken at 146 MHz on 15 mm samples in a superconducting magnet. The use of this technique to follow changes in the concentrations of metabolites or exogenous substances in living animals and humans, or in perfused organs, will be most severely limited by the low sensitivity with which physiologically significant concentrations of such substances can be detected.[15]

Zeugmatographic imaging can extend the usefulness of NMR spectroscopic and dynamical studies to macroscopically heterogeneous objects. In principle, all of the quantities that can be measured in homogeneous samples, such as concentrations, relaxation times (T_1, T_2, $T_{1\varrho}$), spectra, diffusion coefficients, flow,[26] and exchange rates, can be spatially resolved. The most important practical limitation is that the signal-to-noise ratio which must be achieved in order to make such an experiment practical is that for the volume to be resolved, not for the volume of the entire object.

Acknowledgements

This work was supported in part by Grant No. CA-153000, awarded by the National Cancer Institute, DHEW, by Grant No. HL1 985101A1, awarded by the National Heart, Lung and Blood Institute, DHEW, and by the Veterans Administration. This work was also carried out, in part, at Brookhaven National Laboratory under the auspices of the United State Department of Energy.

References and Notes

1. Present address: Intermagnetics General Corporation, P.O. Box 566, Guilderland, New York 12084.
2. Permanent address: Centro de Quimica Estructural da Universidade de Lisboa, Instituto Superior

Technico, Lisboa, Portugal.

3. Lauterbur, P. C., *Nature 242*, 190 (1973).

4. Mansfield, P., and Grannell, P. K., *Phys. Rev. B 12*, 3618 (1975).

5. Mansfield, P., Grannell, P. K., Garroway, A. N., and Stalker, D. C., *Proc. First* Specialized "*Colloque Ampere*," Inst. Nucl. Phys., Krakow, Poland 16 (1973).

6. Mansfield, P., and Grannell, P. K., *J. Phys. C: Solid State Phys. 6*, L422 (1973).

7. Mansfield, P., Grannell, P. K., and Maudsley, A. A., *Magnetic Resonance and Related Phenomena, Proc. of the 18th Ampere Congress*, P. S. Allen, E. R. Andrew and C. A. Bates, Eds., North-Holland, Amsterdam, Vol. 2, p. 431 (1975).

8. Lai, C.-M., House, W. V., Jr., and Lauterbur, P. C., *Proc. IEEE Electro/78 Conf., Session 30,* paper 2 (1978).

9. Mansfield, P., Pykett, I. L., and Morris, P. G., *Brit. J. Radiol. 51*, 921 (1978).

10. Garrett, M. W., *J. Appl. Phys. 22*, 1091 (1951).

11. Franzen, W., *Rev. Sci. Instr. 33*, 933 (1962).

12. Goldsmith, M., Damadian, R., Stanford, M., and Lipkowitz, M., *Physiol. Chem. and Phys. 9*, 105 (1977).

13. Lai, C.-M., Shook, J. W., and Lauterbur, P. C., *Chem. Biomed. Environ. Instr.* (in press).

14. Lauterbur, P. C., *Pure Appl. Chem. 40*, 149 (1974).

15. Lauterbur, P. C., in *NMR in Biology*, R. A. Dwek, I. D., Campbell, R. E. Richards, and R. J. P. Williams, Eds., Academic Press, London, p. 323 (1977).

16. Lauterbur, P. C., *IEEE Trans. Nucl. Sci., NS-26*, No. 2, 2808 (1979).

17. Andrew, E. R., *Proc. IV Ampere Intnl. Summer School*, R. Blinc and G. Lahajuar, Eds., J. Stefan Institute, Ljubljana, Yugoslavia (1977).

18. Mansfield, P., and Pykett, I. L., *J. Mag. Res. 29*, 355 (1978).

19. Lauterbur, P. C., Lai, C.-M., Frank, J. A., and Dulcey, C. S., Jr., *Physics in Canada 32* Abstract 33.11 (July 1977).

20. Hutchison, J. M. S., in *Medical Images: Formation, Perception, and Measurement*, G. A. Hay, Ed., Wiley, p. 135 (1977).

21. House, W. V., and Lauterbur, P. C., *Proc. Med. VII Conf.*, Society of Photo-Optical Instrumentation Engineers, Bellingham, Wash., 1979 (in press).

22. Lauterbur, P. C., Frank, J. A., and Jacobson, M. J., *Physics in Canada 32*, Abstract 33.9 (July 1976).

23. Frank, J. A., Feiler, M. A., House, W. V., Lauterbur, P. C., and Jacobson, M. J., *Clin. Res. 24*, 217A (1976).

24. Lauterbur, P. C., Mendonca Dias, M. H., and Rudin, A. M., in *Frontiers of Biological Energetics*, P. O. Dutton, J. Leigh, and A. Scarpa, Eds., Academic Press, New York, p. 752 (1978).

25. Lauterbur, P. C., Kramer, D. M., House, W. V., Jr., and Chen, C.-N., *J. Am. Chem. Soc. 97*, 6866 (1975).

26. Lauterbur, P. C., and Lai, C-M., Proc. of the NHLBI Div. of Heart and Vascular Diseases Contractors Meeting, 158 (1977).

Index to Subjects